COMPRESSIVE SENSING OF EARTH OBSERVATIONS

Signal and Image Processing of Earth Observations Series

Series Editor

C.H. Chen

Published Titles

COMPRESSIVE SENSING OF EARTH OBSERVATIONS

Edited by
C.H. Chen

CRC Press
Taylor & Francis Group
Boca Raton London New York

CRC Press is an imprint of the
Taylor & Francis Group, an **informa** business

CRC Press

Taylor & Francis Group

6000 Broken Sound Parkway NW, Suite 300

Boca Raton, FL 33487-2742

First issued in paperback 2019

© 2017 by Taylor & Francis Group, LLC

CRC Press is an imprint of Taylor & Francis Group, an Informa business

No claim to original U.S. Government works

ISBN-13: 978-1-4987-7437-6 (hbk)

ISBN-13: 978-0-367-87886-3 (pbk)

Library of Congress Cataloging-in-Publication Data

Names: Chen, C. H. (Chi-hau), 1937- editor.
Title: Compressive sensing of earth observations / edited by C.H. Chen.
Description: Boca Raton, FL : Taylor & Francis, 2017. | Includes
bibliographical references and index.
Identifiers: LCCN 2016050152 | ISBN 9781498774376 (print : alk. paper)
Subjects: LCSH: Earth sciences--Remote sensing. | Compressed sensing
(Telecommunication)
Classification: LCC QE33.2.R4 C66 2017 | DDC 550.28/7--dc23
LC record available at https://lccn.loc.gov/2016050152

Visit the Taylor & Francis Web site at
http://www.taylorandfrancis.com

and the CRC Press Web site at
http://www.crcpress.com

*This book is dedicated to my wife, Wanda,
my sons, Ivan and Christopher,
and grandsons, Jeremy and Benjamin.*

Contents

Preface

Compressive sensing (CS), a new area of signal processing, has been around for 15 years. Its theory has now been mostly developed. With increased high-resolution sensor development for remote sensing, we are dealing with an enormous amount of earth observation data. CS is particularly important for meeting this challenge. CS exploits the sparsity of the remote sensing data and also allows for reduced sampling. It can considerably simplify the data acquisition design for remote sensing systems and speed up the process of extracting the desired information from a large amount of data. It is believed that future remote sensing systems will make extensive use of CS and will include CS as part of the system design. Indeed, CS has enormous potential for many aspects of remote sensing.

Several recent efforts to implement or incorporate CS in remote sensing systems are presented in this book. The book consists, somewhat loosely, of five parts. Part 1 provides a basic review of the theory and its transition to practice in Chapter 1 and an overview of the Slepian functions in Chapter 2. Part 2, containing Chapters 3 through 7, discusses CS for radar and microwave remote sensing. This is the area that has benefited the most from the use of CS. Part 3, with Chapters 8 through 10, covers CS for subsurface sensing. Part 4, with Chapters 11 through 13, examines CS for multispectral and hyperspectral images. Part 5, with Chapters 14 and 15, covers other issues of CS for remote sensing, such as implementation involving hyperspectral remote sensing images. This book was largely inspired by the well-written chapter "Compressive Remote Sensing," by J. Ma et al., published as Chapter 5 of *Signal and Image Processing for Remote Sensing, 2nd edition*, edited by C.H. Chen, CRC Press, 2012, and the well-edited book, *Compressive Sensing for Urban Radar*, by M. Amin, CRC Press, 2014.

Digital signal processing has experienced significant growth since the development (or discovery) of the fast Fourier transform in the early 1960s. Since then, the methodologies of adaptive filtering, spectrum estimation, time–frequency analysis, wavelet transforms, neural networks and machine learning, deconvolutions, and so on have dominated the rapid progress in one-dimensional and two-dimensional (image) signal processing. The success of all of these methodologies was largely built on sound mathematical development and advances. CS was built on the mathematical theory of sparse representation. As with previous major advances in signal processing, the many advances in CS and the meeting of continued challenges was made possible by the close collaboration of mathematicians, engineers, and scientists. Though this volume is more concerned with the application of CS for earth observations, we are very fortunate to have leading mathematician Prof. Dr. Gitta Kutyniok and her colleagues as the authors of Chapter 1, "Compressive Sensing: From Theory to Praxis." This chapter provides a clear and rigorous introduction to CS as a mathematical topic in signal processing. A significant part of the chapter is devoted to some sample applications such as medical imaging, data separation, analog-to-information converters, and sparse channel estimation, among others. Chapter 2, "Compressive Sensing on the Sphere: Slepian Functions for Applications in Geophysics," was authored by Drs. Khalid and Muhammad. As noted by the authors, the Slepian functions are particularly useful for CS in geophysics, which deals with earth observations. The remaining chapters are concerned with practical CS applications for earth observations.

Chapter 3, by Prof. Hoofar et al., titled "Compressive Sensing–Based High-Resolution Imaging and Tracking of Targets and Human Vital Sign Detection behind Walls," outlines their recent works in the application of CS in synthetic aperture radar (SAR), multiple-input multiple-output (MIMO) radar imaging, and simultaneous tracking of multiple targets behind multilayer walls or other layered media. The chapter also presents the use of CS in detection of human vital signs in a cluttered environment and discusses the corresponding modeling of rib-cage breathing. The chapter concludes with a discussion on hardware design of low complexity receivers with reduced analog-to-digital conversion (ADC) requirements suitable for SAR, MIMO, ultra-wideband, or stepped frequency. Chapter 4, by Prof. Ertin and N. Suvaganam, discusses recovery guarantees for high-resolution radar sensing with compressive illumination. The chapter presents a comprehensive radar architecture that combines multitoned linear frequency modulated waveforms on transmit with classical stretch processor and sub-Nyquist sampling on receive. The proposed compressive illumination scheme has fewer random elements than previously proposed compressive radar designs based on stochastic waveforms, resulting in reduced storage and complexity for implementation. Bounds on the operator norm and mutual coherence are presented such that high-resolution recovery is guaranteed for a sparse scene using sampling rates that scale linearly with the scene sparsity. Chapter 5, by Dr. Bacci et al., is on CS for inverse SAR imaging. The applicability of CS to the radar imaging problem has been justified by observing that the radar signal is sparse when represented in the 2D Fourier domain, since it can be represented by a superposition of a few prominent scatter responses in the radar image plane with respect to the pixels in the image. Three different applications of CS to the Inverse Synthetic Aperture Radar (ISAR) imaging problem are discussed, namely image reconstruction from compressed data, resolution enhancement, and image reconstruction from incomplete data. Chapter 6, by Prof. Liu et al., is on a novel CS–based algorithm for space–time signal processing using airborne radars. The chapter presents the general similar sensing matrix pursuit algorithm to reconstruct the sparse radar scene directly using the test cell, which reduces the computational complexity significantly. The algorithm can efficiently cope with the deterministic sensing matrix with high coherence; it can estimate weak elements (targets) as well as strong elements (clutter) in the DOA-Doppler plane accurately and distinguish the targets from clutter successfully.

Chapter 7, by Prof. Bidon et al., covers Bayesian sparse estimation of radar targets in the CS framework. The chapter investigates the use of a wideband waveform jointly with sparse recovery techniques to unambiguously reconstruct the target signal. Several hierarchical Bayes models are presented where sparsity of the target scene is induced via sparse-promoting prior. Advantages of the Bayes approach are clearly presented.

Chapter 8, by Prof. Di Donato et al., on virtual experiments and CS for subsurface microwave tomography, presents a CS-inspired tomographic approach for processing ground-penetrating radar (GPR) data. The approach exploits the recently introduced Virtual Experiments framework, which is applied to surface and borehole GPR surveys for quantitative characterization of buried targets. Chapter 9, by Drs. Rodriguez and Sacchi, on seismic source monitoring with CS, illustrates practical aspects regarding the incorporation of CS into a system for continuous monitoring of seismic activity. Solving the continuous monitoring system as a sparse representation presents an advantage over the least-squares solution in the form of enhanced resolution in event location, particularly when noise is present in the observations. Sparsity makes the problem amenable to the incorporation of CS. Chapter 10, by Drs. Wang and Cao, provides a detailed introduction to seismic data regularization and imaging based on CS and sparse optimization. Numerical experiments based on some reviewed models and methods prove the efficiency of CS theory and sparse optimization.

Chapter 11, by Prof. Melgani et al., on land use classification with sparse models, formulates the land use classification problem within a CS fusion framework. Residuals are generated from the image reconstruction with dictionaries associated with the available set of possible land uses and gathered to form a single-feature image pattern. The patterns obtained from different types of spatial/spectral features are then used to provide the final land use estimate. As demonstrated by experiments run on the basis of a public benchmark database, the presented schemes can achieve significant classification accuracy gains over recent reference methods. With the development of the CS paradigm, sparse representation-based classification and detection have drawn much attention in hyperspectral imagery. With this in mind, Chapter 12, by Prof. Zhang et al., on CS for reconstruction, classification, and detection of hyperspectral images, introduces three reconstruction methods from random projection, three representation-based classifiers and their extensions, and two one-class representation-based target detectors and their improved versions. Chapter 13, by Prof. Rontogiannis et al., proposes a novel semi-supervised algorithm for the problem of abundance matrix estimation in land cover hyperspectral image unmixing. The abundance matrix being sought is constrained to be simultaneously sparse and low rank.

Chapter 14, by Prof. Nascimento et al., covers the parallel coded aperture method for hyperspectral CS on Graphics Processing Unit (GPU). Reduction of the large amount of continuously acquired hyperspectral image data is much needed. Hyperspectral images are often highly compressible owing to the very high spatial and spectral correlation. Therefore, this imaging modality is a perfect candidate for applying CS technology. A simple process can be implemented on the optics of the acquisition system onboard the satellite, while in the CS model the reconstruction is performed at the ground station, where a variety of powerful computational resources may be available. The chapter describes the computationally efficient implementation of two reconstruction algorithms in graphics process units. The hardware is able to exploit data parallelism through a single-instruction, multiple-data computing architecture. Finally, Chapter 15, by Prof. Magli et al., on algorithms and prototyping of a compressed hyperspectral imager, deals with the practical implementation of a compressive hyperspectral imager, highlighting design and data quality issues. Details of the instruments employed as well as image acquisition and recovery are discussed in depth. From Chapters 14 and 15, we may conclude that implementation is a new challenge in designing CS systems for hyperspectral images.

Based on the development thus far, we may conclude that radar remote sensing and hyperspectral image sensing have been the major users of CS. Much continued effort is expected to be driven by the need for system design.

The rapid advances in nearly all areas of science and engineering over the past half century have been the result of research and publications. Although journal publications and textbooks still play a fundamentally important role among publications, conference publications and edited books have also become very important. Since the publication of my first edited book 40 years ago, I have had a particular appreciation for the unique role of edited books in advancing knowledge, which may often be overlooked by readers. Contributors to edited books often do not get proper credit or recognition for making the book possible. Preparing an edited book requires much effort from both editors and contributors. Readers can only appreciate the rich content of an edited book through careful reading of each chapter or certain chapters. I am sure that there will be many more books in the foreseeable future on CS theory and applications, and I hope that readers will give the chapters of this book a careful reading. As in past edited books, my thanks and appreciation go to all contributors of the book for their important role in advancing technical knowledge.

C.H. Chen
University of Massachusetts Dartmouth

MATLAB® is a registered trademark of The MathWorks, Inc. For product information, please contact:

The MathWorks, Inc.
3 Apple Hill Drive
Natick, MA 01760-2098
USA Tel: 508-647-7000
Fax: 508-647-7001
Email: info@mathworks.com
Web: www.mathworks.com

Editor

C.H. Chen was born on December 22, 1937. He received his PhD in electrical engineering from Purdue University, West Lafayette, Indiana, in 1965, his MSEE from the University of Tennessee, Knoxville, in 1962, and his BSEE from the National Taiwan University, Taipei, in 1959.

He is currently the chancellor professor emeriti of electrical and computer engineering at the University of Massachusetts Dartmouth, where he has been a faculty member since 1968. His research areas encompass statistical pattern recognition and signal/image processing with applications to remote sensing, geophysical, underwater acoustics, and nondestructive testing problems, as well as computer vision for video surveillance, time-series analysis, and neural networks.

Dr. Chen has edited and authored 34 books in his area of research. He is the editor of *Digital Waveform Processing and Recognition* (CRC Press, 1982), *Signal Processing Handbook* (Marcel Dekker, 1988), and *Signal and Image Processing for Remote Sensing* (CRC Press, first edition 2006 and second edition 2012). He is the chief editor of *Handbook of Pattern Recognition and Computer Vision*, volumes 1–5 (World Scientific Publishing, 1993, 1999, 2005, 2010, and 2016, respectively). He is the editor of *Fuzzy Logic and Neural Network Handbook* (McGraw-Hill, 1966). In the field of remote sensing, he also edited the books *Information Processing for Remote Sensing* and *Frontiers of Remote Sensing Information Processing* (World Scientific Publishing, 1999 and 2003, respectively).

He served as associate editor of the *IEEE Transactions on Acoustics, Speech, and Signal Processing* for 4 years, associate editor of the *IEEE Transactions on Geoscience and Remote Sensing* for 15 years, and he has been a board member of *Pattern Recognition* since 2008.

Dr. Chen has been a Fellow of the Institute of Electrical and Electronic Engineers (IEEE) since 1988, a Life Fellow of the IEEE since 2003, and Fellow of the International Association of Pattern Recognition (IAPR) since 1996. He is also editor of a book series, Signal and Image Processing of Earth Observations, from CRC Press.

Contributors

Naif Alajlan
College of Computer and Information
Sciences
King Saud University
Riyadh, Saudi Arabia

Alessio Bacci
CNIT Radar and Surveillance System
(RaSS) National Laboratory

and

Department of Information Engineering
University of Pisa
Pisa, Italy

Alessandro Barducci
SOFASI SRL
Viale Alessandro Guidoni
Firenze, Italy

Yakoub Bazi
College of Computer and Information
Sciences
King Saud University
Riyadh, Saudi Arabia

Fabrizio Berizzi
CNIT Radar and Surveillance System
(RaSS) National Laboratory

and

Department of Information Engineering
University of Pisa
Pisa, Italy

Martina Bevacqua
Department of Information Engineering,
Infrastructures and Substainable Energy
University of "Mediterranea" of Reggio
Calabria
Reggio di Calabria, Italy

Stéphanie Bidon
Department of Electronics Optronics and
Signal
University of Toulouse–ISAE-Supaéro
Toulouse, France

José Bioucas-Dias
Instituto de Telecomunicações
Instituto Superior Técnico
Lisboa, Portugal

Jingjie Cao
Hebei GEO University
Shijiazhung, Hebei, China

Davide Cataldo
CNIT Radar and Surveillance System
(RaSS) National Laboratory

and

Department of Information Engineering
University of Pisa
Pisa, Italy

Giulio Coluccia
Dipartimento di Elettronica e
Telecomunicazioni
Politecnico di Torino
Torino, Italy

Lorenzo Crocco
Institute for Electromagnetic Sensing of the
Environment
National Research Council of Italy
Naples, Italy

Loreto Di Donato
Department of Electrical, Electronics and
Computer Engineering (DIEEI)
University of Catania
Catania, Italy

Emre Ertin
Department of Electrical and Computer
Engineering
The Ohio State University
Columbus, Ohio

Aly E. Fathy
Antenna Research Laboratory
Department of Electrical Engineering and
Computer Science
University of Tennessee
Knoxville, Tennessee

Axel Flinth
Institut für Mathematik
Technische Universität Berlin
Berlin, Germany

Lianru Gao
Institute of Remote Sensing and Digital
    Earth
Chinese Academy of Sciences
Beijing, China

Paris V. Giampouras
IAASARS
National Observatory of Athens
Penteli, Greece

Elisa Giusti
CNIT Radar and Surveillance System
    (RaSS) National Laboratory

and

Department of Information Engineering
University of Pisa
Pisa, Italy

Donatella Guzzi
Institute of Applied Physics
    'Nello Carrara' – National Research
    Council of Italy (IFAC-CNR)
C.N.R.-IFAC
Sesto Fiorentino, Italy

Ali Hashemi
Berlin International Graduate School in
    Model and Simulation based Research
    (BIMoS)
Institut für Mathematik
Technische Universitat
Berlin, Germany

Ahmad Hoorfar
Antenna Research Laboratory
Department of Electrical and Computer
    Engineering
Villanova University
Villanova, Pennsylvania

Kaiyu Huang
Department of Automation Science and
    Technology
School of Electronics and Information
    Engineering
Xi'an Jiaotong University
Xi'an, China

Tommaso Isernia
Department of Information Engineering,
    Infrastructures and Substainable Energy
University of "Mediterranea" of Reggio
    Calabria
Reggio di Calabria, Italy

Zubair Khalid
Department of Electrical Engineering
School of Science and Engineering
Lahore University of Management Sciences
Lahore, Pakistan

Ozlem Kilic
Department of Electrical and Computer
    Science
The Catholic University of America
Washington, DC

Konstantinos D. Koutroumbas
IAASARS
National Observatory of Athens
Penteli, Greece

Gitta Kutyniok
Institut für Mathematik
Technische Universität Berlin
Berlin, Germany

Marie Lasserre
Department of Electronics Optronics and
    Signal
University of Toulouse–ISAE-Supaéro
Toulouse, France

Cinzia Lastri
Institute of Applied Physics
    'Nello Carrara' – National Research
    Council of Italy (IFAC-CNR)
C.N.R.-IFAC
Sesto Fiorentino, Italy

François Le Chevalier
Delft University of Technology
Delft, The Netherlands

Wei Li
College of Information Science and
    Technology
Beijing University of Chemical Technology
Beijing, China

Feng Lian
Department of Automation Science and
Technology
School of Electronics and Information
Engineering
Xi'an Jiaotong University
Xi'an, China

Jing Liu
Department of Automation Science and
Technology
School of Electronics and Information
Engineering
Xi'an Jiaotong University
Xi'an, China

Enrico Magli
Dipartimento di Elettronica e
Telecomunicazioni
Politecnico di Torino
Torino, Italy

Marco Martorella
CNIT Radar and Surveillance System
(RaSS) National Laboratory

and

Department of Information Engineering
University of Pisa
Pisa, Italy

Mahendra Mallick
Independent Consultant
Smith River, California

Gabriel Martín
Instituto de Telecomunicações
Lisboa, Portugal

Mohamed L. Mekhalfi
Department of Information Engineering
and Computer Science
University of Trento
Trento, Italy

Farid Melgani
Department of Information Engineering
and Computer Science
University of Trento
Trento, Italy

Abubakr Muhammad
Department of Electrical Engineering
Center for Water Informatics & Technology
(WIT)
SBA School of Science and Engineering,
LUMS
Lahore, Pakistan

José Nascimento
Instituto de Telecomunicações
ISEL–Instituto Superior de Engenharia de
Lisboa
Instituto Politécnico de Lisboa
Lisboa, Portugal

Roberta Palmeri
Department of Information Engineering
Infrastructures and Sustainable Energy
University "Mediterranea" of Reggio
Calabria
Reggio di Calabria, Italy

Valentina Raimondi
Institute of Applied Physics
'Nello Carrara' – National Research
Council of Italy (IFAC-CNR)
C.N.R.-IFAC
Sesto Fiorentino, Italy

Athanasios A. Rontogiannis
IAASARS
National Observatory of Athens
Penteli, Greece

Mauricio D. Sacchi
Department of Physics
University of Alberta
Edmonton, AB, Canada

Nithin Sugavanam
Department of Electrical and Computer
Engineering
The Ohio State University
Columbus, Ohio

Xu Sun
Institute of Remote Sensing and Digital
Earth
Chinese Academy of Sciences
Beijing, China

Konstantinos E. Themelis
IAASARS
National Observatory of Athens
Penteli, Greece

Sonia Tomei
CNIT Radar and Surveillance System
    (RaSS) National Laboratory

and

Department of Information Engineering
University of Pisa
Pisa, Italy

Ismael Vera Rodriguez
Schlumberger Gould Research
Cambridge, UK

Yanfei Wang
Key Laboratory of Petroleum Resources
    Research
Institute of Geology and Geophysics
Chinese Academy of Sciences
Beijing, China

Bing Zhang
Institute of Remote Sensing and Digital
    Earth
Chinese Academy of Sciences
Beijing, China

1

Compressed Sensing: From Theory to Praxis

Axel Flinth, Ali Hashemi, and Gitta Kutyniok

CONTENTS

1.1 Introduction

For a long time, the *Shannon sampling theorem* was the ruling paradigm in the signal processing community when it came to choosing sampling rates. The theorem, proved in 1949 by Claude E. Shannon, states that any function $f : \mathbb{R} \to \mathbb{C}$ having limited bandwidth W (meaning that the support of the Fourier transform $\hat{f}$ is contained in the interval $[-W, W]$) can be exactly reconstructed from its values $\left(f(\frac{n}{2W})\right)_{n \in \mathbb{Z}}$. Put differently, sampling a function at a rate at least two times higher than the bandwidth of the function will provide enough information for perfect reconstruction of the signal of interest. The rate two times higher than the bandwidth of a signal is often referred to as the *Nyquist rate*.

During the course of the last decade, a new sampling theory has emerged, known as *compressive sampling* or *compressed sensing*, which was introduced about 10 years ago as an effective and efficient way of sensing and acquiring data. This theory enables faithful recovery of signals, images, and other types of data from highly sub-Nyquist-rate samples. While Shannon sampling theory only utilizes the bandwidth information of the signal, compressed sensing relies on the crucial observation that data we are interested in acquiring typically are structured. To be concrete, they often possess a sparse or nearly sparse representation in a certain basis or dictionary [5,11,25].

To state the problem mathematically, let $x_0 = ((x_0)_i)_{i=1}^n \in \mathbb{R}^n$ be our signal of interest. As prior information, we first assume that x_0 is sparse by itself, that is, that its support $\operatorname{supp} x_0 = \{i : (x_0)_i \neq 0\}$ contains few elements. Another way of stating this is to say that the ℓ_0-norm of the signal,

$$\|x\|_0 := \# \operatorname{supp} x_0,$$

is small. Furthermore, let A be an $m \times n$ matrix, which is typically called *sensing matrix* or *measurement matrix*. Throughout this chapter, we will always assume that $m < n$. Then the compressed sensing problem can be formulated as follows: Recover x_0 from knowledge of

$$b = Ax_0.$$

The above equation is an underdetermined linear system of equations. The prior information that the signal of interest is sparse makes it possible to still uniquely recover it using compressed sensing techniques.

In this introductory chapter, we will first give an overview of different solution strategies to the above problem. Then we will present the basic results about which properties the measurement matrix has to have in order for those strategies to succeed. In this part, we will also briefly discuss the case of the matrix being chosen at random. In the final part, we will widen our horizon by discussing applications of the theory of compressed sensing. In that part, it will be necessary to briefly discuss some generalizations of compressed sensing.

1.2   Recovery algorithms

Given the prior knowledge that a linear system of equations $Ax = b$ has a sparse solution x_0, and that x_0 is the sparsest among all solutions, it is evident that the ℓ_0-minimization problem,

$$\min \|x\|_0 \text{ subject to } Ax = b, \tag{1.1}$$

will have x_0 as its solution. There is however no general strategy to solve Equation 1.1 efficiently. In fact, one can prove that solving Equation 1.1 for general A and b is NP-hard [60]. Going into detail what this means goes beyond the scope of this introductory chapter. To give the reader some intuition, let us just note that the NP-hardness of Equation 1.1 implies that if we find an algorithm which can solve Equation 1.1 for any matrix A and any vector b in polynomial time, we would have solved the famous millenium problem P versus NP, which certainly is very much to ask for. This motivates the need for formulating tractable relaxations of Equation 1.1.

1.2.1 Convex programming

Since there often exist efficient solvers for convex optimization, it seems reasonable to search for a convex surrogate of Equation 1.1. In some sense, the most natural way (see [20]) of designing a convex program for identifying sparse vectors is to minimize the ℓ_1-norm instead of the ℓ_0-norm, that is, to consider

$$\min \|x\|_1 \text{ subject to } Ax = b. \tag{1.2}$$

The algorithm, or rather strategy, of solving Equation 1.2 for recovering sparse solutions of $Ax = b$ is known as *Basis Pursuit*. Researchers have known for several decades that the ℓ_1-norm promotes sparsity. But it was as late as 2001 that the Basis Pursuit algorithm was formulated [21]. The Basis Pursuit algorithm can be formulated as a linear program, which in particular proves that it can be solved in polynomial time. It is however by no means clear that the solution of the problem defined by Equation 1.2 is equal to the sparse signal of interest x_0. Much of the theory in fact revolves around determining for which type of matrices this is the case. We will discuss this in detail in Section 1.3.

The problem defined by Equation 1.2 assumes that the linear measurements $b = Ax_0$ are exactly known. In applications, they are however often affected by noise; that is, we are only given $b = Ax_0 + n$ for some unknown noise vector n. There exist several ways to handle this situation. Assuming that some appropriate bound for the noise level $\|n\|_2$ is available, say $\|n\|_2 \le \epsilon$, we can use regularized Basis Pursuit, given by

$$\min \|x\|_1 \text{ subject to } \|Ax - b\|_2 \le \epsilon. \tag{1.3}$$

A procedure of similar characteristic is the so-called *Dantzig*[*] *selector* [16]

$$\min \|x\|_1 \text{ subject to } \|A^*(b - Ax)\|_\infty \le \tau.$$

τ is thereby a tuning parameter. If one assumes that the noise is randomly chosen according to the normal distribution with variance σ^2, then $\tau = \sigma\sqrt{2 \log d}$ is a good choice.

Another way to handle noise is to apply the so-called *LASSO* (least absolute shrinkage and selection operator) [71]. The *LASSO* has its origin in statistics, and it again consists of solving a minimization problem, this time given by

$$\min \|Ax - b\|_2 \text{ subject to } \|x\|_1 \le \tau,$$

where τ is a tuning parameter (which has to be chosen carefully).

1.2.2 Thresholding algorithms

For many practical applications such as image processing, linear programming methods face implementation challenges, since the complexity of these algorithms grows superlinearly with the input size. As an example, recall that the required average number of steps in the simplex algorithm for finding a solution grows cubically with the size of the input [12]. In order to decrease the computational complexity of ℓ_1-norm minimization, iterative methods have been proposed, which converge to a very good approximation of the exact solution of the original linear program.

Let us consider the following unconstrained regularized least-squares (LS) problem:

$$\min_x \frac{1}{2}\|b - Ax\|_2^2 + \lambda\|x\|_1.$$

---

[*]The name of the algorithm was chosen to honor George Dantzig, the founding father of linear programming. In particular, it does not directly originate from the German name of the Polish town Gdańsk.

It can be shown that there exists a $\lambda > 0$ such that the solution of the above optimization problem is equal to the one of Equation 1.2. This value of λ although cannot be calculated *a priori*, when it is often empirically determined. Interested readers might want to have a look at [75], where a more systematic approach is described.

Possibilities for efficient solvers are iterative methods such as steepest descend or conjugate gradient approaches. These will converge to the minimizer of Equation 1.2 by selecting an appropriate starting point as an initial value. Several algorithms have been developed for this purpose [6,77]. One of the most prominent is called *iterative hard thresholding* (IHT) [9,24,32] and will be explained in the sequel.

In the IHT algorithm, a sparse solution can be found by refining the estimate of sparse signals by the following iterative procedure:

$$x^{i+1} = \mathcal{H}_\lambda(x^i - A^*(Ax^i - b)), \tag{1.4}$$

where A^* is the (conjugate) transpose of the matrix A and $\mathcal{H}(\cdot)$ is a the hard thresholding operator given by

$$H_\lambda(x)_i = \begin{cases} 0 & : \quad |x_i| \leq \sqrt{\lambda}, \\ x_i & : \quad |x_i| \geq \sqrt{\lambda}. \end{cases}$$

As can be seen from Equation 1.4, the input of the thresholding function only consists of the multiplication of A and A^* with x^i and b. These operations have a significantly smaller computational complexity than linear programming. This feature can be considered as the main advantage of these types of algorithms as compared with the linear programming approaches [65, p. 18].

The threshold in iterative algorithms can be chosen either constant in all iterations such as in Equation 1.4, or adaptively, resulting in iterative shrinkage thresholding algorithms (ISTA). Adaptively choosing the threshold can improve the performance compared with IHT [24,32]. The iterative step then looks as follows:

$$x^{i+1} = \sigma_{\lambda\delta}(x^i - \delta A^*(Ax^i - b)),$$

where δ is a parameter for step size and $\sigma_\lambda(\cdot)$ is the soft shrinkage threshold function given by

$$\sigma_\lambda(x_i) = \max(|x_i| - \lambda, 0) \, \text{sgn}(x_i).$$

As an extended version of the ISTA method, the FISTA algorithm [6] has been proposed by utilizing the solutions of two previous iterations in order to improve the convergence rate of ISTA [6].

1.2.3 Greedy algorithms

The key idea of greedy algorithms is to first identify some subset in support of the original signal and then refine it in a number of steps until a good estimate of the support is found. In compressed sensing problems, the number of steps is typically equal to the sparsity level $s = \|x_0\|_0$. These algorithms have the advantage of being very fast and easy to implement.

One of the simplest algorithms of this class is called matching pursuit (MP) [56]. To explain this method, we first provide an equivalent representation of the compressed sensing problem as follows:

$$b = \sum_{i=1}^{n} a_i x_i, \tag{1.5}$$

where a_i is the ith column of A. By assuming s-sparsity of the signal, the compressed sensing problem in this formulation can be considered as finding the s "active" columns of A (i.e., the support of the vector $x = (x_i)_{i=1}^n$) and the corresponding coefficients x_i [65, p. 19].

MP approximates the source in s steps. In each step, the algorithm finds the column of the sensing matrix A which is most correlated with the current *residual* r (initialized as b). Once the column of A is revealed, the corresponding coefficient x_i is revealed by minimizing the error corresponding to the data fidelity, $\|b - x_i a_i\|_2^2$, through solving a LS problem [10]. In particular, the column a_{i*} with the maximum absolute value of $\langle r, a_i \rangle$ (most correlated with residual) is selected as an active column in Equation 1.5 and $x_i = \langle r, a_{i*} \rangle$, after obtaining the inner products $\langle b, a_i \rangle$ of b. Subsequently, the residual is updated according to

$$r \leftarrow r - x_i a_{i*}.$$

The main disadvantage of this approach is that it assumes that the columns of A are orthogonal, which is not the case for most sensing matrices [65]. If that is the case however, it is very fast and cheap, in particular when the matrix multiplication with A and A^* can be efficiently implemented (that being the main complexity bottleneck).

Orthogonal matching pursuit (OMP) [72] improves MP, making it work for more general measurement matrices, via updating the found x_i's in each step. This approach is explained in detail as follows: At the first stage, the OMP algorithm takes a matrix $A \in \mathbb{R}^{m,n}$, a measurement vector $b \in \mathbb{R}^m$, and an error threshold $\epsilon > 0$ as an input data. In the next step, for starting the algorithm, the initial values for the solution $x^0 = 0$, the residual value $r_0 = b - Ax^0 = b$, and for the support $S^0 = \text{supp } x^0 = \emptyset$ are assigned. If k denotes the index of the iteration, the algorithm updates the support of recovered coefficients at each iteration by choosing a new index i_* such that [46]

$$|\langle a_{i_*}, r_{k-1} \rangle| \leq |\langle a_i, r_{k-1} \rangle| \quad \text{for all } i.$$

Then, based on the updated support $S^k = S^{k-1} \cup \{i_*\}$, OMP computes the residual $r_k = y - Ax^k$ by updating x^k in each step as follows [46]:

$$x^k = \text{argmin}_x \|Ax - b\|_2 \quad \text{subject to} \quad \text{supp } x = S^k.$$

Finally, the algorithm repeats through iterations until the stopping criterion $\|r_k\|_2 < \epsilon$ is met, where k is the index of iteration.

Faster algorithms such as compressive sampling matched pursuit (CoSaMP) [61] improve OMP by choosing more than k columns of the sensing matrix, recovering the signal coefficients based on the projection onto the space of the selected columns, and rejecting those which might not be in the true support.

1.3 Necessary and sufficient conditions of the measurement matrix for sparse recovery

In this section, we will focus on the original problem of compressed sensing: Under which assumptions on the measurement matrix A does a specific algorithm recover all s-sparse vectors x_0 from their respective measurements Ax_0? Note that almost all of the results in this section also hold for complex matrices and signals, but we shall stay in the real setting for the sake of simplicity of the exposition.

For a subset S of $\{1, \ldots, n\}$, let A_S denote the submatrix formed by the columns a_i of A corresponding to indices $i \in S$. Now, let us begin by making a simple observation: If there exists a submatrix of A consisting of $2s$ columns which is not injective, any algorithm whose output only depends on A and $b = Ax_0$ will fail to recover all s-sparse vectors. To see this, let η be a nonzero vector in the kernel of a submatrix A_S with $S = \{i_1, \ldots i_{2s}\}$. Then the two s-sparse vectors x_1 and x_2 defined by

$$x_1(i) = \begin{cases} \eta(i) & : i \in \{i_1, \ldots i_s\}, \\ 0 & : \text{else}, \end{cases} \quad \text{and} \quad x_2(i) = \begin{cases} -\eta(i) & : i \in \{i_{s+1}, \ldots i_{2s}\}, \\ 0 & : \text{else}, \end{cases}$$

will have the same measurements, $b = Ax_1 = Ax_2$. Thus, they cannot be distinguished by any recovery algorithm.

We have derived a first result. To state it formally, let us introduce the *spark* of a matrix A [28]. The spark, denoted by $\text{spark}(A)$, is defined as the cardinality of the smallest set of columns in A which is linearly dependent. Using this naming, we have just proven the following proposition:

Proposition 1.3.1 *If $A \in \mathbb{R}^{m,n}$ is a matrix such that any s-sparse vector x_0 can be recovered from its linear measurements Ax_0, then necessarily $\text{spark}(A) > 2s$.*

Since $\text{spark}(A) \leq m$, we can in particular conclude that recovery of all s-sparse vectors is only possible if the number of measurements, m, is at least equal to $2s$. This will serve as a lower benchmark in our further discussion.

1.3.1   The RIP and the NSP

Spark-based analyses will rarely yield any statements about the success of specific recovery procedures. In particular, it is impossible to prove stability statements solely relying on the spark, since it is possible to construct matrices with arbitrarily ill-conditioned submatrices which are still invertible. Therefore, it is necessary to impose other conditions on A. The arguably most celebrated and well-known condition is the so-called *restricted isometry property* (RIP). It was introduced by Candès and Tao in the pioneer work of [14].

Definition 1.3.2 (restricted isometry property) *Let $A \in \mathbb{R}^{m,n}$. Then A satisfies the (δ, k)-RIP, if all k-sparse vectors x satisfy*

$$(1 - \delta)\|x\|_2^2 \leq \|Ax\|_2^2 \leq (1 + \delta)\|x\|_2^2.$$

The smallest δ such that A satisfies the (δ, k)-RIP is called the kth restricted isometry constant of A and is denoted by $\delta_k(A)$.

Note that despite the popularity of claiming that a matrix "has the RIP," that statement by itself has a vague meaning. Formally, one always has to specify the parameters k and δ to make the statements precise. In reality, however, this is seldom done. The sparsity level is most often clear from the context and δ is implicitly assumed to be sufficiently small. In the following, we will also use this convention.

Next notice that $\delta_k < 1$ implies $\text{spark}(A) > k$. Furthermore, by considering the definition of submatrix, it is not hard to see that using standard theorems about the spectral properties of Hermitian matrices, the inequality $\delta_k(A) \geq \delta$ is equivalent to

$$\|A_S^* A_S - \text{id}_S\| \leq \delta \tag{1.6}$$

for all S with $|S| \leq k$. $\|A\|$ thereby denotes the operator norm of the matrix A. This equivalence will become very useful when proving the RIP, in particular, for random matrices, since there exist a lot of powerful results about their spectral properties.

There exist many proofs of the fact that the RIP-property implies that regularized Basis Pursuit, that is, the problem defined by Equation 1.3, will recover any s-sparse signal in a stable manner. Candès and Tao were in their original article able to construct a so-called *dual certificate*[*] $c \in \mathbb{R}^n$, only using the RIP of the matrix A. We will consider an alternative route, following [33, Chs. 4, 6]. This route takes a detour to another key property, namely the so-called *null space property* (NSP). The term "NSP" originates from the article [23], but it was implicitly known before that.

Definition 1.3.3 *Let $A \in \mathbb{R}^{m,n}$. Then the matrix A has the s-NSP, if for every set $S \subseteq \{1, \ldots, n\}$ with $|S| \leq n$ and $0 \neq \eta \in \ker A$,*

$$\sum_{i \in S} |\eta_i| < \sum_{i \notin S} |\eta_i|.$$

The s-NSP is in fact equivalent to the uniform success of Equation 1.2 at recovering sparse vectors as the following result shows.

Lemma 1.3.4 *Let $A \in \mathbb{R}^{m,n}$. Then the following conditions are equivalent:*

1. *The program Equation 1.2 with $b = Ax_0$ has x_0 as its unique solution for every s-sparse x_0.*

2. *A has the s-NSP.*

Proof. (ii) $\Rightarrow$ (i). Suppose that A has the s-NSP, and let x_0 be an arbitrary s-sparse vector. Suppose further that x_0 is not the unique solution of Equation 1.2, but instead that there exists $x \in \mathbb{R}^n$ with $Ax = Ax_0$ and $\|x\|_1 \leq \|x_0\|_1$. Then, $\eta = x - x_0 \in \ker A$.

We will now prove that

$$\sum_{i \in \operatorname{supp} x_0} |\eta_i| \geq \sum_{i \notin \operatorname{supp} x_0} |\eta_i|,$$

which would be a contradiction to the s-NSP. Due to the triangle inequality,

$$\|x_0 + \theta\eta\|_1 = \|\theta(x_0 + \eta) + (1 - \theta)x_0\|_1 \leq \theta\|x\|_1 + (1 - \theta)\|x_0\|_1 \leq \|x_0\|_1$$

for every $\theta \in [0, 1]$. Hence, for every such θ, we have

$$\sum_{i \in \operatorname{supp} x_0} |x_0(i)| \geq \sum_{i \in \operatorname{supp} x_0} |x_0(i) + \theta\eta(i)| + \sum_{i \notin \operatorname{supp} x_0} \theta|\eta(i)|. \tag{1.7}$$

If we now choose θ sufficiently close to zero, we can ensure that $|x_0(i) + \theta\eta(i)| = |x_0(i)| + \theta \operatorname{sgn}(x_0(i))\eta(i)$ for all $i \in S$. For such θ, inequality Equation 1.7 reads

$$\sum_{i \in \operatorname{supp} x_0} |x_0(i)| \geq \sum_{i \in \operatorname{supp} x_0} |x_0(i)| + \theta \operatorname{sgn}(x_0(i))\eta(i) + \sum_{i \notin \operatorname{supp} x_0} \theta|\eta(i)|$$

$$\Leftrightarrow \sum_{i \notin \operatorname{supp} x_0} \theta|\eta(i)| - \sum_{i \in \operatorname{supp} x_0} \theta \operatorname{sgn}(x_0(i))\eta(i) \leq 0.$$

Dividing by $\theta > 0$ and estimating $\operatorname{sgn}(x_0(i))\eta(i) \leq |\eta_i|$ yields the claim.

---

[*]A dual certificate is, put simply, a vector in the range of A^* with certain properties whose existence implies that a vector x_0 is the solution of a convex optimization problem. The properties of course depend on which problem is considered.

(i) $\Rightarrow$ (ii). Now, suppose that A does not have the s-NSP. We will construct a vector x_0 which is not recovered by Equation 1.2. Due to the definition of the NSP, there exists an $\eta \neq 0$ with $A\eta = 0$ and an S with $|S| \leq s$ such that

$$\sum_{i \in S} |\eta_i| \geq \sum_{i \notin S} |\eta_i|. \tag{1.8}$$

If we define

$$x_0(i) = \begin{cases} \eta(i) & : i \in S, \\ 0 & : \text{else} \end{cases} \quad \text{and} \quad x(i) = \begin{cases} -\eta(i) & : i \notin S, \\ 0 & : \text{else}, \end{cases}$$

then x_0 is s-sparse, $Ax_0 = Ax$, and $\|x\|_1 \leq \|x_0\|$, due to Equation 1.8. Hence, x_0 is not uniquely recovered by Equation 1.2, and the proof is finished.

Let us next analyze the relation between the NSP and the RIP.

Theorem 1.3.5 ([33, Thm. 6.9]) *If the $2s$-th restricted isometry constant of A obeys $\delta_{2s} \leq \frac{1}{3}$, then A satisfies the s-NSP. In particular, Basis Pursuit will uniquely recover any s-sparse vector from its measurements Ax_0.*

There are two important facts about the NSP and the RIP to notice.

Remark 1.3.6

1. The NSP has a nice geometrical interpretation, as can be seen in Figure 1.1. For a function $f : \mathbb{R}^n \to \mathbb{R}$ and a point $x_0 \in \mathbb{R}^n$, define the *descent cone* $\mathcal{D}(f, x_0)$ as the set of descent directions at x_0:

$$\mathcal{D}(f, x_0) = \{v \in \mathbb{R}^n : \exists\, t > 0 \text{ such that } f(x_0 + tv) \leq f(x_0)\}.$$

A careful analysis of the proof of Lemma 1.3.4 shows that it in fact provides an argument for the fact that

$$\mathcal{D}(\|\cdot\|_1, x_0) = \left\{ v \in \mathbb{R}^n : \sum_{i \notin \operatorname{supp} x_0} |v(i)| \leq - \sum_{i \in \operatorname{supp} x_0} \operatorname{sgn}(x_0(i))v(i) \right\}.$$

This formula makes it clear that the NSP is equivalent to the statement that $\mathcal{D}(\|\cdot\|_1, x_0) \cap \ker A = \{0\}$ for every s-sparse x_0. This geometrical characterization of success of programs such as the one given by Equation 1.2 is directly used for many results. Together with methods from high-dimensional geometry, it is possible to provide very elegant arguments about statements on how many Gaussian measurements (see Section 1.3.3) are needed to guarantee unique recovery with high probability. The seminal work in this direction is [2].

2. The RIP is not equivalent to the success of Equation 1.2, see the discussion in [33, p. 169].

As has been mentioned before, it was proven that regularized Basis Pursuit is stable. To be precise, we speak of stability if the solution x_* of the program Equation 1.3 with input $b = Ax_0$ is guaranteed to satisfy

$$\|x_* - x_0\|_2 \leq C_1 \frac{\sigma_s(x_0)_1}{\sqrt{s}} + C_2 \epsilon \tag{1.9}$$

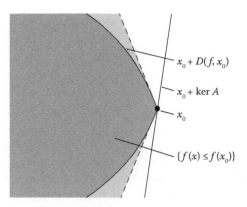

$x_0 + D(f, x_0)$

$x_0 + \ker A$

x_0

$\{f(x) \leq f(x_0)\}$

FIGURE 1.1

An illustration of the situation $\ker A \cap \mathcal{D}(f, x_0) = \{0\}$. Note that there does not exist any vector $x \in x_0 + \ker A$ which satisfies $f(x) < f(x_0)$. Hence, x_0 solves the program $\min f(x)$ subject to $Ax = Ax_0$.

for every $x_0 \in \mathbb{R}^n$, where $\sigma_s(x_0)_1$ denotes the *s-term ℓ_1-approximation error*, that is,

$$\sigma_s(x_0)_1 = \min_{x' \text{ s-sparse}} \|x_0 - x'\|_1.$$

Indeed, a statement like Equation 1.9 holds under the assumption $\delta_{2s} \leq 4/\sqrt{41}$. The strategy resembles the one above: First, it is proven that $\delta_{2s} \leq 4/\sqrt{41}$ implies A to satisfy the so-called *robust null space property* (Robust NSP), namely that

$$\|x_S\|_1 \leq \rho \|x_{S^c}\|_1 + \tau \|Ax\|_2$$

for some $0 < \rho < 1$ and $\tau > 0$, with ρ and τ only dependent on δ_{2s}. It is then argued that the Robust NSP implies stability of Equation 1.3 . It should again be emphasized that one of the first articles on Basis Pursuit [18] derives a similar assertion under the assumption $\delta_{3s} + 3\delta_{4s} < 2$ without taking the detour to the Robust NSP.

1.3.2 Mutual coherence

Another quality measure for matrices of high relevance in the setting of compressed sensing is the so-called *mutual coherence*. It was first introduced in [29] the special case of $A = [\Phi_1, \Phi_2]$, where each matrix Φ_i is an orthogonal matrix, but has then been adopted accordingly for general matrices. For the sake of simplicity of the exposition, we assume that all matrices in this section have ℓ_2-normalized columns. Then, the definition of mutual coherence reads as follows:

Definition 1.3.7 *Let a_i denote the i-th column of the matrix A. Then the* mutual coherence *of A is defined by*

$$\mu(A) = \max_{i \neq j} |\langle a_i, a_j \rangle|.$$

Small mutual coherence of a matrix is a strong property. For instance, a low mutual coherence implies the RIP. The following proposition is a simplification of a theorem from [72].

Proposition 1.3.8 *For every matrix A and positive integer s, we have*

$$\delta_s(A) < (s-1)\mu(A).$$

In particular, $\text{spark}(A) > (1 + \mu(A)^{-1})$.

Sketch of proof. Considering the equivalence of spectral properties of submatrices of A with the RIP (expressed in Equation 1.6), we have to analyze the eigenvalues of the matrices $M^S = A_S^* A_S$ for $S \subseteq \{1, \ldots, n\}$ arbitrary with $|S| \leq s$. The entries of M^S are given by $M_{ij}^S = \langle a_i, a_j \rangle$. Hence, the diagonal entries M_{ii}^S are all equal to 1, due to the assumed unit length of the columns. Besides, all other entries uniformly are bounded in magnitude by $\mu(A)$. Now, one can apply the *Gershgorin Disk Theorem* [35], which states that all eigenvalues λ of a matrix M lie in one of the circles $\mathcal{C}_i \subseteq \mathbb{C}$ with center $M_{ii} = 1$ and radius $R_i = \sum_{j \neq i} |M_{ij}|$. This implies that the eigenvalues λ of M^S all obey

$$|\lambda - 1| \leq \sum_{j \in S \setminus \{i\}} |M_{ij}^S| \leq (s-1)\mu,$$

which is the statement we needed to prove.

The concept of mutual coherence has a major drawback: There is a fundamental lower bound, which is referred to as *Welch Bound* [78]. In fact, the mutual coherence of a matrix $A \in \mathbb{R}^{m,n}$ obeys

$$\mu(A) \geq \sqrt{\frac{n-m}{m(n-1)}}. \tag{1.10}$$

If, via Proposition 1.3.8, we intend to guarantee that $\delta_s(A) \leq c$, we require $\mu(A) \leq c(s-1)^{-1}$. The Welch bound Equation 1.10 then enforces

$$\sqrt{\frac{n-m}{m(n-1)}} \leq \frac{c}{s-1} \iff m\left(1 + \frac{(s-1)^2}{n}\right) \geq \frac{(s-1)^2}{c^2(1 - \frac{1}{n})}.$$

Thus, if we assume a regime where $n \gtrsim s^2$ (i.e., $n \geq Cs^2$ for some constant C), we require an order of s^2 measurements to—via the mutual coherence—ensure that $\delta_s(A)$ is small, which is needed for ensuring success of Basis Pursuit for s-sparse signals. Ideally, we would like to remove the square exponent, so that the number of required measurements essentially scales as the degree of sparsity s. It is fortunately possible to directly prove that certain types of random matrices in $\mathbb{R}^{m,n}$ satisfy the RIP with m in that range. We will return to this issue in Section 1.3.3.

With the last paragraph in mind, one could argue that one should avoid using the notion of mutual coherence for analyzing the performance of algorithms, and instead concentrate on the RIP or NSP directly. There is however one situation when the RIP fails, and that is when analyzing the performance of OMP. Indeed, there exist many matrices having the s-RIP for $s \sim n$ for which there exist s-sparse vectors x_0 such that the first step of OMP fails to pick a correct index (see [27, Sec. 7]). In contrast, we may however prove statements guaranteeing good performance of OMP involving the mutual coherence, such as the following one.

Theorem 1.3.9 (Simplified version of Theorem A and B from [72]) *If $\mu(A)(2s-1) < 1$, then OMP succeeds at recovering every s-sparse signal x_0 from the input $b = Ax_0$ in s steps.*

Proof. Let $S = \operatorname{supp} x_0$. It suffices to prove that, for $k = 1, \ldots, s$, we have

$$\max_{i \in S} |\langle a_i, r_k \rangle| > \max_{i \notin S} |\langle a_i, r_k \rangle|. \tag{1.11}$$

Then, at least one of the indices in S will be chosen ahead of the ones in S^c. Since, due to step (3) of algorithm OMP, r_k will be orthogonal to all columns a_i with $i \in S^k$, we will not choose any index twice. Hence, if Equation 1.11 is true for all $k \geq s$, S^s will equal S.

Since, by assumption, $b \in \operatorname{ran} A_S$ and A_S will be injective (by $\mu(A)(2s-1) < 1$ together with Proposition 1.3.8), we can conclude that $x_0 = \operatorname{argmin}_{\operatorname{supp} x \subseteq S} \|Ax - b\|_2$.

We will now prove Equation 1.11 by induction. Since the argument for $k = 1$ is the same as for the induction argument $k \to k+1$, we will only perform the latter. Let S_k denote the kth support iterate. Due to induction, we can assume that $S_k \subseteq S$. Therefore, we obtain $r_k = Ax_0 - Ax_k \in \operatorname{ran} A_S - \operatorname{ran} A_{S_k} \subseteq \operatorname{ran} A_S$, that is, $r_k = \sum_{i \in S} r_k(i) a_i$ for some $r_k(i) \in \mathbb{R}$. Consequently,

$$|\langle a_i, r_k \rangle| = |\sum_{j \in S} r_k(j) \langle a_i, a_j \rangle| \begin{cases} \geq |r_k(i)| \langle a_i, a_i \rangle - \sum_{j \in S \setminus \{i\}} |r_k(j)| |\langle a_i, a_j \rangle| &: i \in S, \\ \leq \sum_{j \in S} |r_k(j)| |\langle a_i, a_j \rangle| &: i \notin S, \end{cases}$$

$$\begin{cases} \geq |r_k(i)| - (s-1) \mu(A) \max_{j \in S} |r_k(j)| &: i \in S, \\ \leq s \mu(A) \max_{j \in S} |r_k(j)| &: i \notin S. \end{cases}$$

Hence, if we let $i^* := \operatorname{argmax}_{j \in S} |r_k(j)|$, we obtain

$$\max_{i \in S} |\langle a_i, r_k \rangle| \geq |\langle a_{i^*}, r_k \rangle| \geq (1 - (s-1) \mu(A)) \max_{j \in S} |r_k(j)|$$

$$\geq s \mu(A) \max_{j \in S} |r_k(j)| \geq \max_{i \notin S} |\langle a_i, r_k \rangle|,$$

where we in the second to last step used the assumption on μ. The proof is finished.

Theorem 1.3.9 leaves us with the impression that OMP will only work for matrices with low coherence, which inevitably will force us to use $\Omega(s^2)$ measurements. This is however not the entire picture. In fact, if we move away from the uniform regime, that is, drop the goal to construct matrix A so that OMP will reconstruct *any* s-sparse signal, we can indeed circumvent the above s^2-bottleneck and prove significantly more satisfying results. To be precise, it is possible to prove that given some s-sparse x_0, certain random matrix A will have the property that *this specific* x_0 can be reconstructed from the measurements Ax_0 using OMP already when $m \gtrsim s$. The interested reader is referred to [74].

1.3.3 Random matrices

It has turned out to be very hard to deterministically construct matrices in $\mathbb{R}^{m,n}$ which can be proven to satisfy the RIP for reasonable values of m, n, and s. Most arguments in the literature for proving that deterministic constructions have the RIP rely on the mutual coherence, or alternatively the Gershgorin Disk Theorem. As we already discussed, the Welch bound then forces us into an $m \gtrsim s^2$-domain, which is far from desirable. This is known as the "square-root bottleneck," since it implies that we only can certify recovery of $\sqrt{m}$-sparse signals using m measurements.

The square-root bottleneck has probably only been broken once deterministically. This was done by Bourgain et al. [13]. By using very sophisticated arguments, they constructed $(m \times n)$-matrices with $n^{1-\epsilon_0} \leq m \leq n$ that satisfied the s-RIP for $s = O(m^{1/2+\epsilon_0})$ with restricted isometry constant $n^{-\epsilon_0}$. In their approach, ϵ_0 is a positive constant whose numerical value has later been proven to be larger than 4.44×10^{-24} [59]. Hence, although proving the result certainly is an achievement, its practical relevance is limited.

Fortunately, one can achieve considerably better results when going over to random constructions. In this section, we will first consider the important family of *sub-Gaussian* matrices, which already cover most well-known examples. Later, we will also briefly describe the similarly interesting model of *bounded orthonormal systems*, which, for instance, includes random partial Fourier matrices as a special case.

There are several definitions as to what a sub-Gaussian matrix is, although the notion of a sub-Gaussian *variable* is used consistently throughout most of the literature. We choose the easiest one for simplicity reasons.

Definition 1.3.10

1. *A real-valued random variable X is called* sub-Gaussian, *if there exist C, κ called the* sub-Gaussian parameters *of A such that*

$$\mathbb{P}(|X| \geq t) \leq C \exp(-\kappa t^2), \quad t > 0.$$

   It is called subexponential *if there exist K, γ with*

$$\mathbb{P}(|X| \geq t) \leq K \exp(-\gamma t), \quad t > 0.$$

2. *A* sub-Gaussian *matrix A is a matrix whose entries are independent sub-Gaussian.*

Let us start by presenting some examples.

Example 1.3.11

1. Normally distributed variables are sub-Gaussian. In this case, we even speak of Gaussian variables. A *Gaussian matrix* is a matrix with independent identically normally distributed entries.

2. All bounded distributions are sub-Gaussian. In particular, a matrix whose entries are independent Bernoilli distributed variables is sub-Gaussian.

3. If X is sub-Gaussian, X^2 is subexponential, that is,

$$\mathbb{P}(|X|^2 \geq t) = \mathbb{P}\left(|X| \geq \sqrt{t}\right) \leq C \exp(-\kappa t).$$

Sub-Gaussian matrices satisfy the RIP *with high probability* (*whp.*) already when $m \gtrsim s \log\left(\frac{n}{s}\right)$. An exact statement is provided by the following theorem, which is a slight simplification of [76, Thm. 5.65]. For this, the reader might want to recall the definitions of the *mean* $\mu = \mathbb{E}(X)$ and *variance* $\sigma^2 = \mathbb{E}\left((X - \mathbb{E}(X)^2)\right)$ of a random variable.

Theorem 1.3.12 *Let $A \in \mathbb{R}^{m,n}$ be a sub-Gaussian matrix, where each entry is assumed to have mean zero and unit variance, and let $\delta \in (0,1)$. For every $s = 1, \ldots n$, the normalized matrix $\hat{A} = \frac{1}{\sqrt{m}} A$ has an s-th restricted isometry constant $\delta_s(A) \leq \delta$ with probability of at least $1 - \exp(-cm\delta^2)$, provided that*

$$m \geq C\delta^{-2} s \log\left(\frac{en}{s}\right).$$

The constants only depend on the sub-Gaussian parameters of A.

There are many ways to prove Theorem 1.3.12. Let us sketch one variant. The interested reader is referred to [76], in particular, to the proof of Theorem 5.39 in mentioned reference, for details. The argument has three steps three steps (see Figure 1.2).

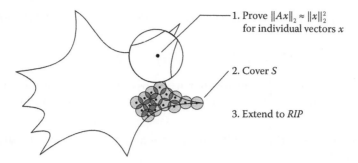

FIGURE 1.2
The structure of the proof of the *RIP* for random matrix A sketched in this section.

1. **Prove that for each (sparse) x_0 we have $\|Ax_0\|_2^2 \approx \|x_0\|_2^2$ with high probability.** Let x_0 be a fixed, unit-norm vector and consider the random variable $Z = \|Ax_0\|_2^2 - \|x_0\|_2^2$. Further, let the rows of A be denoted by A_i. Then, we have

$$Z = \frac{1}{m} \sum_{i=1}^m (\langle A_i, x_0 \rangle^2 - \|x_0\|_2^2) =: \frac{1}{m} \sum_{i=1}^m Y_i,$$

where we defined

$$Y_i = \left(\langle A_i, x_0 \rangle^2 - \|x_0\|_2^2 \right) = \sum_{j=1}^n (A_{ij}^2 - 1) x_0(j)^2 + \sum_{j \neq k} A_{ij} A_{ik} x_0(j) x_0(k).$$

Due to the independence, zero mean, and unit variance of the variables A_{ij}, we can conclude that $\mathbb{E}(Y_i) = 0$. The random variable Y_i is furthermore a subexponential variable as a sum of squares of sub-Gaussians. In our concrete case, the parameters C and κ are in addition only dependent on the sub-Gaussian parameters of A and the Euclidean norm of x_0, which is assumed to be equal to 1. The family $(Y_i)_{i=1}^m$ is also independent, again due to the corresponding properties of the entries of A.

We have hence argued that $\sum_{i=1}^m Y_i$ is a sum of centered, independent subexponential random variables. Such sums obey the *Bernstein inequality*, which can be stated as

$$\mathbb{P}\left(|\sum_{i=1}^m Y_i| \geq t \right) \leq 2 \exp\left(\frac{-(\beta t)^2/2}{2\alpha m + \beta t} \right),$$

where α and β are constants only dependent on the subexponential parameters of the Y_i. Consequently, for every $\tau \in (0,1)$, we have

$$\mathbb{P}\left(|\|Ax_0\|_2^2 - \|x_0\|_2^2| \geq \tau \right) = \mathbb{P}\left(|\sum_{i=1}^m Y_i| \geq m\tau \right) \tag{1.12}$$
$$\leq 2 \exp\left(\frac{-(m\beta\tau)^2/2}{2\alpha m + m\beta\tau} \right) \leq 2 \exp\left(-cm\tau^2 \right),$$

where $c = \frac{\beta^2}{4\alpha + 2\beta}$ only depends on the sub-Gaussian parameters of A. In the last step, we used that $\tau \in (0,1)$.

2. **Covering argument.** Now, consider a set $\mathcal{N}$ consisting of unit vectors with the property that $\text{dist}(x, \mathcal{N}) \leq \rho$ for all s-sparse vectors x with unit norm, where ρ

is some small positive number. Utilizing Equation 1.12 and a union bound, we obtain

$$\mathbb{P}\left(\forall\ x \in \mathcal{N} : |\|Ax\|_2^2 - \|x\|_2^2| \geq \tau\right) \leq 2|\mathcal{N}| \exp\left(-\gamma m\tau^2\right).$$

It is furthermore not hard to prove that it is possible to construct a net $\mathcal{N}$ with the desired property and $|\mathcal{N}| \leq \binom{n}{s}(1 + \frac{2}{\rho})^s$ (see, for instance, [76, Lem. 5.2 and Thm. 5.65]).

3. **Extend to the RIP.** By carefully choosing ρ and τ, we can now finally extend Equation 1.12 to

$$\mathbb{P}\left(\forall\ x \text{ unit-norm, } s\text{-sparse} : |\|Ax\|_2^2 - \|x\|_2^2| \geq \delta\right)$$

$$\leq 2\binom{n}{s}(1 + \tfrac{2}{\rho})^s \exp\left(-\gamma m(1 - 2\rho)^2\delta^2\right),$$

where the latter expression is smaller than ϵ provided that

$$m \geq (\gamma(1 - 2\rho)^2)^{-1}\delta^{-2}s\log(1 + \tfrac{2}{\rho})\log\left(\binom{n}{s}\epsilon^{-1}\right) \asymp C\delta^{-2}s\log\left(\tfrac{en}{s}\right).$$

The special case of Gaussian matrices has a particularly satisfying universality property: If U is an orthogonal matrix and A is Gaussian, AU is also Gaussian. This has the consequence that all guarantees holding for recovery of sparse vectors in the standard basis hold true for sparse vectors in any other orthogonal matrix. Note that although a similar transformation of a sub-Gaussian matrix yields again a sub-Gaussian, the parameters may change considerably.

1.3.3.1   Bounded orthonormal systems

Gaussian random matrices is to some extent the "least structured" random matrices possible. In some applications, it may be a simplified but still reasonable model. But often, there is much more structure to be accounted for. An important example are coefficients $\langle x_0, \phi_i \rangle$ of x_0 which are randomly sampled with respect to an orthonormal system $(\phi_i)_{i=1}^n$ of $\mathbb{R}^n$. A quite general model for such schemes is provided by the concept of *bounded orthonormal systems* [64].

Definition 1.3.13 *Let* $\Omega \subseteq \mathbb{R}^n$ *be endowed with a probability measure* ν. *A family of functions* $\phi_j : \Omega \to \mathbb{C}$ *is called a* bounded orthonormal system *(BOS), if*

1. (ϕ_j) *is orthonormal with respect to* ν, *that is,*

$$\int_\Omega \phi_j(t)\overline{\phi_k(t)}d\nu(t) = \delta_{j,k}.$$

2. (ϕ_j) *is bounded in* $L^\infty(\nu)$; *that is, there exists a constant* $K > 0$ *with*

$$\max_j \sup_{t \in \Omega} |\phi_j(t)| \leq K.$$

Let us next discuss two key examples.

- *Fourier Basis.* For $k \in \mathbb{Z}$, let (φ_k) denote the kth Fourier polynomial on $[0, 1]$, that is,

$$\varphi_k(t) = \exp(2\pi i k t).$$

Then, $(\varphi_k)_{k=-N}^{N}$ is a BOS in the space of trigonometric polynomials of degree at most N with respect to the uniform measure on $[0,1]$, since

$$\int_0^1 \phi_j(t)\overline{\phi_k(t)}dt = \int_0^1 \exp(2\pi i(j-k)t)dt = \delta_{j,k}, \quad |\phi_j(t)| \leq 1, t \in [0,1].$$

- *Discrete Systems.* Let $(\psi_k)_{k=1}^{n}$ be an orthonormal system of $\mathbb{C}^n$ and υ any discrete probability measure on $\Omega = \{1, 2, \ldots, n\}$. Then the set of functions

$$\hat{\psi}_k(t) = \frac{1}{\sqrt{\upsilon(t)}}\psi_k(t)$$

is a BOS with respect to υ, since

$$\int_\Omega \phi_k(t)\overline{\phi_\ell(t)}d\upsilon(t) = \sum_{t=1}^{n} \frac{1}{\sqrt{\upsilon(t)}}\psi_k(t)\frac{1}{\sqrt{\upsilon(t)}}\overline{\psi_\ell(t)}\upsilon(t) = \sum_{t=1}^{n}\psi_k(t)\overline{\psi_\ell(t)} = \delta_{k,l}$$

and

$$\max_k \max_t \frac{1}{\sqrt{\upsilon(t)}}\psi_k(t) := K < \infty.$$

By choosing υ wisely, we can make the constant K small.

The model of BOS is compatible with sparse recovery of signals of the form

$$x(t) = \sum_{k=1}^{n} \upsilon_k\phi_k(t),$$

where ϕ_n is a BOS and $c \in \mathbb{C}^n$ is sparse. Let us now take m samples $x(t_i)$ of x, where t is chosen according to υ. This corresponds to measuring the sparse vector x with the help of the random matrix A, with

$$F_{i,n} = \phi_n(t_i), \quad t_i \sim \upsilon. \tag{1.13}$$

We know that provided the matrix F satisfies the RIP, we can reconstruct c and thus also x. Fortunately, this is the case due to the following result.

Theorem 1.3.14 ([64, Thm. 8.1]) *Let $F \in \mathbb{C}^{m,n}$ be the sampling matrix defined by Equation 1.13 associated with an BOS with constant $K \geq 1$. Further, suppose that for some $\epsilon > 0$ and $\delta \in (0, \frac{1}{2}]$, we have*

$$\frac{m}{\ln(m)} \geq DK^2\delta^{-2}s\log^2(100s)\log(4n)\log(7\epsilon^{-1}),$$

where D is a universal constant. Then, with probability at least $1 - \epsilon$, the sth restricted isometry constant of $\frac{1}{\sqrt{m}}F$ satisfies $\delta_s \leq \delta$.

We would like to stress that the proof of Theorem 1.3.14 uses much more sophisticated methods than the simple argument we presented for Gaussian matrices. It is still possible to prove that $\|Fx\|_2^2 \approx \|x\|_2^2$ for individual, sparse vectors using slightly more sophisticated (matrix) concentration inequalities (an excellent introduction to such inequalites can be found in [73]), but the resulting probability bounds are not small enough to make the covering strategy work. The proof of Theorem 3.14, together with stronger, nonuniform recovery guarantees, can be found in [64].

1.4   Some exemplary applications

A decade after the publication of the pioneering articles of Tao, Candès, Donoho, and their co-authors, compressed sensing is already changing the way engineers and researchers from other application areas are thinking about signal processing. In this final section, we would like to present several prototype applications that make use of compressed sensing ideas. For this purpose, we will first present some generalizations of compressed sensing which are used in the examples.

In the previous part of this chapter, we always considered the problem of recovering a signal x_0 which is sparse by itself, at least when represented in a basis. However, there exists many situations where this is not the case, but the signal still enjoys a certain structure which can be used as prior for compressed sensing–like approaches. It could also be that x_0 is sparse, but that the sparsity pattern exhibits some additional structure. The following list summarizes some of the most important structural constraints:

- *Dictionary sparsity*: In many situations, $x_0 \in \mathbb{R}^n$ can be sparsely represented by a system of vectors, $\Gamma = (\gamma_i)_{i=1}^K$, that is, that

$$x_0 = \sum_{i=1}^K c_i \gamma_i, \tag{1.14}$$

where $c = (c_i)_{i=1}^K$ is a sparse coefficient vector. Thereby, the system $\Gamma = (\gamma_i)_{i=1}^K$ may be overcomplete. This means that although Γ spans the signal space, it does not necessarily need to be linearly independent. Note that we have $K \geq n$ in this scenario. Then, one often speaks of *dictionaries* Γ. The interested reader is referred to [17] for more details on introduction to the theory of compressed sensing using dictionary sparsity.

A particularly interesting choice of the dictionary Γ is a so-called *frame* [22]. A family of vectors is called a *frame* if the spectrum of the *frame operator* $\sum_{i=1}^K \gamma_i \gamma_i^*$ is contained in some interval $[\alpha, \beta]$ with $\alpha, \beta > 0$. If α and β are both close to one, then we have $\sum_{i=1}^K \gamma_i \gamma_i^* \approx \mathrm{id}$ and hence

$$\forall x_0 \in \mathbb{R}^n : x_0 \approx \sum_{i=1}^K \gamma_i \langle \gamma_i, x_0 \rangle.$$

This motivates that although this in general will not give us a sparse representation in the frame in the sense of Equation 1.14, a sparse coefficient vector $(\langle \gamma_i, x_0 \rangle)_{i=1}^K$ is a prior which could be used for a compressed sensing–based strategy[*]. This is indeed the case. One then speaks of *cosparse* signals. A detailed description of this framework which is called *analysis formulation* can be found in [41]. A big advantage of this approach is that it is often easier to guarantee that the sequence $(\langle \gamma_i, x_0 \rangle)_{i=1}^K$ is sparse for x_0 in some signal class $\mathcal{C}$ than to prove that a sparse representation as in Equation 1.14 exists for the same class.

- *Group sparsity*: Suppose that the index set $I = \{1, ..., N\}$ of our vector x_0 is subdivided into K disjoint subsets I_k of size ℓ with the property that provided one entry $x_0(i)$ with $i \in I_k$ is nonzero, then probably the other entries of the group I_k

---

[*]A more precise motivation is given by the concept of *dual frames* [22].

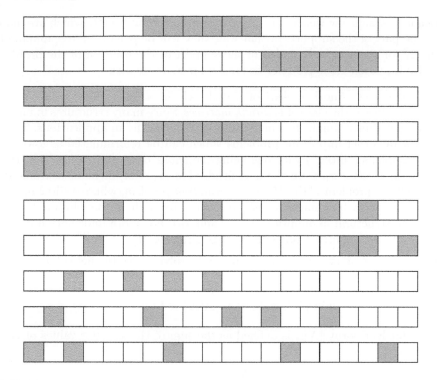

FIGURE 1.3
The difference between group sparsity (above) and classical sparsity (below).

are as well—but most entries of x_0 are still equal to zero. One could think of a set of users in a transmission network, each having ℓ antennas which they all exploit while actively transmitting, but they do this very sparsely in time. This type of sparsity is known as *group sparsity*. Figure 1.3 illustrates this scenario.

One approach to recover such types of signals is to view the vector x_0 as a matrix $X^0 \in \mathbb{R}^{\ell,K}$ with jth column X_j^0 equal to $x_0|_{I_j}$ and to solve the $\ell_{1,2}$-minimization problem given by

$$\min \sum_{j=1}^{k} \|X_j\|_2 \quad \text{subject to } \mathcal{A}(X) = b,$$

where $\mathcal{A}$ thereby denotes the linear map corresponding to the map $x \mapsto Ax$ when applied to the vectors written in matrix form. For an introduction to this approach and some initial results, see [31].

- *Nonuniform sparsity*: Suppose again that we can subdivide our index set into K subsets I_k, but now with the property that it is known to be more likely for indices in certain sets I_k to be nonzero. Again, we may think of a transmission network: It could happen that some of the users transmit more often than others. A convex programming approach to recover these signals is to solve a weighted ℓ_1-minimization problem of the form

$$\min_{w_k} \sum_{k=1}^{K} \sum_{i \in I_k} w_k |x(i)| \text{ subject to } Ax = b. \tag{1.15}$$

The idea is to assign low weights $w_k > 0$ to the groups of indices I_k which are more likely to be part of the support. In that way, they will be penalized less, making it more likely for them to be included in the support of the solution of the problem defined by Equation 1.15. We refer to [42,62] for more details.

- *Low-rank matrices*: Another class of signals which is quite different from the previous examples is the class of *low-rank matrices*. Since many structures in nature tend to be low dimensional, such matrices appear naturally in applications. There also exist some other important problems which can be reformulated as low-rank recovery problems ("lifted") with one being the *phase retrieval problem* [19], that is, recovering a vector x_0 from the magnitudes $|\langle a_i, x_0 \rangle|$ of linear measurements. Another related problem is the *blind deconvolution problem*, where a filter h_0 and a signal x_0 need to be simultaneously recovered from their convolution $h_0 * x_0$ (under some structural assumptions on h_0 and x_0). For more information on the latter, as well as the "lifting procedure" in general, see [1].

The low-rank matrix recovery problem can be solved with the help of a *nuclear-norm* minimization

$$\min \|X\|_* \text{ subject to } \mathcal{A}(X) = b,$$

where the nuclear norm $\|X\|_*$ is defined as the sum of the singular values of X:

$$\|X\|_* = \sum_{i=1}^{n} \sigma_i(X).$$

The foundations of the theory of low-rank matrix recovery using this program can be found in [15].

With the above generalized versions of compressed sensing, we are now ready to discuss some exemplary applications.

1.4.1   Compressed medical imaging

Compressed sensing is being actively pursued for medical imaging, particularly in magnetic resonance imaging (MRI). Put very simply, an MRI machine exposes the object to be studied to a strong, varying magnetic field. This causes the spin of the nuclei of (primarily) hydrogen atoms present in the object to align with the external magnetic field. By measuring to which extent this happens, one can conclude in which region much hydrogen, thus much water, is present.

From a mathematical perspective, we can regard the MRI machine as a device which samples the Fourier transform

$$\mathscr{F}(f)(\xi) = \frac{1}{(2\pi)^{n/2}} \int f(x) e^{2\pi i \langle \xi, x \rangle} dx, \quad \xi \in \mathbb{R}^N$$

of the object of interest f at certain frequencies ξ_k. If we discretize f as an n-dimensional vector, this corresponds to taking samples of the discrete Fourier transform (DFT) Ff, that is,

$$(Ff)_k = \langle f, \varphi_k \rangle, k = 1, \ldots, n,$$

with $(\varphi_k)_{k=1,\ldots,n}$ being the Fourier basis of $\mathbb{C}^n$. If we could sample all values of the DFT, it would be trivial to reconstruct the object f from the measurements Ff. This is unfortunately not always the case, since an MRI measurement process is very expensive and time consuming. During the scan, the patient is required to retain his or her body in a fixed position, which is not easy, especially for children or patients suffering from claustrophobia. This naturally leads us to considering compressed sensing as a mean of reducing the number of samples needed. Several compressed sensing algorithms have in fact been specifically designed for this problem [34,40,53,54,79].

In order to apply the tools of compressed sensing, we have to take some care. An MRI image of, for instance, a brain is in fact not sparse in the space domain—most pixels in the image are not equal to zero. Its wavelet transform is however at least approximately sparse, since an image is typically smooth with the exception of a few line singularities. We can therefore assume that

$$f = \sum_{\ell=1}^{n} c_\ell \psi_\ell,$$

where $(\psi_\ell)_{\ell=1}^n$ is the wavelet basis and the vector $c \in \mathbb{R}^n$ is sparse. The Fourier samples then take the form

$$\langle f, \varphi_k \rangle = \sum_{\ell=1}^{n} c_\ell \langle \psi_\ell, \varphi_k \rangle, \quad k = 1, \ldots, n.$$

Hence, we need to recover the (approximately) sparse vector c from measurements $y = Ac$, where

$$A_{k,\ell} = \langle \psi_\ell, \varphi_k \rangle =: \phi_\ell(k).$$

This leads to the following ℓ_1-minimization problem:

$$\min_c \|c\|_1 \text{ subject to } \|y - Ac\|_2 \leq \epsilon.$$

The system of functions $(\phi_\ell(\cdot))_{\ell=1}^N$ is an orthonormal basis of $\mathbb{C}^n$, since

$$\langle \phi_\ell, \phi_j \rangle = \sum_{k=1}^{n} \langle \psi_\ell, \varphi_k \rangle \langle \varphi_k, \psi_j \rangle = \langle \psi_\ell, \psi_j \rangle = \delta_{\ell,j}$$

for which we used the orthogonality of the discrete Fourier system and the wavelet basis. Hence, the theory of bounded orthonormal systems (Subsection 1.3.3.1) applies. It turns out that in order to get a good constant K, one should sample the k not uniformly, but instead sample low frequencies with a higher probability. One typical example of a sampling density is

$$\nu(k) = \frac{C}{k_1^2 + k_2^2}.$$

Since it is very important that a large part of the low frequency part is fully sampled, Krahmer and Ward [45] have proposed using a version of the above probability distribution which is "leveled" close to the origin;

$$\eta(k_1, k_2) = C_N \min\left(C, \frac{1}{k_1^2 + k_2^2}\right),$$

where C is an absolute constant and C_N is chosen such that η is a probability distribution [45].

In [45,48,54], the authors have proposed theoretical formulations based on randomly sampling points. In practice, however, the samples have to follow continuous lines which makes random isolated measurements infeasible. It is possible to analyze compressed sensing reconstructions from random sampling of lines in the k-space using so-called block models (see [8]).

The reason why compressed sensing strategies should work for MRI is the wavelet sparsity prior. However, wavelets are not optimal to detect curvilinear singularities. In fact, images of the type we mentioned above, that is, which are smooth with the exception of a few singularities of curvilinear type, become optimally sparse when mapped by a so-called *shearlet transform* [49]. Shearlets are a recent type of directional representation systems— an overview can be found in the book [47]. Indeed, it is possible to achieve significantly better recovery results with shearlets than with wavelets, since the transformed signals are sparser in the former case [55]. Since shearlets do not form orthonormal systems, the theory of Section 3 is not applicable in this situation. It is however possible to generalize the results also to this setting, by carefully adjusting the sampling pattern (see Figure 1.4). For more details, we refer the interested reader to [48].

1.4.2   Data separation

Given a composed signal x of the form $x = x_1 + x_2$, the data separation problem can be defined as extracting the unknown components x_1 and x_2 from it. Solving this problem seems to be impossible, since there are infinitely many choices of x_1 and x_2 for a problem with one known data and two unknowns. However, by leveraging the concept of compressed sensing and appropriately incorporating the sparsity property, this challenge becomes feasible.

To extract the two components from x, some knowledge of the structure of those components should be given as a prior information. This additional information allows us to have sparse expansions of x_1 and x_2, by choosing two representation systems, Φ_1 and Φ_2, respectively. Such representation systems might be chosen from the collection of well-known systems such as wavelets. A different possibility is to design the systems adaptively via dictionary-learning procedures. In this section, we focus on the data separation problem with orthonormal bases and frames as sparsifying representation systems [25, Chp. 11; 46].

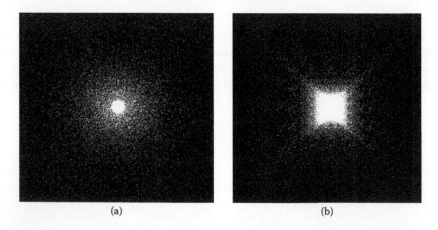

(a)   (b)

FIGURE 1.4

A figure demonstrating the difference between the (a) variable sampling density scheme to be used with wavelets (proposed by Krahmer and Ward [45]) and (b) directional sampling scheme to be used with shearlets (proposed by Kutyniok and Lim [48]).

1.4.2.1 Orthonormal bases

At first, we consider the case where Φ_1 and Φ_2 form orthonormal bases. Let us assume that we are given two such representation systems Φ_1 and Φ_2 for $\mathbb{R}^n$ such that the ground-truth coefficient vectors $\tilde{c}_i = \Phi_i^* x_i$ ($i = 1, 2$) are sparse, meaning that $\|\tilde{c}_1\|_0$ and $\|\tilde{c}_2\|_0$ are *sufficiently small*. This leads to the following underdetermined linear system of equations

$$x = [\Phi_1|\Phi_2] \begin{bmatrix} \tilde{c}_1 \\ \tilde{c}_2 \end{bmatrix}.$$

Hence, it is quite natural to encourage sparsity for both coefficient vectors. This can be realized by the following ℓ_1-minimization problem:

$$\min_{c_1,c_2} \|c_1\|_1 + \|c_2\|_1 \text{ subject to } x = [\Phi_1|\Phi_2] \begin{bmatrix} \tilde{c}_1 \\ \tilde{c}_2 \end{bmatrix}. \tag{1.16}$$

Then the data separation problem can be solved by computing $x_1 = \Phi_1(c_1)$ and $x_2 = \Phi_2(c_2)$, provided that the sparse vector $[c_1, c_2]^* = [\Phi_1^* x_1, \Phi_2^* x_2]^*$ can be truly recovered [25, Chp. 11; 46]. In addition to the sparsity of $[\Phi_1^* x_1, \Phi_2^* x_2]^*$, separation can only be achieved provided that the components x_1 and x_2 are in some sense *morphologically* distinct. Notice that this property is indeed encoded in the problem through the required incoherence of the matrix $[\Phi_1|\Phi_2]$. The reader is referred to [26,29,30] for more details.

1.4.2.2 Frames

Choosing redundant frames as representation systems for the components of interest, x_1 and x_2, is sometimes beneficial or even necessary. Note that some well-known representation systems are in fact redundant and typically constitute Parseval frames, such as curvelets or shearlets. In addition, systems generated by dictionary learning are typically highly redundant as well. Finally, we may not be able to find sparsifying orthonormal bases in a specific situation. In these cases, the minimization problem we stated in Equation 1.16 might cause problems.

More precisely, for each possible separation, $x = x_1 + x_2$, there exist infinitely many coefficient sequences $[c_1, c_2]^T$ satisfying $x_1 = \Phi_1 c_1$ and $x_2 = \Phi_2 c_2$, due to the redundancy of the frames. This particularly causes numerical instabilities if the redundancy of either frame is too high.

As discussed at the beginning of this chapter, we can circumvent this problem by applying the ℓ_1-norm *analysis* formulation, rather than the *synthesis* formulation. In other words, we select a particular coefficient sequence, called analysis coefficients, for each separation. Since Φ_1 and Φ_1 are assumed to be Parseval frames, that is, that $\Phi_i \Phi_i^* = I$ ($i = 1, 2$), we can write:

$$x = x_1 + x_2 = \Phi_1(\Phi_1^* x_1) + \Phi_2(\Phi_2^* x_2).$$

This leads to the following alternative ℓ_1-minimization problem [25, Chp. 11; 46]:

$$\min_{x_1,x_2} \|\Phi_1^* x_1\|_1 + \|\Phi_2^* x_2\|_1 \text{ subject to } x = x_1 + x_2$$

1.4.3 Analog-to-information converters

Analog-to-digital converters (ADC) are devices which use sampling to convert analog signals defined on a continuous domain to digital signals in a discrete time representation. Based on Shannon theory, the sampling frequency should be at least two times larger than the bandwidth of the signal, in order to avoid aliasing effects. Therefore, analog to digital communication systems face practical challenges when sensing wide-band spectrums in the order of a few GHz. The reason is that the Nyquist rate might exceed the sampling rate of current ADC devices. Besides, sampling at this high frequency results in huge power consumption [36].

However, in most applications such as cognitive radio (CR) networks, ultra-wideband communications, and radar signal detection, the part of the signal containing the information is much smaller than its bandwidth. In other words, the signal may be sparse in the frequency domain. Therefore, it seems conceivable that one can efficiently sample the wide-band signal by leveraging the concept of compressed sensing, without using highly complicated resources [44].

Analog-to-information conversion (AIC) [44] is a promising solution to this practical challenge. Let us illustrate this approach by explaining the mathematical model of this problem. Assume that an orthogonal basis $\Psi = \{\psi_1(t), \psi_2(t), ..., \psi_N(t)\}$ is given, under which the analog signal $x(t)$ can be sparsely represented by

$$x(t) = \sum_{i=1}^{N} c_i \psi_i(t).$$

AIC is then a combination of three components, namely a wide-band pseudo-random signal demodulator $p_c(t)$, a filter $h(t)$ (typically a low-pass filter), and a low-rate ADC which can be considered as uniform sampling [44,50]. The input of AIC is a signal $x(t), t \in \mathbb{R}$. At first stage, $x(t)$ is convolved by a wide-band pseudo-random signal demodulator $p_c(t)$. Then, it is passed through a low-pass filter $h(t)$, and finally sampled with a much slower rate than the Nyquist rate [44]. Put together, the output is the low-rate measurement y_i [4, p. 58]:

$$y_i = \int_{-\infty}^{\infty} x(\tau) p_c(\tau) h(t - \tau) d\tau |_{t=i\Delta}, \tag{1.17}$$

where Δ is the sampling period.

By plugging $x(t)$ into Equation 1.17, we obtain

$$y_i = \sum_{j=1}^{N} c_j \int_{-\infty}^{\infty} \psi_j(\tau) p_c(\tau) h(i\Delta - \tau) d\tau.$$

This shows that we can solve the ADC problem by using compressed sensing in the following way [39]:

$$\min \|c\|_1 \text{ subject to } y = \Phi c,$$

with the measurement matrix Φ, chosen as

$$\phi_{i,j} = \int_{-\infty}^{\infty} \psi_j(\tau) p_c(\tau) h(i\Delta - \tau) d\tau.$$

1.4.4   Communication and networking domain

Compressed sensing is also a promising tool to exploit sparsity properties of network and communication systems. In the following sections, we will provide a few prototype examples that are associated with this domain of applications.

1.4.4.1 Error correction

Data corruption is an ubiquitous issue in every realistic data transmission process due to distortions in the transmission channel. Therefore, it is essential to design mechanisms to detect and correct the error in the received data.

Assume that we aim to send a data vector $z \in \mathbb{R}^n$. One of the standard approaches is to use redundancy for sending the data. Hence, the encoded data $v = Bz \in \mathbb{R}^N$ with $N = n + m$ is transmitted with redundancy m compared to the data vector z itself. The receiving device measures the vector $w = v + x \in \mathbb{R}^N$, consisting of the transmitted signal v plus the transmission error x. By assuming that transmission errors happen infrequently, that is, that $\|x\|_0 < s$, we can exploit compressed sensing for reconstructing the data vector z as follows [33].

For decoding, a generalized checksum matrix $A \in \mathbb{R}^{m,N}$ has to be constructed such that $AB = 0$. Then, the standard compressed sensing measurement rule with the matrix A and the sparse error vector x can be achieved as follows [33]:

$$y = Aw = A(v + x) = ABz + Ax = Ax.$$

Under suitable conditions which have been described in the previous sections, one may recover the vector x as well as the originally transmitted vector v by $v = w - x$. Finally, by solving an overdetermined system $v = Bz$, the data vector z can be computed.

1.4.4.2 Sparse channel estimation

Wireless channels play a significant role in the performance of many wireless communication systems. Due to a number of environmental effects, such as scattering and fading, two or more delayed versions of the transmitted signal with severe fluctuations of the amplitudes and phases will arrive at the receiver. In general, these effects can be categorized as reflection, scattering, and attenuation, which causes the so-called fading phenomena and can be modeled as a multipath propagation channel response. To obtain the spatial and frequency diversity gain provided by multipath property of the wireless channels and to alleviate the side effect of the channel fading, knowledge of channel state information (CSI) is of significant importance for the transceiver design [36].

Channel estimation is referred to as the problem of system identification in which the channel H as a system should be estimated with sufficient accuracy, given the known transmitted signal $s(t)$ and the received signal $x(t)$ at time t. Many practical factors such as long signaling duration, large number of antennas, and operation at large bandwidths, enable sparse representation of multiple-input multiple-output (MIMO) channels in appropriate bases. This sparsity feature can also be understood from the fact that only a relatively small number of paths are dominant in rich multipath behavior of MIMO channels. Because of the sparsity in compressed sensing can be used in MIMO channels to improve the performance of channel estimation, wheres conventional techniques of training-based estimation, using LS methods, may not be an optimal choice.

Various studies have employed compressed sensing for sparse channel estimation. Compressed channel estimation (CCS) has been shown to achieve a much smaller reconstruction error while utilizing significantly less energy and, in some cases, less latency and bandwidth as well [3,63]. CCS gives much better reconstruction using its non-linear reconstruction algorithm as opposed to linear reconstruction of LS-based estimators. In addition to nonlinearity, the CCS framework also enables a scaling analysis [63].

For introducing the problem mathematically, we consider a MIMO channel with N_T transmitters and N_R receivers. Further, assume that the channel has a two-sided bandwidth of W and the signaling has a duration of T. Denoting $s(t)$ as the transmitted signal at time t, the received signal is given by the following equation without considering noise [3,36,39]:

$$x(t) = \int_{-W/2}^{W/2} h(t,f)\tilde{s}(f)e^{j2\pi ft}df,$$

where $\tilde{s}(f)$ is the element-wise Fourier transform of the transmitted signal $s(t)$ and $h(t,f)$ denotes the time-varying frequency response matrix, whose dimension is $N_R \times N_T$.

For a multipath channel, the frequency response can be usually represented in two different ways: The nonlinear physical model in Equation 1.18 and the linear virtual model [66,67] in Equation 1.19, which approximates the physical model expressed in Equation 1.18 for path n by uniformly sampling the physical parameter space $[\beta_n, \theta_{R,n}, \theta_{T,n}, \tau_n, \nu_n]$ at a resolution of $(\Delta\theta_{R,n}, \Delta\theta_{T,n}, \Delta\tau_n, \Delta\nu_n) = (1/N_R, 1/N_T, 1/W, 1/T)$. The parameters N_R, N_T, $L = \lceil W\tau_{max} \rceil + 1$, and $M = \lceil T\nu_{max} \rceil$ denote the maximum number of resolvable *angle of arrival* (AoA), *the angle of departure* (AoD), delays, and one-sided Doppler shifts, where $\theta_{R,n}$ is the AoA at the receiver, $\theta_{T,n}$ is the AoD at the transmitter, ν_n is the Doppler shift, and τ_n is the relative delay.

The physical channel response and the approximated virtual channel response are given as follows [36,39,66,67]:

$$h(t,f) = \sum_{n=1}^{N_P} \beta_n a_R(\theta_{R,n})a_T^H(\theta_{T,n})e^{j2\pi\nu_n t}e^{j2\pi\tau_n f}. \tag{1.18}$$

In Equation 1.18, all parameters can take on any value in their respective continuous ranges. To simplify the problem, the problem is discretized on a uniform grid with fixed spatial, temporal and spectral resolution, resulting in a finite-dimensional space-time signal space [36]. The virtual channel representation after this discretization can be written as follows:

$$h(t,f) \approx \sum_{i=1}^{N_R}\sum_{k=1}^{N_T}\sum_{l=1}^{L-1}\sum_{m=-M}^{M} h_v(i,k,l,m)a_R(i/N_R)a_T^H(k/N_T)e^{j2\pi\frac{m}{T}t}e^{j2\pi\frac{l}{W}f}, \tag{1.19}$$

where N_P in Equation 1.18 denotes the number of paths. Thus, the frequency response $h(t,f)$ is the superposition of contributions from all N_P paths. The N_R-dimensional vector $a_R(\theta_{R,n})$ in Equation 1.18 is the array response vector at the receiver, the N_T-dimensional vector $a_T(\theta_{T,n})$ is the array steering vector at the transmitter, and the superscript H denotes the matrix conjugate transpose. In Equation 1.18, β_n for path n denotes the complex path gain, and its linear model, $h_v(i,k,l,m)$ in Equation 1.19, can be observed in equation Equation 1.20. The summation in Equation 1.20 is over all paths that contribute to the sampling point, and a phase and attenuation factor has been absorbed in β_n [39]:

$$h_v(i,k,l,m) \simeq \sum_{n\in[sampling\ point]} \beta_n \tag{1.20}$$

The tensor h_v in Equation 1.19, which has the dimension of $D = N_R \times N_T \times L \times (2M+1)$ plays a significant role for characterizing the physical channel. Due to the noise level in practice, there are only very few components that are large. Therefore, the size of the tensor, D, is often much larger than the number of its nonzero elements d [36,39]. Therefore, we can formulate the received training signal as the linear equation to the characteristic of the virtual model

$$y_{tr} = \Phi H_v,$$

where is a D-dimensional column vector constructed by concatenation of all the elements in h_v, ordered according to the index set (i, k, l, m). The measurement matrix Φ with size $M \times D$ is a function of the transmitted training signal, the array steering and the response vector.

Finally, utilizing those considerations, we arrive at a compressed sensing formulation given by [39]

$$H_v^* = \mathrm{argmin}_{H_v} \|H_v\|_1 \text{ subject to } y_{tr} = \Phi H_v.$$

Bajwa et al. [3] show that for sparse-MIMO channels, significant savings in training resources can be achieved by only using d instead of D measurements. They also show that d measurements are sufficient for recovering the D-dimensional channel vector H_v with high probability. For more details, the reader is referred to [3]. The results and analysis for radio frequency (RF) signals and for underwater acoustic signals can be found in [37] and [7], respectively.

1.4.4.3 Spectrum sensing in CR networks

Rapidly increasing data rates on one hand, and limitation of the natural frequency on the other hand results in spectrum scarcity problems. In order to address this issue, an emerging approach called cognitive radio (CR) was proposed in [58], which can sense the environment and adjust the transmitting behavior of secondary users in order to prevent the interference to other primary users. Since we aim to sense the entire frequency band in the order of GHz, a well-developed wide-band spectrum sensing method is highly desirable. Therefore, wide-band spectrum sensing is a vital component in CR networks, that if realized by using conventional methods [80] (e.g., narrow-band methods) can lead to some practical difficulties such as high computational cost. This problem is mainly due to the reliance on the Nyquist sampling rate [68]. On the other hand, considering the electromagnetic spectrum ranging from a few megahertz to a few gigahertz shows that a high portion of the frequency range is idle due to spectrum underutilization, suggesting the frequency spectrum of signal to be highly sparse [57]. Hence, one can make the wide-band spectrum sensing practically feasible by using compressed sensing techniques to reduce the sampling rate, if the signal exhibits sparsity properties in some domain.

Now, we assume that $x(t)$ denotes the signal sensed by a secondary CR user. Further, let W be the frequency range of the whole wide-band spectrum which should be sensed. If B denotes the set of frequency bands currently used by other primary users, one can assume $|B| \ll W$ due to the spectrum underutilization. The condition $|B| \ll W$ also represents the sparsity feature of $x(t)$ in the frequency domain. Denoting f_N as the Nyquist rate, f as the signal representation in the frequency domain, F^{-1} as the inverse Fourier transform, S as the reduced-rate sampling matrix operating at the rate close to $|B|f_N/W$, and finally x_t as the reduced rate measurements, the compressed sensing formulation of spectrum sensing can be written as follows [39]:

$$\min \|f\|_1 \text{ subject to } x_t = Sx(t) = SF^{-1}f.$$

Hence, compressed sensing is applicable by sampling at a much lower rate, that is, $|B|f_N/W$, compared with the Nyquist sampling rate f_N.

Compressed sensing can be also applied directly to analog signals, which has the advantage of saving the ADC resources, especially in cases where the sampling rate is high [69]. Let $x = [x(T_s), ..., x(nT_s)]^T$ denote a vector composed of the full sample set of the received analog wide-band signal $x(t)$ at each secondary user terminal, which is acquired by exploiting the Nyquist rate, $1/T_s$, where T_s is the sampling period equal to the inverse of Nyquist rate. If we assume that $t \in [0, nT_S]$, then at least n samples of $x(t)$ are required in order to fulfill the Nyquist condition and recover $x(t)$ without aliasing.

Now, assume that Φ denotes a measurement matrix of size $m \times n$, which satisfies $m < n$ in order to apply the sub-Nyquist sampling rate. The possible choices of Φ can be considered as a matrix obtained by randomly selecting m rows of an $n \times n$ identity, Gaussian, or Bernoulli matrix. Therefore, the discrete time-domain sampling process for obtaining sub-Nyquist linear measurements can be modeled by

$$y = \Phi x.$$

If D denotes the unitary DFT matrix, whose (l, m)-element is given by $\frac{1}{\sqrt{n}} e^{-i \frac{2\pi}{n} lm}$, the standard compressed sensing problem can be formalized as follows [38,69]:

$$c^* = \operatorname{argmin}_c \|c\|_1 \text{ subject to } y = \Phi D^* c,$$

where c is the discrete spectrum of $x(t)$ and can be calculated by $c = Dx$. ΦD^* is regarded as the multiplication of sensing matrix and dictionary, and c is an unknown sparse vector. One can recover the vector c, if the discrete spectrum of the primary signal occupies the frequency band sparsely enough.

In [69], the authors propose a method which utilizes the sparsity of the primary signal itself in the frequency domain as well as the sparsity of subband edges. The edge spectrum can then be calculated by

$$z_s = \Gamma D \Psi_s x,$$

where Γ is the differentiation matrix, defined as follows [69]:

$$\Gamma := \begin{bmatrix} 1 & 0 & \cdots & \cdots & 0 \\ -1 & 1 & \ddots & & \vdots \\ 0 & \ddots & \ddots & \ddots & \vdots \\ \vdots & \ddots & \ddots & \ddots & 0 \\ 0 & \cdots & 0 & -1 & 1 \end{bmatrix}$$

and Ψ_s is a matrix representation of a wavelet smoothing function with scale factor s. Tian and Giannakis [69] assume that a total of W Hz in the frequency range $[f_0, f_k]$ is available for a wide-band wireless network. Assume that we receive the signal $x(t)$ that occupies k consecutive spectrum bands, with their frequency boundaries located at $f_0 < f_1 < ... < f_k$. Based on the sparsity of z_s, by assuming the piecewise smooth power spectrum density (PSD), the following optimization problem can be considered as another compressed sensing problem in CR application [38,69]:

$$z_s^* = \operatorname{argmin}_{z_s} \|z_s\|_1 \text{ subject to } y = \Phi(\Gamma D \Psi_s)^{-1} z_s.$$

Since sparse reconstruction methods show poor performance in low SNR regimes due to the loss of sparsity properties, which results from increasing noise power, we cannot utilize the compressed sensing approach directly in the low SNR regimes. To overcome this problem, Kim et al. [43] and Lundén et al. [52] proposed spectrum sensing based on cyclostationary features. Exploiting cyclostationary features introduces robustness against noise uncertainty; however, as in [43,52], the cyclostationary feature extraction block comes after the compressed sensing block, the reconstruction error of compressed sensing propagates through the cyclostationary features extraction, and therefore leads to limited performance. To overcome this limitation, Leus and Tian [51] proposed to combine these two blocks, which was then improved in [70]. In [70], the spectral correlation density function (SCD) has been stated in a matrix formulation and directly related to variable time covariance matrix of compressed measurements utilizing the Kronecker product.

Acknowledgments

A. Hashemi acknowledges support from Berlin International Graduate School in Model and Simulation based Research (BIMoS). He also likes to thank Jackie Ma for providing the MATLAB® codes and figures for generating the sampling schemes in MRI application.

A. Flinth acknowledges support from Deutsche Forschungsgemeinschaft (DFG) Grant KU 1446/18-1 and Berlin Mathematical School (BMS).

G. Kutyniok acknowledges partial support by the Einstein Foundation Berlin, the Einstein Center for Mathematics Berlin (ECMath), the European Commission-Project DEDALE (contract no. 665044) within the H2020 Framework Program, DFG Grant KU 1446/18, DFG-SPP 1798 Grants KU 1446/21 and KU 1446/23, the DFG Collaborative Research Center TRR 109 Discretization in Geometry and Dynamics, and by the DFG Research Center MATHEON "Mathematics for Key Technologies" in Berlin.

References

[1] A. Ahmed, B. Recht, and J. Romberg. Blind deconvolution using convex programming. *IEEE Trans. Inform. Theory*, 60(3):1711–1732, 2014.

[2] D. Amelunxen, M. Lotz, M. B. McCoy, and J. A. Tropp. Living on the edge: Phase transitions in convex programs with random data. *Inf. Inference*, 3:224–294, 2014.

[3] W. U. Bajwa, J. Haupt, A. M. Sayeed, and R. Nowak. Compressed channel sensing: A new approach to estimating sparse multipath channels. *Proc. IEEE*, 98(6):1058–1076, 2010.

[4] R. Baraniuk, M. A. Davenport, M. F. Duarte, C. Hegde, et al. *An Introduction to Compressive Sensing*. Connexions e-textbook, 2011. OpenStax-CNX http://legacy.cnx.org/content/col11133/1.5/.

[5] R. G. Baraniuk, E. Candès, R. Nowak, and M. Vetterli. Compressive sampling. *IEEE Sign. Process. Mag.*, 25(2):12–13, 2008.

[6] A. Beck and M. Teboulle. A fast iterative shrinkage-thresholding algorithm for linear inverse problems. *SIAM J. Imaging Sci.*, 2(1):183–202, 2009.

[7] C. R. Berger, S. Zhou, J. C. Preisig, and P. Willett. Sparse channel estimation for multicarrier underwater acoustic communication: From subspace methods to compressed sensing. *IEEE Trans. Signal Process.*, 58(3):1708–1721, 2010.

[8] J. Bigot, C. Boyer, and P. Weiss. An analysis of block sampling strategies in compressed sensing. *IEEE Trans. Inf. Theory*, 62(4):2125–2139, 2016.

[9] T. Blumensath and M. E. Davies. Iterative hard thresholding for compressed sensing. *Appl. Comput. Harmon. Anal.*, 27(3):265–274, 2009.

[10] T. Blumensath, M. E. Davies, and G. Rilling. Greedy algorithms for compressed sensing. In *Compressed Sensing: Theory and Applications*, Eds. Y. Eldar, G. Kutyniok, pages 348–393. Cambridge University Press, UK, 2012.

[11] H. Boche, R. Calderbank, G. Kutyniok, and J. Vybíral. A survey of compressed sensing. In *Compressed Sensing and Its Applications*, Eds. Y. Eldar, G. Kutyniok, pages 1–39. Basel: Springer, 2015.

[12] K.-H. Borgwardt. The average number of pivot steps required by the simplex-method is polynomial. *Math. Method. Oper. Res.*, 26(1):157–177, 1982.

[13] J. Bourgain, S. Dilworth, K. Ford, S. Konyagin, D. Kutzarova, et al. Explicit constructions of RIP matrices and related problems. *Duke Math. J.*, 159(1):145–185, 2011.

[14] E. Candès and T. Tao. Decoding by linear programming. *IEEE Trans. Inf. Theory*, 51:4203–4215, 2005.

[15] E. J. Candès and B. Recht. Exact matrix completion via convex optimization. *Found. Comput. Math.*, 9(6):717–772, 2009.

[16] E. Candès and T. Tao. The Dantzig selector: Statistical estimation when p is much larger than n. *Ann. Stat.*, 16:2313–2351, 2007.

[17] E. J. Candès, Y. C. Eldar, D. Needell, and P. Randall. Compressed sensing with coherent and redundant dictionaries. *Appl. Comput. Harmon. Anal.*, 31(1):59–73, 2011.

[18] E. J. Candès, J. K. Romberg, and T. Tao. Stable signal recovery from incomplete and inaccurate measurements. *Commun. Pure Appl. Math.*, 59(8):1207–1223, 2006.

[19] E. J. Candès, T. Strohmer, and V. Voroninski. Phaselift: Exact and stable signal recovery from magnitude measurements via convex programming. *Commun. Pur. Appl. Math.*, 66(8):1241–1274, 2013.

[20] V. Chandrasekaran, B. Recht, P. A. Parrilo, and A. S. Willsky. The convex geometry of linear inverse problems. *Found. Comput. Math.*, 12(6):805–849, 2012.

[21] S. S. Chen, D. L. Donoho, and M. A. Saunders. Atomic decomposition by basis pursuit. *SIAM Rev.*, 43(1):129–159, 2001.

[22] O. Christensen. *An Introduction to Frames and Riesz Bases*. Boston: Birkhäuser, 2013.

[23] A. Cohen, W. Dahmen, and R. DeVore. Compressed sensing and best k-term approximation. *J. Am. Math. Soc.*, 22(1):211–231, 2009.

[24] I. Daubechies, M. Defrise, and C. De Mol. An iterative thresholding algorithm for linear inverse problems with a sparsity constraint. *Appl. Comput. Harmon. Anal.*, 57(11):1413–1457, 2004.

[25] M. Davenport, M. Duarte, Y. Eldar, and G. Kutyniok. *Introduction to Compressed Sensing*, pages 1–64. Cambridge: Cambridge University Press, 2012.

[26] D. Donoho and G. Kutyniok. Microlocal analysis of the geometric separation problem. *Appl. Comput. Harmon. Anal.*, 66(1):1–47, 2013.

[27] D. L. Donoho. For most large underdetermined systems of linear equations the minimal ℓ_1-norm solution is also the sparsest solution. *Appl. Comput. Harmon. Anal.*, 59(6): 797–829, 2006.

[28] D. L. Donoho and M. Elad. Optimally sparse representation in general (nonorthogonal) dictionaries via ℓ_1-minimization. *Proc. Natl. Acad. Sci. U S A.* 100(5):2197–2202, 2003.

[29] D. L. Donoho and X. Huo. Uncertainty principles and ideal atomic decomposition. *IEEE Trans. Inf. Theory*, 47(7):2845–2862, 2001.

[30] M. Elad and A. M. Bruckstein. A generalized uncertainty principle and sparse representation in pairs of bases. *IEEE Trans. Inf. Theory*, 48(9):2558–2567, 2002.

[31] Y. Eldar and M. Mishali. Robust recovery of signals from a structured union of subspaces. *IEEE Trans. Inf. Theory*, 55(11):5302—5316, 2009.

[32] M. Fornasier and H. Rauhut. Iterative thresholding algorithms. *Appl. Comput. Harmon. Anal.*, 25(2):187–208, 2008.

[33] S. Foucart and H. Rauhut. *A Mathematical Introduction to Compressed Sensing*. New York: Birkhäuser, 2013.

[34] U. Gamper, P. Boesiger, and S. Kozerke. Compressed sensing in dynamic MRI. *Magn. Reson. Med.*, 59(2):365–373, 2008.

[35] S. Gerschgorin. Über die Abgrenzung der Eigenwerte einer Matrix. *Bull. Acad. Sci. USSR.*, 6:749–754, 1931.

[36] Z. Han, H. Li, and W. Yin. *Compressive Sensing for Wireless Networks*. Cambridge University Press, 2013.

[37] J. Haupt, W. U. Bajwa, G. Raz, and R. Nowak. Toeplitz compressed sensing matrices with applications to sparse channel estimation. *IEEE Trans. Inf. Theory*, 56(11):5862–5875, 2010.

[38] K. Hayashi, M. Nagahara, and T. Tanaka. A user's guide to compressed sensing for communications systems. *IEICE Trans. Commun.*, 96(3):685–712, 2013.

[39] H. Huang, S. Misra, W. Tang, H. Barani, and H. Al-Azzawi. *Applications of Compressed Sensing in Communications Networks. Preprint.* Available at: arxiv.org. Identifier: 1305.3002, 2013.

[40] H. Jung, K. Sung, K. S. Nayak, E. Y. Kim, and J. C. Ye. k-t FOCUSS: A general compressed sensing framework for high resolution dynamic MRI. *Magn. Reson. Med.*, 61(1):103–116, 2009.

[41] M. Kabanava and H. Rauhut. Cosparsity in compressed sensing. In *Compressed Sensing and Its Applications*, Eds. Boche, H., Calderbank, R., Kutyniok, G., Vybral, J., pages 315–339. Basel: Springer, 2015.

[42] M. A. Khajehnejad, W. Xu, A. S. Avestimehr, and B. Hassibi. Weighted ℓ_1 minimization for sparse signals with prior information. In *IEEE International Symposium on Information Theory*, pages 483–487. IEEE, 2009.

[43] K. Kim, I. A. Akbar, K. K. Bae, J.-S. Um, C. M. Spooner, and J. H. Reed. Cyclostationary approaches to signal detection and classification in cognitive radio. In *2nd IEEE International Symposium on New Frontiers in Dynamic Spectrum Access Networks*, pages 212–215. IEEE, 2007.

[44] S. Kirolos, J. Laska, M. Wakin, M. Duarte, D. Baron, T. Ragheb, Y. Massoud, and R. Baraniuk. Analog-to-information conversion via random demodulation. In *IEEE Dallas/CAS Workshop on Design, Applications, Integration and Software*, pages 71–74. IEEE, 2006.

[45] F. Krahmer and R. Ward. Stable and robust sampling strategies for compressive imaging. *IEEE Trans. Image Process.*, 23(2):612–622, 2014.

[46] G. Kutyniok. Theory and applications of compressed sensing. *GAMM-Mitt.*, 36(1): 79–101, 2013.

[47] G. Kutyniok and D. E. Labate. *Shearlets: Multiscale Analysis for Multivariate Data.* Basel: Springer, 2012.

[48] G. Kutyniok and W.-Q. Lim. *Optimal Compressive Imaging of Fourier Data.* Preprint. Available at: arXiv.org, identifier 1510.05029, 2015.

[49] G. Kutyniok, W.-Q. Lim, and R. Reisenhofer. Shearlab 3D: Faithful digital shearlet transforms based on compactly supported shearlets. *ACM Trans. Math. Softw.*, 42(1):5, 2016.

[50] J. N. Laska, S. Kirolos, M. F. Duarte, T. S. Ragheb, R. G. Baraniuk, and Y. Massoud. Theory and implementation of an analog-to-information converter using random demodulation. In *2007 IEEE International Symposium on Circuits and Systems*, pages 1959–1962, 2007.

[51] G. Leus and Z. Tian. Recovering second-order statistics from compressive measurements. In *4th IEEE International Workshop on Computational Advances in Multi-Sensor Adaptive Processing (CAMSAP)*, pages 337–340, 2011.

[52] J. Lundén, V. Koivunen, A. Huttunen, and H. V. Poor. Collaborative cyclostationary spectrum sensing for cognitive radio systems. *IEEE Trans. Signal Process.*, 57(11):4182–4195, 2009.

[53] M. Lustig, D. Donoho, and J. M. Pauly. Sparse MRI: The application of compressed sensing for rapid MR imaging. *Magn. Reson. Med.*, 58(6):1182–1195, 2007.

[54] M. Lustig, D. L. Donoho, J. M. Santos, and J. M. Pauly. Compressed sensing MRI. *IEEE Sign. Proc. Mag.*, 25(2):72–82, 2008.

[55] J. Ma. Generalized sampling reconstruction from Fourier measurements using compactly supported shearlets. *Appl. Comput. Harmon. Anal.*, 47(2):294–318, 2015.

[56] S. G. Mallat and Z. Zhang. Matching pursuits with time-frequency dictionaries. *IEEE Trans. Signal Process.*, 41(12):3397–3415, 1993.

[57] M. A. McHenry, P. A. Tenhula, D. McCloskey, D. A. Roberson, and C. S. Hood. Chicago spectrum occupancy measurements & analysis and a long-term studies proposal. In *Proceedings of the First International Workshop on Technology and Policy for Accessing Spectrum*, pages 1–12, 2006.

[58] J. Mitola and G. Q. Maguire. Cognitive radio: Making software radios more personal. *IEEE Pers. Commun.*, 6(4):13–18, 1999.

[59] D. G. Mixon. Explicit matrices with the restricted isometry property: Breaking the square-root bottleneck. In *Compressed Sensing and Its Applications*, Eds. Boche, H., Calderbank, R., Kutyniok, G., Vybral, J., pages 389–417. Switzerland: Springer, International Publishing, 2015.

[60] B. K. Natarajan. Sparse approximate solutions to linear systems. *SIAM J. Comput.*, 24(2):227–234, 1995.

[61] D. Needell and J. A. Tropp. Cosamp: Iterative signal recovery from incomplete and inaccurate samples. *Appl. Comput. Harmon. Anal.*, 26(3):301–321, 2009.

[62] S. Oymak, M. Khajehnejad, and B. Hassibi. Recovery threshold for optimal weight ℓ_1-minimization. In *IEEE International Symposium on Information Theory*, pages 2032–2036, 2012.

[63] S. Qaisar, R. M. Bilal, W. Iqbal, M. Naureen, and S. Lee. Compressive sensing: From theory to applications, a survey. *J. Commun. Netw.*, 15(5):443–456, 2013.

[64] H. Rauhut. Compressive sensing and structured random matrices. In *Theoretical Foundations and Numerical Methods for Sparse Recovery*. Berlin: de Gruyter, 2010.

[65] M. Rostami. Compressed sensing with side information on feasible region. In *Compressed Sensing with Side Information on the Feasible Region*, pages 23–31. Springer, 2013.

[66] A. M. Sayeed. Deconstructing multiantenna fading channels. *IEEE Trans. Signal Process.*, 50(10):2563–2579, 2002.

[67] A. M. Sayeed. A virtual representation for time-and frequency-selective correlated MIMO channels. In *Proceedings of 2003 IEEE International Conference on Acoustics, Speech, and Signal Processing*, vol. 4, pages IV–648–651. IEEE, 2003.

[68] H. Sun, A. Nallanathan, C.-X. Wang, and Y. Chen. Wideband spectrum sensing for cognitive radio networks: A survey. *IEEE Wireless Commun.*, 20(2):74–81, 2013.

[69] Z. Tian and G. B. Giannakis. Compressed sensing for wideband cognitive radios. In *2007 IEEE International Conference on Acoustics, Speech and Signal Processing*, vol. 4, pages IV–1357–1360, 2007.

[70] Z. Tian, Y. Tafesse, and B. M. Sadler. Cyclic feature detection with sub-Nyquist sampling for wideband spectrum sensing. *IEEE J. Sel. Top. Signal Process.*, 6(1):58–69, 2012.

[71] R. Tibshirani. Regression shrinkage and selection via the lasso. *J. Roy. Stat. Soc. B Methodol.*, 58(1):267–288, 1996.

[72] J. Tropp. Greed is good: Algorithmic results for sparse approximation. *IEEE Trans. Inf. Theory*, 50:2231–2242, 2004.

[73] J. A. Tropp. User-friendly tail bounds for sums of random matrices. *Found. Comput. Math.*, 12(4):389–434, 2012.

[74] J. A. Tropp and A. C. Gilbert. *Signal Recovery from Random Measurements via Orthogonal Matching Pursuit*. Technical report, University of Michigan, 2006.

[75] E. Van den Berg and M. P. Friedlander. Probing the Pareto frontier for basis pursuit solutions. *SIAM J. Sci. Comput.*, 31(2):890–912, 2008.

[76] R. Vershinyn. Introduction to the non-asymptotic analysis of random matrices. In *Compressed Sensing, Theory and Applications*, Eds. Y. Eldar, G. Kutyniok, pages 210–268, Cambridge: Cambridge University Press, 2012.

[77] C. Vonesch and M. Unser. A fast iterative thresholding algorithm for wavelet-regularized deconvolution. In *Proceedings of the SPIE Optics and Photonics 2007 Conference on Mathematical Methods*, pages 1–5. International Society for Optics and Photonics, 2007.

[78] L. R. Welch. Lower bounds on the maximum cross correlation of signals (corresp.). *IEEE Trans. Inf. Theory*, 20(3):397–399, 1974.

[79] W. Yin, S. Osher, D. Goldfarb, and J. Darbon. Bregman iterative algorithms for ℓ_1-minimization with applications to compressed sensing. *SIAM J. Imaging Sci.*, 1(1): 143–168, 2008.

[80] T. Yucek and H. Arslan. A survey of spectrum sensing algorithms for cognitive radio applications. *IEEE Commun. Surveys Tutorials*, 11(1):116–130, 2009.

2

Compressive Sensing on the Sphere: Slepian Functions for Applications in Geophysics

Zubair Khalid and Abubakr Muhammad

CONTENTS

2.1 Introduction

Signals are inherently defined on the sphere $\mathbb{S}^2$ in a large number of applications found in various branches of science and engineering, including, but not limited to, geophysics [1–3], planetary science [4], medical imaging [5], cosmology [6], and acoustics [7]. In these applications, the measurements of the signal are either unavailable over the whole sphere, or the estimate of the signal is required to be localized to a spatially limited region $R \subset \mathbb{S}^2$. In this context, the problem that arises is to estimate the signal on the sphere from noisy measurements acquired over the spatially limited region. This estimation (inversion) problem is inherently ill-posed (ill-conditioned) and therefore, in general, impossible to stably estimate the signal. Conventionally, the estimation is carried out in spherical harmonic domain employing regularized inversion such as damped least-squares approach [2], where the damping is introduced to stabilize the estimation of the signal but at the cost of added bias. Spherical harmonic functions (or spherical harmonics) serve as complete orthonormal basis for functions defined on the sphere. The ill-conditioning of the estimation problem in spherical harmonic domain is because the spherical harmonics are global (defined on the whole sphere) basis and are not orthogonal over the spatially limited region. Alternatively, the use of functions that are orthogonal simultaneously over the spatially limited region and whole sphere and serve as complete basis is more suitable for modeling of the signal and/or measurements over the spatially limited region. Such basis functions arise as a solution for Slepian spatial spectral concentration problem on the sphere [1,8,9] and are designed to have their energy maximally (or optimally) concentrated inside the region R. These basis functions, referred to as *Slepian basis* or *Slepian functions*, are band-limited in spherical harmonic domain, orthogonal over the whole sphere $\mathbb{S}^2$ and region R, and serve as complete basis for the representation of band-limited signals on the sphere. Exploiting these characteristics, Slepian functions on the sphere have been used for localized spectral and spatial analyses [10–12], signal estimation from incomplete measurements [13,14], and sparse and efficient representations of spherical signals in a wide range of applications found in geophysics [8,12,15], cosmology and planetary studies [16,17], optics [14], and computer graphics [18], to name a few.

In this chapter, we present the Slepian concentration problem [19–22] on the sphere published elsewhere [1,2,5,8–10]. After reviewing the necessary mathematical background, we present the Slepian concentration problem on the sphere, formulate the problem as an eigenvalue problem, and analyze the properties of the Slepian functions. Furthermore, we present the analytical expression for the computation of matrix defining the concentration problem for polar cap region, rotationally symmetric region, and limited colatitude-longitude region. We also show that the Slepian basis enables the sparse representation of spatially concentrated band-limited signals. We provide illustrations to support the developments.

Moreover, we present the construction of dictionary of Slepian functions based on the sequential division of the region into two approximately equal-sized regions. The dictionary of Slepian functions is similar to the wavelets in nature with elements that are band-limited, multiscale, and mostly localized within the given spatial region. Finally, we provide a framework to estimate the signal from its incomplete and noisy measurements. The presented framework exploits the sparsity of the signal representation in Slepian basis and dictionary elements, enables the employment of compressive sensing methods, and consequently, allows the better estimation, in terms of average reconstruction error, when compared with the traditional methods.

2.2   Mathematical preliminaries—signals on the sphere

For signals defined on the sphere, we here review the necessary mathematical background: coordinate systems, measures, inner products, subspaces, and harmonic analysis.

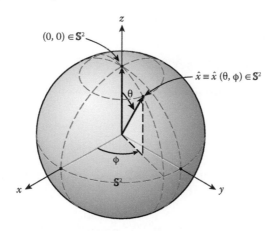

FIGURE 2.1

2-Sphere, $\mathbb{S}^2$, representation—spherical coordinate system consisting of colatitude $\theta \in [0, \pi]$ and longitude $\phi \in [0, 2\pi]$ parameterizes a point unit vector $\hat{\boldsymbol{x}} \equiv \hat{\boldsymbol{x}}(\theta, \phi) \triangleq (\sin\theta \cos\phi, \ \sin\theta \sin\phi, \ \cos\theta)' \in \mathbb{R}^3$ on the 2-sphere $\mathbb{S}^2$.

2.2.1 Signals on the sphere

The spherical domain, also referred to as sphere, 2-sphere, or unit sphere, is denoted by $\mathbb{S}^2$ and is defined as $\mathbb{S}^2 \triangleq \{\boldsymbol{x} \in \mathbb{R}^3 : |\boldsymbol{x}|_2 = 1\} \subset \mathbb{R}^3$, where $|\cdot|_2$ represents Euclidean norm [23]. A point on $\mathbb{S}^2$ is given by the unit vector $\hat{\boldsymbol{x}} \equiv \hat{\boldsymbol{x}}(\theta, \phi) \triangleq (\sin\theta \cos\phi, \ \sin\theta \sin\phi, \ \cos\theta)' \in \mathbb{R}^3$, where $(\cdot)'$ denotes the vector transpose operation, $\theta \in [0, \pi]$ is the colatitude that is measured with respect to the positive z-axis, and $\phi \in [0, 2\pi)$ is the longitude that is measured with respect to the positive x-axis in the x–y plane (see Figure 2.1).

We consider the complex-valued square-integrable functions defined on the sphere. The set of such functions forms a Hilbert space denoted by $L^2(\mathbb{S}^2)$ equipped with the inner product given by [23]

$$\langle f, g \rangle \triangleq \int_{\mathbb{S}^2} f(\hat{\boldsymbol{x}}) \overline{g(\hat{\boldsymbol{x}})} \, ds(\hat{\boldsymbol{x}}), \tag{2.1}$$

where f, g are functions defined on $\mathbb{S}^2$, $ds(\hat{\boldsymbol{x}}) = \sin\theta \, d\theta \, d\phi$ represents the differential area element on $\mathbb{S}^2$, $\overline{(\cdot)}$ denotes complex conjugation, and the integration is carried out over the whole unit sphere. The inner product in Equation 2.1 induces norm $\|f\| \triangleq \langle f, f \rangle^{1/2}$. Throughout this chapter, the functions with finite energy (finite induced norm), that is, $\|f\|^2 < \infty$, belonging to the space are referred to as signals on the sphere or signals for short. For a region $R \subset \mathbb{S}^2$ and $f, g \in L^2(\mathbb{S}^2)$, we also define $\langle f, g \rangle_R \triangleq \int_R f(\hat{\boldsymbol{x}}) \overline{g(\hat{\boldsymbol{x}})} \, ds(\hat{\boldsymbol{x}})$ as the norm over the region R and $\|f\|_R^2 \triangleq \langle f, f \rangle_R$ as the energy of the signal f in the region R. In the sequel, we use $|\cdot|$ to denote the absolute value.

2.2.2 Spherical harmonic domain representation

2.2.2.1 Spherical harmonics

The Hilbert space $L^2(\mathbb{S}^2)$ is separable, and the spherical harmonic functions (or spherical harmonics for short) form the archetype complete orthonormal set of basis functions. The spherical harmonic functions, which arise as angular solutions to the Helmholtz's equation in spherical coordinates, are denoted by $Y_\ell^m(\hat{x}) = Y_\ell^m(\theta, \phi)$, for integer degree $\ell \geq 0$ and integer order $|m| \leq \ell$, and are defined by [23]

$$Y_\ell^m(\theta, \phi) = N_\ell^m P_\ell^m(\cos\theta)\, e^{im\phi}, \qquad (2.2)$$

where N_ℓ^m is the normalization factor given by

$$N_\ell^m \triangleq \sqrt{\frac{2\ell+1}{4\pi}\frac{(\ell-m)!}{(\ell+m)!}}, \qquad (2.3)$$

which ensures that $\|Y_\ell^m\| = 1$. P_ℓ^m in Equation 2.2 is the associated Legendre function defined for degree ℓ and order $0 \leq m \leq \ell$ as [23]

$$P_\ell^m(u) = \frac{(-1)^m}{2^\ell \ell!}(1-u^2)^{m/2}\frac{d^{\ell+m}}{du^{\ell+m}}(u^2-1)^\ell \qquad (2.4)$$

$$P_\ell^{-m}(u) = (-1)^m \frac{(\ell-m)!}{(\ell+m)!} P_\ell^m(u), \qquad (2.5)$$

for $|u| \leq 1$. The zero-order associated Legendre polynomials are referred to as Legendre polynomials $(P_\ell(u))$, that is, $P_\ell^0(u) = P_\ell(u)$.

2.2.2.2 Spectral domain representation

By completeness of spherical harmonics [23], any signal $f \in L^2(\mathbb{S}^2)$ can be expanded as

$$f(\hat{x}) = \sum_{\ell=0}^{\infty} \sum_{m=-\ell}^{\ell} (f)_\ell^m Y_\ell^m(\hat{x}), \qquad (2.6)$$

where

$$(f)_\ell^m \triangleq \langle f, Y_\ell^m \rangle = \int_{\mathbb{S}^2} f(\hat{x})\overline{Y_\ell^m(\hat{x})}\, ds(\hat{x}), \qquad (2.7)$$

denotes the spherical harmonic coefficient of degree ℓ and order m, which constitute the spectral (spherical harmonic) domain representation of a signal. The equality in Equation 2.6 is understood in terms of convergence in the mean (strong convergence in the norm)

$$\lim_{L\to\infty}\left\| f(\hat{x}) - \sum_{\ell=0}^{L}\sum_{m=-\ell}^{\ell}(f)_\ell^m Y_\ell^m(\hat{x})\right\| = 0. \qquad (2.8)$$

The signal $f \in L^2(\mathbb{S}^2)$ is defined to be band-limited at degree L if $(f)_\ell^m = 0$ for $\ell \geq L$. The set of band-limited signals forms an L^2 dimensional subspace of $L^2(\mathbb{S}^2)$, which is denoted by $\mathcal{H}_L$. For the spectral domain representation of a band-limited signal $f \in \mathcal{H}_L$, we define the column vector containing spherical harmonic coefficients as $\mathbf{f} \triangleq \left((f)_0^0, (f)_1^{-1}, (f)_1^0, (f)_1^1, (f)_2^{-2}, \cdots, (f)_{L-1}^{L-1}\right)'$ of size L^2.

2.3   Slepian concentration problem on the sphere

For signals on the sphere, the Slepian concentration problem [19–22], to find the band-limited (or space-limited) functions with optimal energy concentration in the spatial (or spectral) domain, has been extensively investigated [1,2,5,8–10,23,24]. To maximize the spatial concentration of a band-limited signal $h \in \mathcal{H}_L$ within the spatial region $R \subset \mathbb{S}^2$, we seek to maximize the spatial concentration (energy) ratio λ given by [9]

$$\lambda = \frac{\|h\|_R^2}{\|h\|^2} = \frac{\int_R |h(\widehat{\boldsymbol{x}})|^2 ds(\widehat{\boldsymbol{x}})}{\int_{\mathbb{S}^2} |h(\widehat{\boldsymbol{x}})|^2 ds(\widehat{\boldsymbol{x}})}, \quad 0 \leq \lambda < 1, \tag{2.9}$$

which can be equivalently expressed in spectral domain as

$$\lambda = \frac{\sum_{\ell=0}^{L-1} \sum_{m=-\ell}^{\ell} \sum_{p=0}^{L-1} \sum_{q=-p}^{p} \overline{(h)_\ell^m} (h)_p^q K_{\ell m, pq}}{\sum_{\ell=0}^{L-1} \sum_{m=-\ell}^{\ell} (h)_\ell^m \overline{(h)_\ell^m}}, \tag{2.10}$$

where

$$K_{\ell m, pq} \triangleq \int_R \overline{Y_\ell^m(\widehat{\boldsymbol{x}})} Y_p^q(\widehat{\boldsymbol{x}}) ds(\widehat{\boldsymbol{x}}). \tag{2.11}$$

Using the spectral domain version of the concentration ratio Equation 2.10, the Slepian concentration problem to maximize the concentration ratio λ can be solved as an algebraic eigenvalue problem given by

$$\sum_{p=0}^{L-1} \sum_{q=-p}^{p} K_{\ell m, pq} (h)_p^q = \lambda (h)_\ell^m, \tag{2.12}$$

with matrix formulation

$$\mathbf{K}\mathbf{h} = \lambda \mathbf{h}, \tag{2.13}$$

where the matrix $\mathbf{K}$ contains elements $K_{\ell m, pq}$ with similar indexing adopted for $\mathbf{h}$ and has dimension $L^2 \times L^2$.

2.3.1 Analysis

Once the matrix $\mathbf{K}$ is computed for the spatial region R and band-limit L, the concentration problem can be solved using its formulation as an algebraic eigenvalue problem given in Equation 2.13, the solution of which gives L^2 band-limited eigenvectors. Each eigenvector represents the spectral domain representation (spherical harmonic coefficients) of the band-limited eigenfunction (in spatial domain) and the eigenvalue associated with each eigenvector represents the energy concentration of the associated eigenfunction in the spatial region R. Since $\mathbf{K}$ is complex valued and Hermitian symmetric, by definition, the eigenvalues are real and the eigenvectors are orthogonal; we choose eigenvectors to be orthonormal. Furthermore, the eigenvalues are non-negative as $\mathbf{K}$ is positive semidefinite which follows from the numerator in Equation 2.10 that represents the energy of the band-limited function in some spatial region.

Let the eigenvectors of $\mathbf{K}$ and the corresponding eigenfunctions be denoted by $\mathbf{h}_\alpha$ and $h_\alpha(\theta, \phi)$, respectively, for $\alpha = 1, 2, \ldots, L^2$, where we index the eigenfunctions such that $0 \leq \lambda_{\alpha+1} \leq \lambda_\alpha < 1$. With this indexing, the eigenfunction $h_1(\theta, \phi)$ is most concentrated in R, while $h_{L^2}(\theta, \phi)$ is most concentrated in $\mathbb{S}^2 \backslash R$.

2.3.1.1   Orthogonality of eigenfunctions

The eigenvectors, by definition, are orthogonal, and are chosen here as orthonormal, that is,

$$\mathbf{h}_\alpha^H \mathbf{h}_\beta = \sum_{\ell=0}^{L-1} \sum_{m=-\ell}^{\ell} \overline{(h_\alpha)_\ell^m} \, (h_\beta)_\ell^m = \delta_{\alpha,\beta}, \tag{2.14}$$

$$\mathbf{h}_\alpha^H \mathbf{K} \mathbf{h}_\beta = \sum_{\ell=0}^{L-1} \sum_{m=-\ell}^{\ell} \sum_{p=0}^{L-1} \sum_{q=-p}^{p} \overline{(h_\alpha)_\ell^m} \, (h_\beta)_p^q K_{pq,\ell m} = \lambda_\alpha \, \delta_{\alpha,\beta}, \tag{2.15}$$

where $(\cdot)^H$ denotes the Hermitian of a vector or matrix. The orthogonality relations in Equations 2.14 and 2.15 can be equivalently expressed in terms of the associated eigenfunctions as

$$\|h\|^2 = \int_{\mathbb{S}^2} h_\alpha(\widehat{\boldsymbol{x}}) \overline{h_\beta(\widehat{\boldsymbol{x}})} \, ds(\widehat{\boldsymbol{x}}) = \delta_{\alpha,\beta}, \tag{2.16}$$

$$\|h\|_R^2 = \int_R h_\alpha(\widehat{\boldsymbol{x}}) \overline{h_\beta(\widehat{\boldsymbol{x}})} \, ds(\widehat{\boldsymbol{x}}) = \lambda_\alpha \, \delta_{\alpha,\beta}. \tag{2.17}$$

Remark 2.3.1 (On the orthogonality of eigenfunctions) The orthogonality relations in Equations 2.14–2.17 indicate that the eigenfunctions are not only orthonormal over the sphere but also orthogonal over the spatial region R. This double orthogonality is one of the important features of these eigenfunctions which makes them useful in the analysis of the signal over the spatial region R [10,13,16,25].

2.3.1.2   Number of concentrated eigenfunctions

We also note that the sum of the eigenvalues of $\mathbf{K}$ for the spatial region R is given by the trace of $\mathbf{K}$ [9,23] with,

$$N = \sum_{\alpha=1}^{L^2} \lambda_\alpha = \text{tr}(\mathbf{K}) = \frac{A}{4\pi} L^2, \quad A = \int_R ds(\widehat{\boldsymbol{x}}), \tag{2.18}$$

where $\text{tr}(\cdot)$ denotes the trace of the matrix and A is the area of the region R. If the spectrum of eigenvalues has a narrow transition from significant (near unity) to insignificant (near zero) eigenvalues, the sum of the eigenvalues, given by N, well approximates the number of significant eigenvalues and the number of well-concentrated eigenfunctions.

2.3.1.3   Slepian basis

Since we obtain a set of L^2 band-limited orthonormal eigenfunctions as a solution of the (Slepian) spatial-spectral concentration problem, these eigenfunctions span the L^2-dimensional subspace $\mathcal{H}_L$ and therefore serve as a complete basis, referred to as the *Slepian basis* or *Slepian functions* [9], for the representation of any band-limited signal in the space $\mathcal{H}_L$. Any band-limited signal $f \in \mathcal{H}_L$ can be expressed in the Slepian basis as

$$f(\widehat{\boldsymbol{x}}) = \sum_{\alpha=1}^{L^2} (f)_\alpha \, h_\alpha(\widehat{\boldsymbol{x}}), \tag{2.19}$$

where

$$(f)_\alpha \triangleq \langle f, h_\alpha \rangle = \int_{\mathbb{S}^2} f(\widehat{\boldsymbol{x}}) \, \overline{h_\alpha(\widehat{\boldsymbol{x}})} ds(\widehat{\boldsymbol{x}}) = \frac{1}{\lambda_\alpha} \int_R f(\widehat{\boldsymbol{x}}) \overline{h_\alpha(\widehat{\boldsymbol{x}})} ds(\widehat{\boldsymbol{x}}) \tag{2.20}$$

denotes the Slepian coefficient of index α. The signal $f(\widehat{\boldsymbol{x}})$ within the spatial region R can be well approximated by excluding the basis functions with almost zero concentration within the region in the expansion of the signal given in Equation 2.19; that is, the summation in Equation 2.19 can be truncated at J such that $\lambda_{J+1} \approx 0$ as

$$f(\widehat{\boldsymbol{x}}) \approx \sum_{\alpha=1}^{J} (f)_\alpha \, h_\alpha(\widehat{\boldsymbol{x}}), \quad \widehat{\boldsymbol{x}} \in R. \tag{2.21}$$

The quality of approximation of the signal given in Equation 2.21 within the spatial region R can be measured by defining the quality measure as a ratio of the energy concentration of the approximate representation and the energy of the exact representation within the spatial region, that is,

$$Q(J) = \frac{\displaystyle\int_R \left| \sum_{\alpha=1}^{J} (f)_\alpha h_\alpha(\widehat{\boldsymbol{x}}) \right|^2 ds(\widehat{\boldsymbol{x}})}{\displaystyle\int_R \left| \sum_{\alpha=1}^{L^2} (f)_\alpha h_\alpha(\widehat{\boldsymbol{x}}) \right|^2 ds(\widehat{\boldsymbol{x}})}$$

$$= \frac{\displaystyle\sum_{\alpha=1}^{J} \lambda_\alpha |(f)_\alpha|^2}{\displaystyle\sum_{\alpha=1}^{L^2} \lambda_\alpha |(f)_\alpha|^2}, \tag{2.22}$$

where we have used the orthogonality of Slepian basis functions over the spatial region R, given in Equation 2.17, in obtaining the second equality.

Remark 2.3.2 (Truncation at N) Since the number of Slepian basis functions that is well concentrated in the region is approximately represented by the sum of the eigenvalues, N, given in Equation 2.18, the truncation level in Equation 2.21 can be chosen as $J = N$. We note that such truncation at $J = N$ is based on the assumption that the eigenvalue spectrum has sharp transition from 1 to 0. If for some cases this assumption is not fairly supported, N can be used to estimate the truncation level $J > N$ such that $\lambda_{J+1} \approx 0$.

The representation of the signal within the region R using N basis functions and the computation of Slepian coefficients as an integral over different spatial regions have also been adopted and studied for multidimensional Euclidean domains and various geometries [9,10,13,26,27].

2.3.2 Computation of Slepian functions

Following the formulation of Slepian concentration problem presented above, the computation of Slepian functions for a given band-limit L and an arbitrary-shaped region R is a two-step procedure: (1) calculate the $L^2 \times L^2$ matrix $\mathbf{K}$ composed of inner products of spherical harmonic functions Equation 2.11 over the spatial region R and (2) carry out the eigenvalue decomposition of $\mathbf{K}$ to compute eigenvalues λ_α and eigenvectors $\mathbf{h}_\alpha$ for $\alpha = 1, 2, \ldots, L^2$. Eigenvectors represent the Slepian functions in the spectral (spherical harmonic) domain.

For an arbitrary region R on the sphere, the inner product of spherical harmonics is carried out via numerical integration as there does not exist any (exact) quadrature rule to evaluate the integral of the function on the sphere over arbitrary region, and therefore $K_{\ell m, pq}$, given in Equation 2.11, is computed numerically by employing the approximate quadrature rule for the discretization of integral over arbitrary region R. However, analytical expressions have been presented in the literature for the polar cap region [9,23], polar gap region (equatorial belt) [2], and the limited colatitude-longitude region [28].

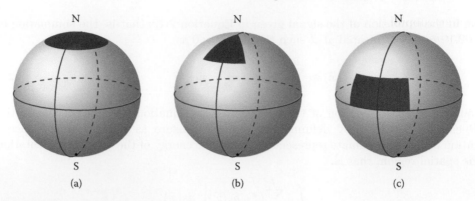

FIGURE 2.2
Different regions on the sphere (shaded in red): (a) Polar cap region R_Θ for $\Theta = \pi/6$. Limited colatitude-longitude region $\tilde{R}$ on the sphere, defined in Equation 2.30, for different parameters; (b) $[\theta_1, \theta_2] = [0, \pi/4]$, $[\phi_1, \phi_2] = [-\pi/6, \pi/6]$; and (c) $[\theta_1, \theta_2] = [\pi/3, \pi/2]$, $[\phi_1, \phi_2] = [-\pi/6, \pi/6]$. Here the periodicity of ϕ implies $-\phi = 2\pi - \phi$.

2.3.2.1   Polar cap region

We first review the computation of Slepian functions for a polar cap region $R_\Theta \triangleq \{\widehat{\boldsymbol{x}}(\theta, \phi) \in \mathbb{S}^2 \,|\, 0 \leq \theta \leq \Theta\}$, parameterized by central angle Θ formed by the boundary of the polar cap with the positive z-axis. As an example, the region R_Θ is shown in Figure 2.2(a) for $\Theta = \pi/6$.

To solve the Slepian concentration problem Equation 2.12, we are first required to evaluate $K_{\ell m,pq}$, given by the integral over R in Equation 2.11. For the special case of a polar cap region R_Θ, analytic expressions have been devised in the literature to compute $K_{\ell m,pq}$ [9,26] in terms of Wigner-3j symbols [23], that is,

$$
\begin{aligned}
K_{\ell m,pq} &= 2\pi \delta_{m,q} N_\ell^m N_p^m \int_0^\Theta P_\ell^m(\cos\theta) P_p^m(\cos\theta) \, \sin\theta \, d\theta \\
&= (-1)^m \frac{\sqrt{(2\ell+1)(2p+1)}}{2} \sum_{n=|\ell-p|}^{\ell+p} \begin{pmatrix} \ell & n & p \\ 0 & 0 & 0 \end{pmatrix} \begin{pmatrix} \ell & n & p \\ m & 0 & -m \end{pmatrix} \\
&\quad \times [P_{n-1}(\cos\Theta) - P_{n+1}(\cos\Theta)],
\end{aligned}
\tag{2.23}
$$

which implies that $K_{\ell m,pq} = 0$ for $m \neq q$ and $K_{\ell m,pq} = K_{\ell(-m),p(-q)}$.

Remark 2.3.3 (On the computation of Slepian functions for polar cap region)
The formulation in Equation 2.23 implies $K_{\ell m,pq} = 0$ for $m \neq q$ and $K_{\ell m,pq} = K_{\ell(-m),p(-q)}$, which, by appropriate switching of rows and columns of the matrix $\mathbf{K}$, enable us to formulate $\mathbf{K}$ as a block diagonal matrix, where nonzero elements with a fixed order m appear next to each other forming submatrices $\mathbf{K}^{(m)}$ of size $(L-m) \times (L-m)$ [2] with

$$
\mathbf{K}^{(m)} = \begin{pmatrix}
K_{mm,mm} & K_{mm,(m+1)m} & \cdots & K_{mm,(L-1)m} \\
K_{(m+1)m,mm} & K_{(m+1)m,(m+1)m} & \cdots & K_{(m+1)m,(L-1)m} \\
\vdots & \vdots & \ddots & \vdots \\
K_{(L-1)m,mm} & K_{(L-1)m,(m+1)m} & \cdots & K_{(L-1)m,(L-1)m}
\end{pmatrix}
\tag{2.24}
$$

for $0 \leq m < L$ and $\mathbf{K}^{(m)} = \mathbf{K}^{(-m)}$. Due to the block diagonal structure of $\mathbf{K}$ for the polar cap region, rather than solving the $L^2 \times L^2$ eigenvalue problem in Equation 2.13, we can solve L smaller subproblems of size $(L-m) \times (L-m)$ for $0 \leq m < L$, the largest being of size $L \times L$ for $m = 0$, of the form

$$\mathbf{K}^{(m)}\mathbf{h}^{(m)} = \lambda\,\mathbf{h}^{(m)}, \qquad (2.25)$$

where $\mathbf{h}^{(m)} = \left((h)^m_{|m|}, (h)^m_{|m|+1}, \cdots, (h)^m_{L-1}\right)'$ contains the spherical harmonic coefficients of order m. For each eigenvector obtained from the problem for fixed m, the associated Slepian functions can be obtained using Equation 2.6 in a rather simplified formulation given by

$$h(\widehat{\boldsymbol{x}}) \equiv h(\theta, \phi) = e^{im\phi} \sum_{\ell=|m|}^{L-1} (h)^m_\ell Y^m_\ell(\theta, 0). \qquad (2.26)$$

The number of Slepian functions that are well concentrated in the polar cap is found by substituting its area $A_\Theta \triangleq 2\pi(1 - \cos\Theta)$ into Equation 2.18 to give the number of well-concentrated Slepian functions in R_Θ as

$$N_\Theta = \frac{(1 - \cos\Theta)}{2} L^2. \qquad (2.27)$$

2.3.2.2 Rotationally symmetric region

We define a rotationally symmetric region centered at $\widehat{\boldsymbol{x}}_c = \widehat{\boldsymbol{x}}_c(\theta_c, \phi_c)$ as $R_\Theta(\widehat{\boldsymbol{x}}_c) = \{\widehat{\boldsymbol{x}}(\theta_c, \phi_c) \in \mathbb{S}^2 \,|\, \Delta(\widehat{\boldsymbol{x}} \cdot \widehat{\boldsymbol{x}}_c) \leq \Theta\}$, where $\Delta(\widehat{\boldsymbol{x}} \cdot \widehat{\boldsymbol{x}}_c)$ denotes the great circle distance between $\widehat{\boldsymbol{x}}$ and $\widehat{\boldsymbol{x}}_c$. The region is rotationally symmetric around its center $\widehat{\boldsymbol{x}}_c$. We note that $R_\Theta = R_\Theta(\widehat{\boldsymbol{x}}_c(0,0))$; that is, a polar cap region is a special case of rotationally symmetric region with $\widehat{\boldsymbol{x}}_c = \widehat{\boldsymbol{x}}_c(0,0)$ (North pole).

For a given band-limit and rotationally symmetric region $R_\Theta(\widehat{\boldsymbol{x}}_c)$, we denote the Slepian functions by $g_\alpha(\widehat{\boldsymbol{x}})$, $\alpha = 1, 2, \ldots, L^2$. Noting that the polar cap region R_Θ, when rotated around the y-axis by θ_c and then by ϕ_c around the z-axis, becomes rotationally symmetric region $R_\Theta(\widehat{\boldsymbol{x}}_c)$, we compute Slepian functions for rotationally symmetric region $R_\Theta(\widehat{\boldsymbol{x}}_c)$ by first computing the Slepian functions $h_\alpha(\widehat{\boldsymbol{x}})$, $\alpha = 1, 2, \ldots, L^2$ for the polar cap region R_Θ followed by the rotation of polar cap Slepian functions as

$$g_\alpha(\widehat{\boldsymbol{x}}) = (\mathcal{D}(\phi_c, \theta_c, 0)h_\alpha)(\widehat{\boldsymbol{x}}), \quad \alpha = 1, 2, \ldots, L^2. \qquad (2.28)$$

The rotation of $h_\alpha(\widehat{\boldsymbol{x}})$ in the spatial domain using the rotation operator $\mathcal{D}(\phi_c, \theta_c, 0)$ in Equation 2.28 is carried out in spectral (spherical harmonic) domain as a linear transformation given by [23]

$$(g_\alpha)^m_\ell = e^{-im\phi_c} \sum_{m'=-\ell}^{\ell} d^{m,m'}_\ell(\theta_c) \left(h_\alpha\right)^{m'}_\ell, \qquad (2.29)$$

where $d^{m,m'}_\ell$ is the Wigner-d function of degree ℓ and orders m, m' [23]. We also note that the number of Slepian functions that are well concentrated in the region $R_\Theta(\widehat{\boldsymbol{x}}_c)$ is N_Θ, given in Equation 2.27.

2.3.2.3 Limited colatitude-longitude region

A limited colatitude-longitude spatial region $\tilde{R}$ is defined as a Cartesian product of a range of limited colatitudes and limited longitudes, that is,

$$\tilde{R} \triangleq \{(\theta, \phi) \colon \theta_1 \leq \theta \leq \theta_2, \phi_1 \leq \phi \leq \phi_2\}. \qquad (2.30)$$

We note that the region $\tilde{R}$ is parameterized by four parameters: θ_1, θ_2, ϕ_1, and ϕ_2. As an example, we have indicated the region $\tilde{R}$ in Figure 2.2(b) and (c) for two different choices of parameters.

For a limited colatitude-longitude spatial region $\tilde{R}$ defined in Equation 2.30, the elements of the matrix $\mathbf{K}$ given in Equation 2.11, have the following analytical expression [28]

$$K_{\ell m, pq} = \sum_{m'=-\ell}^{\ell} \sum_{q'=-p}^{p} F_{m',m}^{\ell} F_{q',q}^{p} Q(m'+q') S(q-m), \tag{2.31}$$

where

$$Q(m) = \begin{cases} \frac{1}{4}\left(2im(\theta_2 - \theta_1) + e^{2im\theta_1} - e^{2im\theta_2}\right), & |m| = 1 \\ \frac{1}{m^2-1}\left(e^{im\theta_1}(-\cos\theta_1 + im\sin\theta_1) \right. \\ \left. + e^{im\theta_2}(\cos\theta_2 - im\sin\theta_2)\right), & |m| \neq 1, \end{cases} \tag{2.32}$$

$$S(m) = \begin{cases} \phi_2 - \phi_1, & m = 0 \\ \frac{i}{m}\left(e^{im\phi_1} - e^{im\phi_2}\right), & m \neq 0, \end{cases} \tag{2.33}$$

and

$$F_{m',m}^{\ell} = (-i)^m \sqrt{\frac{2\ell+1}{4\pi}} \Delta_{m',m}^{\ell} \Delta_{m',0}^{\ell}, \tag{2.34}$$

with $\Delta_{m,n}^{\ell} = d_{\ell}^{m,m'}(\pi/2)$.

We also note that the sum of the eigenvalues of $\mathbf{K}$ for the spatial region $\tilde{R}$ can be computed analytically as [28]

$$N = \sum_{\alpha=1}^{L^2} \lambda_\alpha = \sum_{\ell=0}^{L-1} \sum_{m=-\ell}^{\ell} K_{\ell m, \ell m}$$

$$= \frac{L^2}{4\pi} \int_{\tilde{R}} \sin\theta \, d\theta d\phi = \frac{L^2}{4\pi}(\phi_2 - \phi_1)(\cos\theta_1 - \cos\theta_2). \tag{2.35}$$

2.3.3 Illustrations

2.3.3.1 Eigenvalue spectrum and eigenfunctions

We show the eigenvalue spectrum and Slepian functions for band-limit at $L = 20$ for limited colatitude-longitude regions shown in Figure 2.2(b) and (c), which we refer to as Example 1 and Example 2, respectively. Figures 2.3 and 2.4 shows the magnitude of first 12 most concentrated Slepian functions on the sphere $|h_\alpha(\hat{\boldsymbol{x}})|$, $\alpha = 1, 2, \ldots, 12$ for Example 1 and Example 2, respectively. Figure 2.5(a) and (b) show the first 60 eigenvalues in the eigenvalue spectrum for Example 1 and Example 2, respectively, where we have also plotted the trace of the matrix $\mathbf{K}$ given by Equation 2.18 and computed using Equation 2.35, which serve as a good approximation of the number of well-concentrated eigenfunctions in the region. Since Example 1 has region of smaller area than Example 2, it has a smaller number of well-concentrated eigenfunctions; that is, we have $N = 11$ and $N = 18$ for Examples 1 and 2, respectively.

2.3.3.2 Sparse representation of spatially concentrated signal

Now, we present an illustration to show that the use of Slepian basis enables sparse representation of a spatially concentrated band-limited signal. We take earth's topology

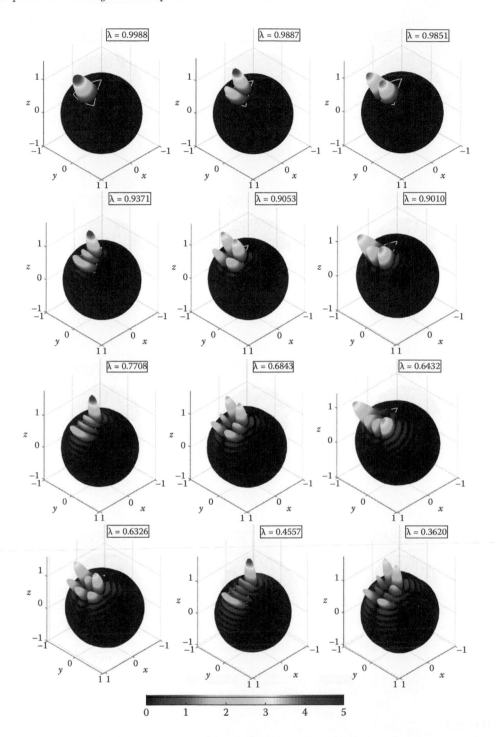

FIGURE 2.3
Magnitude of Slepian functions $|h_\alpha(\widehat{\boldsymbol{x}})|$, $\alpha = 1, 2, \ldots, 12$ concentrated in $\tilde{R}$ in Example 1, shown with their corresponding eigenvalue λ_α, with band-limit $L = 20$ ($N = 11$). The ordering of concentration is left to right, top to bottom.

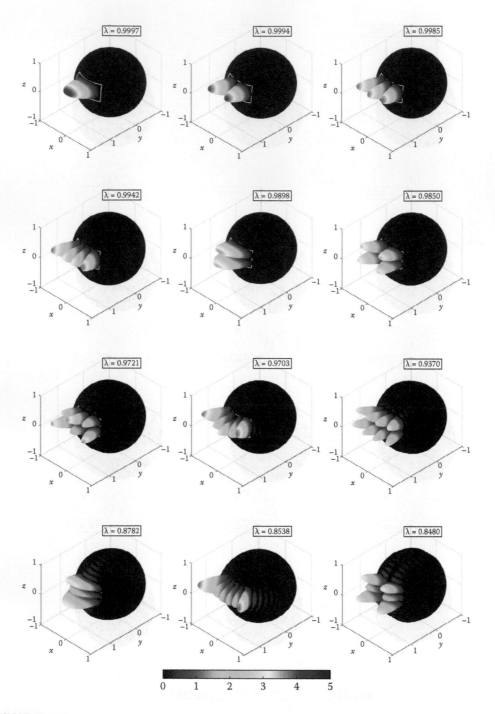

FIGURE 2.4
Magnitude of Slepian functions $|h_\alpha(\widehat{\boldsymbol{x}})|$, $\alpha = 1, 2, \ldots, 12$ concentrated in $\tilde{R}$ in Example 2, shown with their corresponding eigenvalue λ_α, with band-limit $L = 20$ ($N = 18$). The ordering of concentration is left to right, top to bottom.

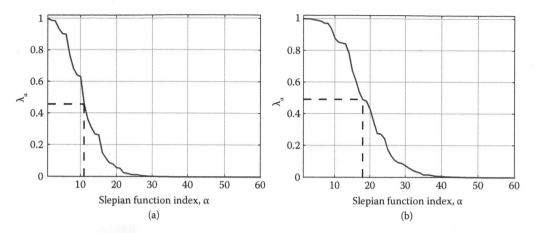

FIGURE 2.5

Eigenvalue spectrum λ_α, $\alpha = 1, 2, \ldots, 60$ for the Slepian functions with band-limit $L = 20$ concentrated in the limited colatitude-longitude regions shown in (a) Figure 2.2(b) with $N = 11$ and (b) Figure 2.2(c) with $N = 18$. The number of well-concentrated eigenfunctions is well approximated by N indicted with the dashed line.

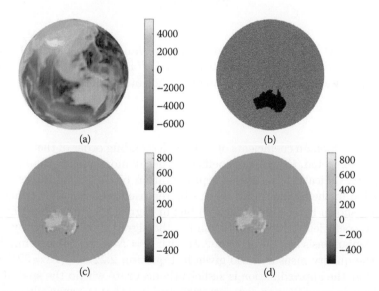

FIGURE 2.6

(a) Earth's topological map and (b) spacial mask for mainland Australia which is used to obtain (c) test signal $f(\hat{x})$ band-limited at $L = 100$. (d) The reconstructed test signal used is obtained using $N = 321$ well-concentrated Slepian functions. The quality measure for the reconstructed signal is $Q(N = 321) = 99.4\%$.

from the EMM2008 model band-limited as a signal shown in Figure 2.6(a) for $L = 100$. Next, we obtain the test signal $f(\hat{x})$ as topology for mainland Australia region R, which we obtain by masking $f(\hat{x})$ with a binary mask of mainland Australia shown in Figure 2.6(b), followed by the band-limiting of the signal at $L = 100$. The test signal is shown in Figure 2.6(c). Since Slepian functions form a complete basis for the subspace of band-limited signals, $f(\hat{x})$ can be represented in the Slepian basis using Equation 2.19. We plot the

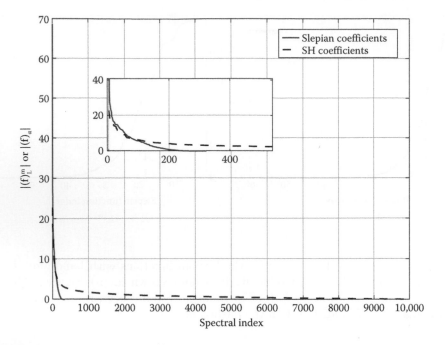

FIGURE 2.7
Spherical harmonic (SH) (black dashed line) and Slepian (blue solid line) coefficients of the band-limited spatially concentrated test signal $f(\hat{\boldsymbol{x}})$ shown in Figure 2.6(c). The magnitude of the spherical harmonic and Slepian coefficients is plotted in descending order of magnitude.

spherical harmonic and Slepian coefficients of $f(\hat{\boldsymbol{x}})$ in descending order of their magnitude in Figure 2.7, where, as expected, the Slepian coefficients decay more quickly than the spherical harmonic coefficients, indicating the sparse representation of the signal. Since $f(\hat{\boldsymbol{x}})$ has a sparse representation in the Slepian basis, it can be represented accurately using $N = 321$ Slepian coefficients, shown by the blue (dashed) line in Figure 2.7, rather than $L^2 = 10{,}000$ spherical harmonic coefficients. Figure 2.6(d) shows the signal reconstructed by expansion in the truncated Slepian basis using Equation 2.21 with $J = N = 321$ Slepian functions. For such truncation, the quality measure ratio given in Equation 2.22 is $Q(N = 321) = 99.4\%$, which illustrates that the approximation is sufficiently accurate within the spatial region of interest. This is the essence of Slepian concentration problem that the spatially concentrated and band-limited signal can be efficiently represented in N number of Slepian basis functions as compared to L^2 spherical harmonic basis.

2.4   Construction of multiscale dictionaries of Slepian functions

We here review the construction of overcomplete (redundant) dictionary of Slepian functions that are band-limited, mostly localized within the given spatial region and multiscale in nature. The dictionary of Slepian functions is similar to the wavelets but allowing for band-limits. The construction of such dictionary allows us to employ the compressive sensing methods to solve ill-posed inverse problems in geophysics and consequently enable the sparse representation of band-limited signals and accurate reconstruction and estimation of band-limited signals from their noisy measurements acquired over the spatially limited region.

As we highlighted earlier, Slepian functions exhibit orthogonality over the region of interest due to which they have been found useful in signal or spectrum estimation in applications [2,10,13], where the data sets or measurements are available over the spatially limited region. With an expectation that the construction of dictionary of Slepian functions that are concentrated in both spatial and spectral domains and enable multiscale analysis is of significant use to solve ill-posed inverse problems, an algorithm has been devised in [29,30] for the construction of dictionary of multiscale Slepian functions. We first present the algorithm for the construction of such dictionary, denoted by $\mathcal{D}$, before analyzing the properties of dictionary elements (Slepian functions).

The construction of dictionary of Slepian functions is based on the sequential division of the region of interest into two (approximately) equal area regions. For a given connected region R and band-limit L, the dictionary $\mathcal{D}$ is composed of functions band-limited at L. The construction of dictionary $\mathcal{D}$ is based on a binary tree shown in Figure 2.8 which requires explanation. Let n be the node capacity of each node of the tree and H be the height of the tree. Each node of the tree corresponds to the first n Slepian functions with band-limit L and concentrated on a subset $R' \subseteq R$, where region for each node is indicated in the tree. Since there are $2^{H+1} - 1$ nodes in a tree of height H, the dictionary $\mathcal{D}$ is composed of $n\left(2^{H+1} - 1\right)$ Slepian functions, that is,

$$\mathcal{D}_{R,L,n} = \{d^{(1,1)}, d^{(1,2)}, \cdots, d^{(1,n)}, \cdots, d^{(2^{H+1}-1,1)}, \cdots, d^{(2^{H+1}-1,n)}\}, \qquad (2.36)$$

for a given band-limit L, node capacity n and spatial region R. Here $d^{(i,j)}$ denotes the jth most concentrated Slepian function obtained as a solution of concentration problem for band-limit L and spatial region $R^{(i)}$.

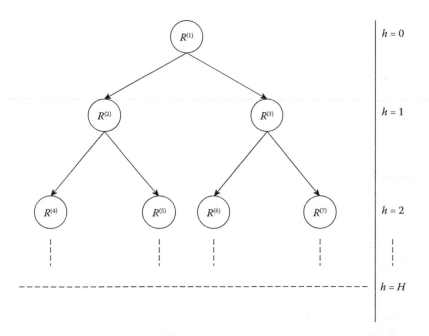

FIGURE 2.8
The binary tree illustrating the sequential division of the region $R = R^{(1)}$ into approximately equal area subregions for the construction of dictionary $\mathcal{D}_{R,L,n}$ of Slepian functions.

To construct the dictionary, we first obtain the Slepian functions for the region $R^{(1)} = R$ (top node) and band-limit L and choose first (most concentrated) n Slepian functions. We then divide the region into two approximately equal area regions $R^{(2)}$ and $R^{(3)}$, which correspond to the next level in the tree, and choose the n most concentrated Slepian functions for each of the two regions in the current level of the tree as elements of the dictionary. Following this, the construction of the dictionary is completed by the subdivision of each region at the current level of the tree into two subregions and obtaining n most concentrated eigenfunctions for each of the subregions. Since the child nodes will be concentrated in disjoint subsets of the parent region, all of their corresponding functions and children are effectively incoherent (which will be explained shortly).

The height H of the tree is chosen as the minimum integer such that we can obtain n well-concentrated Slepian functions when solved for each of the region at the lowest level (H) of the tree and band-limit L. Since we divide area into approximately equal area two subregions as we traverse to lower level in the tree, the area of each of the subregion at lowest level H is $2^{-H} A$, where A is the area of the region R. Noting that the number of well-concentrated functions are given by Equation 2.18, we can determine the height H of the tree for a given region R, band-limit L and node capacity n as

$$H = \left\lceil \log_2 \left(\frac{A L^2}{4 \pi n} \right) \right\rceil, \quad A = \int_R ds(\widehat{\boldsymbol{x}}).$$

2.4.1 Properties of dictionary

The dictionary $\mathcal{D}_{R,L,n}$ of Slepian functions exhibits desirable properties. We refer the readers to [29,30] for the analysis of these properties. The functions in the dictionary are well concentrated in the spatial region R, which follows from the fact the height H of the binary tree used in the construction of dictionary is chosen such that we only include well-concentrated functions at each level of the tree. The dictionary is multiscale in nature, by definition, and is therefore suitable for carrying out localized spectral analysis and efficient representation of data. Furthermore, the dictionary is incoherent; that is, the pairwise inner products of its elements have almost always low amplitude. This is a useful property in signal estimation, which requires the dictionary elements to be mutually incoherent. Consequently, the dictionary of Slepian functions is inherently suitable to solve ill-posed problems of inversion in geophysics. Exploiting these useful properties exhibited by the dictionary of Slepian functions and employing the compressed sensing tools as we formulate in the next section, the use of dictionary of Slepian functions to solve inverse problems yields lower reconstruction error in comparison to the use of conventional inversion techniques.

2.5 Signal estimation on the sphere

We now consider the classical geophysical problem of signal estimation on the sphere from incomplete and/or noisy measurements. We assume that the noisy measurements of the function f are available over the region R, where f is given by

$$f(\widehat{\boldsymbol{x}}) = s(\widehat{\boldsymbol{x}}) + z(\widehat{\boldsymbol{x}}). \tag{2.37}$$

Here $s(\widehat{\boldsymbol{x}})$ denotes the signal to be estimated contaminated with the noise $z(\widehat{\boldsymbol{x}})$. Let $X = \{\widehat{\boldsymbol{x}}_k\}_{k=1}^M \in R$ be a set of M sampling points on which the measurements are made. Given the measurements of f over some sampling points X for spatially limited region R (e.g., the polar gap region [2]), we seek to estimate the signal s from these incomplete and/or noisy measurements.

2.5.1 Problem formulation

Now, we express the measured signal using expansion of the function f in spherical harmonic basis, Slepian basis, and Slepian dictionary functions, present the estimation of signal in these different basis, and review the potential advantages enabled by these different representations. By defining the vector

$$\tilde{\boldsymbol{f}} = [f(\widehat{\boldsymbol{x}}_1),\, f(\widehat{\boldsymbol{x}}_2),\, \ldots,\, f(\widehat{\boldsymbol{x}}_M)], \tag{2.38}$$

containing the measurements made over the sampling points in X, we can linearly model the measurements in spherical harmonic basis, Slepian basis, and Slepian dictionary.

Using spherical harmonic domain representation for the signal $s(\widehat{\boldsymbol{x}})$, we can express the measured signal as

$$f(\widehat{\boldsymbol{x}}_k) = n(\widehat{\boldsymbol{x}}_k) + \sum_{\ell=0}^{L-1} \sum_{m=-\ell}^{\ell} \left(s\right)_\ell^m Y_\ell^m(\widehat{\boldsymbol{x}}_k). \tag{2.39}$$

Since Slepian basis serves as an alternative basis for the representation of band-limited signal, we can express the measured signal, following Equation 2.21, as

$$f(\widehat{\boldsymbol{x}}_k) = z(\widehat{\boldsymbol{x}}_k) + \sum_{\alpha=1}^{J} (s)_\alpha \, h_\alpha(\widehat{\boldsymbol{x}}_k), \tag{2.40}$$

where we have used the truncated expansion in Slepian basis. For $J = L^2$, the formulation in Equations 2.39 and 2.40 is equivalent. Since we are interested in the estimation of the signal over the region R (region of observation), we choose $J = N$ (following remark 2.3.2); that is, we only include the well-concentrated Slepian functions (basis) in the expansion in Equation 2.40. With this truncation and using the expansion of Slepian functions in spherical harmonic domain, we can write Equation 2.40 as

$$f(\widehat{\boldsymbol{x}}_k) = z(\widehat{\boldsymbol{x}}_k) + \sum_{\alpha=1}^{N} (s)_\alpha \sum_{\ell=0}^{L-1} \sum_{m=-\ell}^{\ell} \left(h_\alpha\right)_\ell^m Y_\ell^m(\widehat{\boldsymbol{x}}_k). \tag{2.41}$$

In addition to the formulation of measurements in spherical harmonic basis Equation 2.39 and truncated Slepian basis Equation 2.41, the measurements can also be modeled using the dictionary $\mathcal{D}_{R,L,n}$ of Slepian functions presented in the previous section as

$$f(\widehat{\boldsymbol{x}}_k) = z(\widehat{\boldsymbol{x}}_k) + \sum_{\alpha=1}^{n} \sum_{j=1}^{2^{H+1}-1} (s)^{(j,\alpha)} \, d^{(j,\alpha)}(\widehat{\boldsymbol{x}}_k), \tag{2.42}$$

where

$$(s)^{(j,\alpha)} = \int_{\mathbb{S}^2} s(\widehat{\boldsymbol{x}}) \, \overline{d^{(j,\alpha)}(\widehat{\boldsymbol{x}})} \, ds(\widehat{\boldsymbol{x}}) \tag{2.43}$$

is referred to as the dictionary coefficient that represents the contribution of dictionary element $d^{(k,\alpha)}(\widehat{\boldsymbol{x}})$ in the signal s. We next review the estimation of signal using three different representations of the measured signal, that is, in spherical harmonic basis, truncated Slepian basis, and dictionary representation.

2.5.2 Estimation—inverse problem

All of the three models of the measured signal f, given in Equations 2.39, 2.40, and 2.42, are of the form

$$\tilde{\boldsymbol{f}} = \mathbf{A}\,\tilde{\boldsymbol{s}} + \tilde{\boldsymbol{z}}, \tag{2.44}$$

which requires explanation: $\tilde{\boldsymbol{f}}$ and $\tilde{\boldsymbol{z}}$ are column vectors of length M denoting the samples of the measured signal and unknown noise, respectively, $\mathbf{A}$ denotes the (sensing) matrix of spherical harmonic basis, (complete or truncated) Slepian basis, or dictionary elements, and $\tilde{s}$ is the vector containing spherical harmonic coefficients, Slepian coefficients, or dictionary coefficients which we wish to estimate.

For the representation in spherical harmonic basis, the matrix $\mathbf{A}$ of spherical harmonic basis is of size $M \times L^2$ and depends on the sampling set X. The representation in spherical harmonic basis has an advantage that the inversion gives an estimate of the signal, not only over the region R (region of observation) but also anywhere on the sphere. In practice, $\mathbf{A}$ is highly rank deficient and therefore the inversion of the model in Equation 2.39 to estimate the signal s is ill-conditioned and requires regularization [2,31].

Instead of employing regularization, the alternative approach is to restrict the estimation of the signal over the region R only. In comparison to the spherical harmonics, which form global basis on the sphere, the Slepian basis functions are orthogonal over the region R and enable sparse representation of the signal over the localized region; the truncated representation of the measured signal in Slepian basis is more suitable for signal estimation over the localized region. The use of Slepian basis for signal estimation on sphere has been extensively investigated in the literature [2,29,31]. The signal estimation requires to adopt the model of the measured signal given in Equation 2.41 and employ least-squares approach. We refer to such estimation method as Slepian least-squares method.

Instead of employing least squares for the signal estimation, an alternative approach is to include all Slepian functions in the signal representation, that is, use the representation given in Equation 2.40, assume κ-sparsity of the signal s with respect to the Slepian basis, and solve the following minimization problem [29]

$$\underset{\tilde{s}}{\text{minimize}} \quad \|\tilde{f} - \mathbf{A}\tilde{s}\|_2 \quad \text{subject to} \quad \|\tilde{s}\|_0 \leq \kappa, \tag{2.45}$$

where $\mathbf{A}$ is a matrix of size $M \times L^2$ and composed of L^2 Slepian basis functions evaluated at M sampling points of the set X, $\|\cdot\|_2$ denotes the 2-norm, and $\|\tilde{s}\|_0 = |\text{supp}(\tilde{s})| = |\{\alpha : s_\alpha \neq 0\}|$.

For the representation of the measured signal, given in Equation 2.42, in terms of multiscale and spatially localized dictionary elements, we note that the sparsity is enabled by the overcompleteness of the dictionary. Again, assuming κ-sparsity, the estimate of the signal can be obtained using the model in Equation 2.42 by solving the minimization problem of the form given in Equation 2.46. For the dictionary representation, the matrix $\mathbf{A}$ is of size $M \times n \left(2^{H+1} - 1\right)$ with columns representing the dictionary elements evaluated at M sampling points of the set X. For both the representation of the measured signal in Slepian basis and dictionary elements, we note that the κ can be chosen anywhere between 1 and N as N number of well-concentrated Slepian functions provide almost complete support in the region R for any band-limited signal.

The problem in Equation 2.45 is non-convex and therefore becomes intractable. Provided that the matrix A fulfills one of a number of special properties, e.g., the well-known restricted isometry property (RIP) [32], the problem in Equation 2.45 can be solved as an equivalent convex problem given by

$$\underset{\tilde{s}}{\text{minimize}} \quad \|\tilde{s}\|_1 \quad \text{subject to} \quad \|\tilde{f} - \mathbf{A}\tilde{s}\|_2 \leq \epsilon. \tag{2.46}$$

In practice, RIP and/or other properties may not be checked as the matrix $\mathbf{A}$ depends on the sampling set X. In such cases when the equivalent formulation of the problem (Equation 2.45) in Equation 2.46 is not valid, an estimate of the signal can be obtained as debiased ℓ_1 solution [29] which includes the LASSO (ℓ_1) estimation of support using a sparsity-inducing prior, followed by the debiasing using least squares on the support set.

For the ill-posed problems in geophysics, such as the estimation of signal from incomplete and noise measurements, it has been demonstrated that the debiased ℓ_1 estimation using the models of the measured signal in Slepian basis and dictionary elements enable superior performance in terms of average reconstruction error over the region R over the standard techniques (regularized inversion).

2.6 Future considerations

Slepian functions on the sphere form an alternative basis for the representation of band-limited signals. For spatially localized/concentrated signals/data sets on the sphere, the Slepian functions enable sparse representation and employment of compressed sensing tools for signal estimation from noisy and/or incomplete measurements. The construction of dictionary of Slepian functions reviewed here is a pioneering step in this direction and has been demonstrated to be of significant use to determine the solution of ill-posed problems in geophysics and beyond. However, the construction of the dictionary can be further customized to suit the needs of applications. We here indicate the possible future research directions. For example, the division of each region into two equal area regions in the current construction of dictionary using the binary tree needs to be addressed. Currently, the division of the region into subregions is based on the zero crossing of the second most concentrated Slepian function for the region. Such division may be suitable for circularly connected regions; however, it needs to be improved for the division of arbitrary-shaped region. Furthermore, since the dictionary of Slepian functions shares most of the properties of third generation wavelets, the concept of the construction presented here can be used to construct the wavelet-like transform on the whole sphere, that is, beginning the construction with $R = \mathbb{S}^2$ and employing the equal area division into limited colatitude-longitude regions. Such a reconstruction of the wavelets enables the equal area division of the region into two subregions and allows us to compute the Slepian functions exactly, using the formulation presented in Section 2.3.2.1 up to the computation of matrix. As also indicated in [29], another research direction is to construct an alternative difference dictionary by orthogonalizing each node with its parent and sibling using Gram–Schmidt orthogonalization.

The computation of Slepian functions for an arbitrary region is currently carried out employing numerical integration techniques and therefore the analytic computation remains an open problem. Currently, we can compute the matrix defining the Slepian concentration problem for some special cases which we have reviewed here. For an arbitrary region, the triangulation scheme can be used to partition the region into spherical triangles and then the integration is carried out over each spherical triangle. Analytic integration of spherical harmonic may be explored in future in this direction to support the computation of Slepian functions for arbitrary spatial region. Furthermore, the computation of Slepian functions of band-limit L requires the eigen-decomposition of the matrix of size $L^2 \times L^2$ which becomes infeasible even for moderate band-limits ($L \approx 100-200$). Consequently, there is a need to develop computationally efficient algorithms for the computation of Slepian functions on the sphere for the analysis of modern data sets which support large band-limits ($L \approx 2000-4000$).

Acknowledgment

We gratefully acknowledge the support provided by Alice P. Bates in generating some of the illustrations included in this chapter. We also acknowledge the use of software available on Frederik Simons' webpage: http://www.frederik.net.

References

[1] A. Albertella, F. Sansò, and N. Sneeuw, Band-limited functions on a bounded spherical domain: The Slepian problem on the sphere, *J. Geodesy*, vol. 73, no. 9, pp. 436–447, 1999.

[2] F. J. Simons and F. A. Dahlen, Spherical Slepian functions and the polar gap in geodesy, *Geophys. J. Int.*, vol. 166, no. 3, pp. 1039–1061, 2006.

[3] C. Harig, K. W. Lewis, A. Plattner, and F. J. Simons, A suite of software analyzes data on the sphere, *Eos*, vol. 96, pp. 18–22, 2015.

[4] P. Audet, Directional wavelet analysis on the sphere: Application to gravity and topography of the terrestrial planets, *J. Geophys. Res.*, vol. 116, E01003, 1–16, 2011.

[5] A. P. Bates, Z. Khalid, and R. A. Kennedy, An optimal dimensionality sampling scheme on the sphere with accurate and efficient spherical harmonic transform for diffusion MRI, *IEEE Signal Process. Lett.*, vol. 23, pp. 15–19, 2016.

[6] J.-L. Starck, Y. Moudden, P. Abrial, and M. Nguyen, Wavelets, ridgelets and curvelets on the sphere, *Astron. Astrophys.*, vol. 446, no. 3, pp. 1191–1204, 2006.

[7] A. P. Bates, Z. Khalid, and R. A. Kennedy, Novel sampling scheme on the sphere for head-related transfer function measurements, *IEEE/ACM Trans. Audio Speech Language Process.*, vol. 23, pp. 1068–1081, 2015.

[8] A. Albertella and N. Sneeuw, The analysis of gradiometric data with Slepian functions, *Phys. Chem. Earth A*, vol. 25, no. 9–11, pp. 67–672, 2000.

[9] F. J. Simons, F. A. Dahlen, and M. A. Wieczorek, Spatiospectral concentration on a sphere, *SIAM Rev.*, vol. 48, no. 3, pp. 504–536, 2006.

[10] M. A. Wieczorek and F. J. Simons, Localized spectral analysis on the sphere, *Geophys. J. Int.*, vol. 162, pp. 655–675, 2005.

[11] C. Beggan, J. Saarimäki, K. Whaler, and F. Simons, Spectral and spatial decomposition of lithospheric magnetic field models using spherical Slepian functions, *Geophys. J. Int.*, vol. 193, no. 1, pp. 136–148, 2013.

[12] M. A. Sharifi and S. Farzaneh, Regional TEC dynamic modeling based on Slepian functions, *Adv. Space Res.*, vol. 56, no. 5, pp. 907–915, 2015.

[13] A. M. Wieczorek and J. F. Simons, Minimum-variance multitaper spectral estimation on the sphere, *J. Fourier Anal. Appl.*, vol. 13, no. 6, pp. 665–692, 2007.

[14] K. Jahn and N. Bokor, Vector Slepian basis functions with optimal energy concentration in high numerical aperture focusing, *Optic. Commun*, vol. 285, no. 8, pp. 2028–2038, 2012.

[15] C. Harig and F. J. Simons, Ice mass loss in Greenland, the Gulf of Alaska, and the Canadian Archipelago: Seasonal cycles and decadal trends, *Geophys. Res. Lett.*, vol. 43, no. 7, pp. 3150–3159, 2016.

[16] F. A. Dahlen and F. J. Simons, Spectral estimation on a sphere in geophysics and cosmology, *Geophys. J. Int.*, vol. 174, no. 3, pp. 774–807, 2008.

[17] A. Plattner and F. J. Simons, High-resolution local magnetic field models for the Martian South Pole from Mars Global Surveyor data, *J. Geophys. Res. Planets*, vol. 120, no. 9, pp. 1543–1566, 2015.

[18] C. Lessig and E. Fiume, On the effective dimension of light transport, in *Proceedings of Eurographics Symposium on Rendering 2010*, vol. 29, no. 4, Saarbrücken, Germany, 2010, pp. 1399–1403.

[19] D. Slepian and H. O. Pollak, Prolate spheroidal wave functions, Fourier analysis and uncertainty—I, *Bell Syst. Tech. J.*, vol. 40, no. 1, pp. 43–63, 1961.

[20] H. J. Landau and H. O. Pollak, Prolate spheroidal wave functions, Fourier analysis and uncertainty—II, *Bell Syst. Tech. J.*, vol. 40, no. 1, pp. 65–84, 1961.

[21] H. J. Landau and H. O. Pollak, Prolate spheroidal wave functions, Fourier analysis and uncertainty—III: The dimension of the space of essentially time- and band-limited signals, *Bell Syst. Tech. J.*, vol. 41, no. 4, pp. 1295–1336, 1962.

[22] D. Slepian, Prolate spheroidal wave functions, Fourier analysis and uncertainty—IV: Extensions to many dimensions; generalized prolate spheroidal functions, *Bell Syst. Tech. J.*, vol. 43, no. 6, pp. 3009–3057, 1964.

[23] R. A. Kennedy and P. Sadeghi, *Hilbert Space Methods in Signal Processing*, Cambridge, UK: Cambridge University Press, 2013.

[24] Z. Khalid and R. A. Kennedy, Maximal multiplicative spatial-spectral concentration on the sphere: Optimal basis, in *Proceedings of IEEE International Conference on Acoustics, Speech and Signal Processing, ICASSP'2015*, Brisbane, Australia, April 2015, pp. 4160–4164.

[25] J. Mathews, J. Breakall, and G. Karawas, The discrete prolate spheroidal filter as a digital signal processing tool, *IEEE Trans. Acoust., Speech, Signal Process.*, vol. 33, no. 6, pp. 1471–1478, 1985.

[26] Z. Khalid, R. A. Kennedy, and J. D. McEwen, Slepian spatial-spectral concentration on the ball, *Appl. Comput. Harmon. Anal.*, vol. 40, no. 3, pp. 470–504, 2016.

[27] F. J. Simons and D. V. Wang, Spatiospectral concentration in the Cartesian plane, *Int. J. Geomath.*, vol. 2, no. 1, pp. 1–36, 2011.

[28] A. P. Bates, Z. Khalid, and R. A. Kennedy, Slepian Spatial-Spectral Concentration. Problem on the Sphere: Analytical Formulation for Limited Colatitude-longitude Spatial Region, *IEEE Trans. Signal Process.*, vol. 65, no. 6, pp. 1527–1537, 2017.

[29] E. Brevdo, Efficient representations of signals in nonlinear signal processing with applications to inverse problems, PhD dissertation, Princeton University, 2011.

[30] F. J. Simons, I. Loris, E. Brevdo, and I. C. Daubechies, Wavelets and wavelet-like transforms on the sphere and their application to geophysical data inversion, in *Wavelets and Sparsity XIV*, SPIE, 2011, p. 81380X.

[31] F. J. Simons, Slepian functions and their use in signal estimation and spectral analysis, in *Handbook of Geomathematics*, Eds. W. Freeden, M.Z. Nashed, and T. Sonar, Springer: New York, pp. 891–923, 2010.

[32] D. L. Donoho, Compressed sensing, *IEEE Trans. Inf. Theory*, vol. 52, pp. 1289–1306, 2006.

3

Compressive Sensing–Based High-Resolution Imaging and Tracking of Targets and Human Vital Sign Detection behind Walls

Ahmad Hoorfar, Ozlem Kilic, and Aly E. Fathy

CONTENTS

3.1 Introduction

The high-resolution and noninvasive imaging of stationary and moving human and its vital signatures through clutter, such as visually opaque obstacles (e.g. walls, doors, ground) has sparked a growing interest in through-the-wall radar imaging (TWRI) and ground-penetrating radar (GPR) in both military and civilian applications, such as homeland security, urban counterterrorism, search and rescue missions, and monitoring of the sick and elderly [1,2]. In many TWRI situations, exterior and/or interior building walls induce shadowing effects, which may result in image degradation, errors in geolocation, or complete masking of targets within the building. All of these effects are attributed to direct and multiple reflections from the scene, as well as amplitude and phase distortions of the electromagnetic waves as they penetrate the medium [3–8]. Furthermore, the effects are exacerbated for multilayered and composite walls [9,10]. In order to aid in mitigating these adverse wall effects and enhance the capability for imaging and classification of targets behind walls, various advanced synthetic aperture radar (SAR) and multiple-input multiple-output (MIMO) radar techniques have been developed over the years to accurately estimate the wall constituent parameters. We note that in addition to TWRI, intrawall imaging is also important in many scenarios. For example, detecting and imaging of metallic reinforcement, or locating embedded water pipes and other construction materials, may assist law enforcement personnel in operational planning of counterterrorism or other hostile situations, firefighters in civilian search and rescue operations, and utility companies in planning and delivering electricity, gas, and water services.

In order to achieve high target imaging and tracking resolution in the cross and down ranges in all of the above applications, a long aperture must be synthesized and an ultra-wideband (UWB) signal must be properly transmitted. This results in a large amount of space–time or space–frequency data and a long data acquisition time, together with large storage and memory requirements, which may be expensive and complex to implement. Hence, smart reduction of the data volume using compressive sensing (CS) is important in TWRI applications, as it accelerates processing and, subsequently, allows prompt actionable intelligence. It also relaxes constraints on system aperture and bandwidth and creates different design and deployment paradigms that are more flexible than those underlying conventional SAR operations. The concept of CS indeed relinquishes the dogma of the Nyquist theorem or the need for a large number of measurements, which has ruled the entire electronics industry for a very long time. It is a relatively new concept, with possibilities for application in a wide variety of areas. The capability of CS to reconstruct a sparse signal from far fewer nonadaptive measurements provides a new perspective for meeting these objectives in TWRI and GPR. Luckily, there have been several approaches reported for application of CS to radar imaging [11–15]. In general, CS for through-the-wall imaging and tracking is a hybrid of CS and urban sensing, where it enables reliable high-resolution images of indoor targets using a very small percentage of the entire data volume [16–20].

In this chapter, we outline some of our recent works in application of CS in SAR and MIMO radar imaging and tracking of multiple targets simultaneously behind multilayer or multiple walls. We also discuss the use of CS in detection of human vital signatures in such a cluttered environment. In our proposed approach, wall effects, including wave propagation through the wall and the multiple reflections within the wall, are fully accounted for by employing the full Green's functions of layered media in image formation and target tracking and by combining them with CS for near real-time data acquisition and postprocessing of collected data. In addition to the standard CS using l_1-based minimization, we also present the use of total variation minimization (TVM) in a sparse reconstruction of through-the-wall and GPR targets. TVM minimizes the gradient magnitude of the image and can result in better edge preservation and shape reconstruction than standard CS. Finally, the chapter concludes with a discussion on the design of a low complexity receiver system with reduced analog to digital converter (ADC) requirements suitable for SAR, MIMO, UWB, or stepped frequency.

3.2   CS-based radar imaging of through-the-wall and GPR targets

Over the last decade, various beamforming, diffraction tomography (DT), and other imaging techniques have been developed for TWRI [1–10,21], together with associated techniques for wall parameter estimations [22–25]. Efficient DT-based techniques have also been reported for intrawall imaging, that is, the characterization of hidden objects within the wall [26]. Even though the DT-based imaging results in near real-time image processing, it requires the full frequency measurements at each receiving antenna element and therefore does not contribute to reductions in data acquisition time or complexity and cost of the radar system. More recently, several research groups have investigated the application of CS to TWRI for scenarios when the target space behind the wall is sparse [16–20]. Most of the reported methods of CS for TWRI were developed for monostatic SAR and are only capable of imaging a target behind a single-layer wall. In urban sensing applications, however, we often encounter situations requiring us to detect and identify targets inside a building with multilayered inner walls or walls separated by a hallway. This is challenging and beyond the capability of the delay-and-sum beamforming approach based on solving a nonlinear equation in order to find the wave propagation time as it travels through the wall [3]. Moreover, one of the main drawbacks of TWRI with monostatic SAR is the long data acquisition time to synthesize an aperture. The target in TWRI is not always stationary. During a long data collection time, the target may have moved such that the imaging result is no longer valid, as the target is not in its previous position anymore. The MIMO radar system does not need aperture synthesis; thus, its data acquisition is in real time. In the following, we outline a generalized CS-based formulation of the 2D imaging of targets behind single or multilayered building walls as well as imaging of intrawall targets for both SAR and MIMO radar by collecting measurements at random frequency points and antenna locations. In addition, we discuss the application of TVM to TWRI and GPR. TVM is based on minimizing the gradient magnitude of the image and can result in better edge preservation and shape reconstruction than the standard l_1 minimization-based CS.

3.2.1 Forward model for imaging through layered walls

Figure 3.1 shows a typical scenario of TWRI with MIMO radar. The wall region may consist of a single layer or multilayered walls. The MIMO radar system consists of P transmitting antennas located at $\boldsymbol{r}_{tp} = (x_{tp}, z_{tp})$ and Q receiving antennas located at $\boldsymbol{r}_{rq} = (x_{rq}, z_{rq})$. The operating frequency covers the range from $f_{\min}$ to $f_{\max}$ with a frequency step Δf, resulting in M frequency bins. For monostatic SAR, the transmitter and receiver are at the same location, that is, $\boldsymbol{r}_{rp} = \boldsymbol{r}_{tq}$, $p = 1, 2, \ldots, Q$.

Under the point target model, which ignores the multiple scattering effects, the received signal can be written as follows [20]:

$$E_s(\boldsymbol{r}_{rq}, \boldsymbol{r}_{tp}, k_m) = \int G(\boldsymbol{r}_{rq}, \boldsymbol{r}, k_m) G(\boldsymbol{r}, \boldsymbol{r}_{tp}, k_m) \sigma(\boldsymbol{r}) d\boldsymbol{r} \tag{3.1}$$

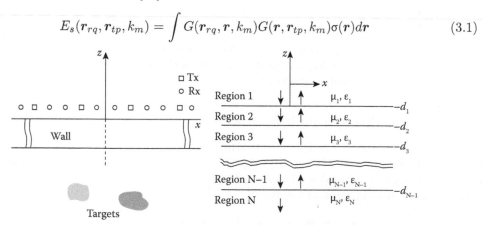

FIGURE 3.1
Configuration of through-the-wall imaging with multiple-input multiple-output (MIMO) radar and model of the multilayered wall (Left figure after Zhang, W., and Hoorfar, A., *IEEE Antennas Wireless Propag. Lett.*, 14, 10521055, 2015).

and the through-the-wall image can be reconstructed as follows:

$$I(\boldsymbol{r}) = \int_{k_{\min}}^{k_{\max}} dk_m \int_{L_{t\min}}^{L_{t\max}} d\boldsymbol{r}_{tp} \int_{L_{r\min}}^{L_{r\max}} d\boldsymbol{r}_{rq} E_s(\boldsymbol{r}_{rq}, \boldsymbol{r}_{tp}, k_m)$$
$$\cdot\, G^{-1}(\boldsymbol{r}_{rq}, \boldsymbol{r}, k_m) G^{-1}(\boldsymbol{r}, \boldsymbol{r}_{tp}, k_m), \tag{3.2}$$

where $E_s(\boldsymbol{r}_{rq}, \boldsymbol{r}_{tp}, k_m)$ is the received scattered field at the qth receiver location due to the illumination of the pth transmitter; σ is the reflectivity of the target; $\boldsymbol{r}$ is the position vector of the target; $\boldsymbol{r} = (x, z)$; k_m is the freespace wavenumber of the mth frequency; and $G(\boldsymbol{r}, \boldsymbol{r}_{tp}, k_m)$ and $G(\boldsymbol{r}_{rq}, \boldsymbol{r}, k_m)$ are the layered medium Green's functions, which characterize the wave propagation process from the transmitter to the target and from the target to the receiver in the presence of wall layers. The freespace wavenumbers of the minimum and maximum operating frequencies are denoted by $k_{\min}$ and $k_{\max}$, respectively, and $L_{t\min}$, $L_{t\max}$, $L_{r\min}$, and $L_{r\max}$ are the minimum and maximum extents of the transmitter and receiver elements. Given the received signal at all antenna locations and all frequency points, the image in Equation 3.2 can then be reconstructed through discretization as follows:

$$I(\boldsymbol{r}) = \sum_{m=1}^{M} \sum_{q=1}^{Q} \sum_{p=1}^{P} E_s(\boldsymbol{r}_{rq}, \boldsymbol{r}_{tp}, k_m) G^{-1}(\boldsymbol{r}_{rq}, \boldsymbol{r}, k_m) G^{-1}(\boldsymbol{r}, \boldsymbol{r}_{tp}, k_m) \tag{3.3}$$

Computation of the above imaging formula in Equation 3.3 requires an efficient evaluation of the layered medium Green's function, which involves a time-consuming computation of Sommerfeld-type integrals [27]. To compromise between the computation time and the exact evaluation of the Green's function, the far-field approximation for Green's function may be used [4,10]. This is valid in most TWRI scenarios as the target is often located in the far field of the antenna aperture. The reconstructed image in Equation 3.3 can then be written as follows [20]:

$$I(\boldsymbol{r}) = \sum_{m=1}^{M} \sum_{q=1}^{Q} \sum_{p=1}^{P} E_s(\boldsymbol{r}_{rq}, \boldsymbol{r}_{tp}, k_m) T_t^{-1}(\boldsymbol{r}, \boldsymbol{r}_{tp}, k_m) T_r^{-1}(\boldsymbol{r}, \boldsymbol{r}_{rq}, k_m) e^{jk_m(R_{tp}+R_{rq})}, \tag{3.4}$$

where T_t and T_r are the transmission coefficients from the transmitter and receiver to the target, respectively, which can be derived by applying the boundary conditions at the interface of each layer or by using an equivalent cascaded transmission-line model of the layered medium [27]. R_{tp} and R_{rq} are the distances from the pth transmitter and qth receiver to the target behind the wall, respectively.

3.2.2   Imaging through multilayered walls with CS

For the above P-transmitter, Q-receiver MIMO step-frequency radar with total of M transmitted frequency bins, the received signal at the qth receiver due to the illumination of the pth transmitter, after discretization of received scattered field in Equation 3.1, can be written in the matrix form as follows:

$$\mathbf{y}_{p,q} = \boldsymbol{\Psi}_{p,q}\mathbf{s}; \quad [\mathbf{y}_{p,q}]_m = E_s(\boldsymbol{r}_{rq}, \boldsymbol{r}_{tp}, k_m), \tag{3.5}$$

where $\mathbf{s}$ is a weighted indicator vector defining the target space behind the wall, which is a $K \times L$ pixel image, vectorized into a $KL \times 1$ column vector. In essence, each element of vector $\mathbf{s}$ represents the target reflectivity, σ, on a given image pixel. The target space and measured data are related through the wall's "dictionary" $\boldsymbol{\Psi}_{p,q}$, which is an $M \times KL$ matrix, encompassing the two-way wave propagation through the multilayered wall. Utilizing the far-field representation of the Green's functions, the (m, n)th element of $\mathbf{\Psi}_{p,q}$ may be expressed as follows:

$$[\mathbf{\Psi}_{p,q}]_{m,n} = T_t(\mathbf{r}_{rq}, \mathbf{r}_n, k_m) T_r(\mathbf{r}_n, \mathbf{r}_{tp}, k_m) e^{jk_m(R_{tp}+R_{rq})}, \qquad (3.6)$$

where R_{tp} and R_{rq} are the distances from the pth transmitter and qth receiver to the target, respectively.

As opposed to the standard beamforming technique using data measured at all antenna locations for all frequencies, in order to reduce the amount of data in CS-based TWRI, we measure at a random subset of J frequencies at each receiver location, where $J \ll M$. Then the measurement data at the qth receiving antenna can be written as follows:

$$\mathbf{y}_{p,q} = \Phi_{p,q} \mathbf{\Psi}_{p,q} \mathbf{s}, \qquad (3.7)$$

where $\Phi_{p,q}$ is a measurement matrix in which each row has only one nonzero element, which is equal to one. The location of this valued-one element corresponds to the index of the measured frequency bin in the transmitting frequency sequence. In essence, the above measurement matrix is formed by randomly choosing J rows from $\mathbf{\Psi}_{p,q}$. In addition to the random frequency sampling, to further reduce the MIMO radar measurement data, we randomly select Q_r receivers from the total Q receivers, $Q_r < Q$, and P_r transmitters from the total P transmitters, $P_r < P$. These will further reduce the complexity and cost of the MIMO radar system for TWRI.

For coherent processing of the received data, we superpose the matrix equation in Equation 3.8 for all Q_r receivers and P_r transmitters to form a large matrix: $\mathbf{y} = \mathbf{\Phi}\mathbf{\Psi}\mathbf{s} = \mathbf{\Theta}\mathbf{s}$. The reconstruction of $\mathbf{s}$ is then a sparse constraint optimization problem. From [28,29], assuming $\mathbf{\Theta}$ satisfies the restricted isometry property, robust reconstruction of a sparse image under noise-corrupted data can be achieved by solving the convex optimization problem in Equation 3.8, which is also referred to as the *Dantzig selector*:

$$\hat{s} = \arg\min \|s\|_1, \; s.t. \; \|\Theta^T(y - \Theta s)\|_\infty < \delta, \qquad (3.8)$$

where δ represents a small tolerance error, which can be determined using the cross-validation strategy in [15,29] for an automatic selection of the error in the optimization process. In (3.8), $\|\cdot\|$ is the l_1-norm.

In the following, we present numerical examples for different wall-target scenarios for SAR and MIMO radar imaging through single-layer and multilayered walls. The Dantzig selector solver in the sparse constraint optimization package l_1-magic [30] is employed to solve Equation 3.8. For calculation of transmission coefficients in Equations 3.4 and 3.6, we assume that the wall parameters are known and used in the imaging. We refer the reader to [22–25] for the estimation of parameters for single-layer or multilayered walls using time and frequency domain methods.

In the first example, we investigate the imaging of a rectangular Perfect Electric Conductor (PEC) target behind a single-layer homogeneous wall in Figure 3.2a using SAR. The scattered data are simulated using the 2D finite-difference time-domain method and then Fast Fourier Transform (FFT) is performed to get the frequency domain data. The target is 0.3×0.2 m and is 1.5 m away from the front boundary of the wall. The permittivity, conductibility, and thickness of the wall are $\varepsilon_r = 6$, $\sigma = 0.03$ S/m, and $d = 0.2$ m, respectively. The SAR scans the region of interest at a standoff distance of 0.3m, synthesizing a 2 m aperture with 0.05 m interelement spacing. The operating frequency ranges from 1–3 GHz with 61 equally spaced frequency bins. Figure 3.2b shows the imaging result using the back-projection through-wall beamformer in Equation 3.4 with full data, that is, data

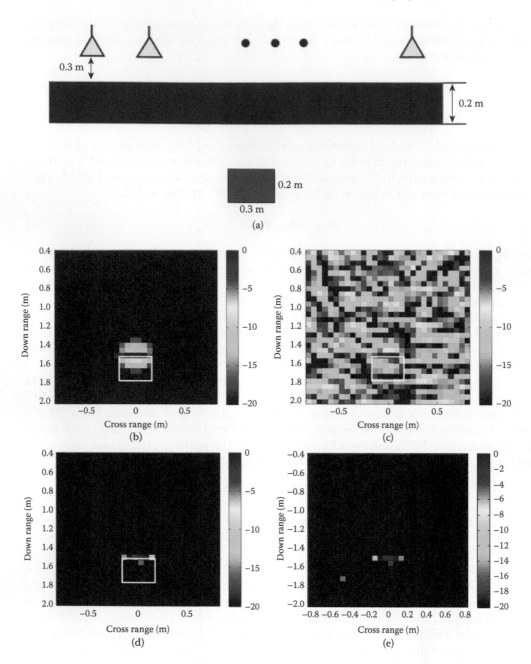

FIGURE 3.2
Imaging of target behind a single-layer wall with synthetic aperture radar (SAR): (a) simulation geometry; (b) beamforming result with full data; (c) beamforming result with limited data; (d) compressive sensing (CS) imaging result with limited data; (e) CS imaging result with 20% random noise (Figure 3.2b through d). (From Zhang, W., and Hoorfar, A., *IEEE Antennas Wireless Propag. Lett.*, 14, 1052–1055, 2015. With permission.)

measured at all frequencies and all antenna locations. The true region of the target is indicated with a white rectangle. If only two frequencies are randomly measured at each antenna location, the through-wall back-projection beamformed image appears as shown in Figure 3.2b; the image is blurred and distorted and has higher sidelobe levels. By exploiting the sparsity of the image space and solving the sparse constraint optimization problem in Equation 3.8 with the same two randomly measured frequencies at each antenna location, we obtain the reconstructed image shown in Figure 3.2c and d for noiseless and noisy data. As seen from the CS imaging result in Figure 3.2d, with only 3.3% of the full set of data, the CS method gives an even sharper and less cluttered image than the beamforming approach using the full set of data in Figure 3.2b.

The next example is the imaging of multiple targets behind external and interior walls separated by a hall wall using MIMO radar, as shown in Figure 3.3a. The dielectric constant, conductivity, and thickness of the exterior and interior walls are each $\varepsilon_b = 6$, $\sigma_b = 0.03\,\text{S/m}$, and $d = 20\,\text{cm}$, respectively. The width of the hallway between the exterior and interior walls is 1.2 m. The targets under investigation are rectangular and circular cylinders [20]. The MIMO radar consists of 4 transmitters equally spaced from -1 to 0.7 m and 18 receivers equally spaced from -1.22 to 1.16 m. The operating frequency covers the range from 1 to 3 GHz with 51 equally spaced frequency bins. Figure 3.3b shows the imaging result using

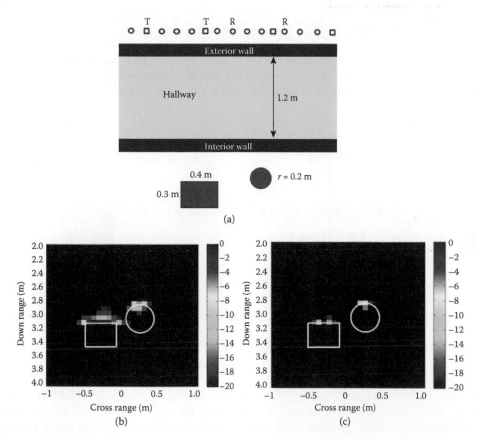

FIGURE 3.3

Imaging of targets behind walls separated by a hallway with MIMO radar: (a) simulation geometry; (b) beamforming result with full data; (c) CS imaging result with limited data. (From Zhang, W., and Hoorfar, A., *IEEE Antennas Wireless Propag. Lett.*, 14, 1052–1055, 2015. With permission.)

Equation 3.4 with full data, that is, data collected at all 18 receivers and all 51 frequencies. To reduce the complexity and cost of the MIMO radar system, we randomly collected data at five receivers, and measured five random frequencies at each receiver and applied CS according to Equation 3.8. From Figure 3.3c, it is clear that the above Green's function–based CS approach is successful for accurately geolocating and imaging the two targets behind multilayered walls using MIMO radar with a significantly reduced number of both receivers and frequencies. This will drastically reduce the complexity and cost of the MIMO radar system for TWRI.

3.2.3   Intrawall and GPR imaging using CS

The far-field Green's function approximation for the multilayered wall model used in the above section does not lend itself to the intrawall imaging of objects inside a wall or target imaging in the GPR case. For those configurations, it is necessary to employ the Green's function of a half-space medium in Equations 3.1 through 3.3:

$$G(\boldsymbol{r}_{rp}, \boldsymbol{r}, k_m) = \frac{jk_m}{4\pi} \int_{-\infty}^{\infty} F(\alpha, k_m) \exp(jk_m\alpha(x_{rp} - x) + jk_{1z}z_{rp} - jk_{2z}z)d\alpha$$

$$F(\alpha, k) = \frac{2}{k_{1z} + k_{2z}}, \quad k_{1z} = k\sqrt{1 - \alpha^2}, \quad k_{2z} = k\sqrt{\tilde{\varepsilon}_r - \alpha^2},$$

(3.9)

where $\boldsymbol{r}_{rp} = (x_{rp}, z_{zp})$ and $\tilde{\varepsilon}_r$ is the complex permittivity of the wall for interwall imaging or the ground for GPR imaging. A similar equation also holds for $G(\boldsymbol{r}, \boldsymbol{r}_{tp}, k_m)$ with $\boldsymbol{r}_{tp} = (x_{tp}, z_{tp})$. The integral in Equation 3.9 can be efficiently evaluated using the stationary phase method (SPM):

$$G_{SPM}(r, r', k) \simeq \frac{j}{4k} F(\alpha_0, k)e^{jk\Phi(\alpha_0)} \sqrt{\frac{2}{\pi k|\Phi''(\alpha_0)|}} e^{j\pi/4},$$

(3.10)

where

$$\Phi(\alpha) = \alpha|x_r - x| - z\sqrt{\tilde{\varepsilon}_r - \alpha^2} + z_r\sqrt{1 - \alpha^2}$$

(3.11)

and α_0 is the saddle point that satisfies $\Phi'(\alpha_0) = 0$. The corresponding half-space's medium dictionary can then be written as follows:

$$[\Psi_{m,n}]_{p,q} = G_{SPM}(\mathbf{r}_{rp}, \mathbf{r}_q, k_m)G_{SPM}(\mathbf{r}_q, \mathbf{r}_{tp}, k_m)$$

(3.12)

We note that for a lossless wall, α_0, ϕ, ϕ', and ϕ'' are not functions of frequency and can be precomputed and stored for a given wall scenario. This would significantly speed up the CS-based imaging, especially when dealing with a large target space.

Figure 3.4 shows the application of CS to SAR imaging of a 1×30 cm metallic plate hidden inside of a 20 cm dielectric wall with $\varepsilon_r = 6$. The receiving antenna aperture is 2 m long with an interelement spacing of 5 cm. The frequency range is 1–2.6 GHz with 55 frequency bins. Standard beamforming results using the full data set and the random number of frequency points together with the sparse reconstructed images using CS are shown in the figure. The image with the full data set of 55 frequency bins shows the target inside the wall as well as the front and back of the wall. As shown, CS-based images using only three frequency bins accurately localize the target and remove the wall clutter seen in conventional beamformed images.

We note the above CS-based imaging using the SPM evaluation of the half-space Green's function can also be used in GPR target imaging.

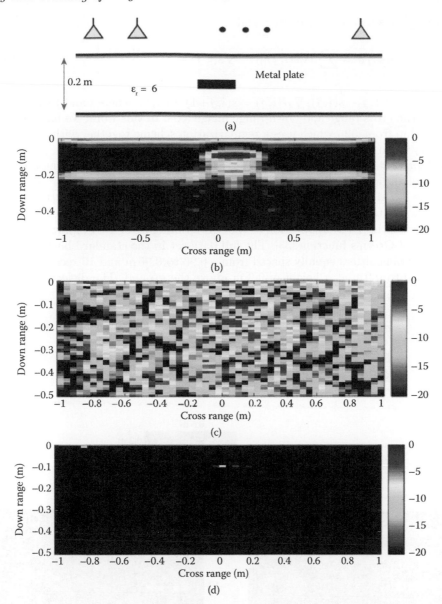

FIGURE 3.4
Intrawall imaging with SAR: (a) simulation geometry; (b) beamforming result with full data; (c) beamforming result with limited data; (d) CS imaging result with limited data.

3.2.4 Through-the-wall and GPR imaging using TVM

The standard CS techniques are mainly effective in detecting the presence of targets, but they cannot accurately reconstruct the target shape and/or differentiate closely spaced targets from an extended target. However, the TVM is based on minimizing the gradient magnitude of the image and can result in better edge preservation and shape reconstruction than standard l_1 minimization-based CS. It has been previously applied in image compression [31]. Unlike the standard CS, where $\|s\|_1$ is minimized, the TVM technique minimizes the total gradient of s:

$$\hat{s} = \arg\min \|s\|_{TV}, \quad \text{s.t.}, \quad \|(y - \Theta s)\|_2 < \delta, \tag{3.13}$$

where δ represents a small tolerance error, and

$$||\mathbf{s}||_{TV} = \sum_{i,j} ||\nabla \mathbf{s}(i,j)||_1; \quad \nabla \mathbf{s}(i,j) = \begin{bmatrix} \nabla_1 s(i,j) \\ \nabla_2 s(i,j) \end{bmatrix} \tag{3.14}$$

with $\nabla_1 s(i,j) = s(i+1,j) - s(i,j)$, $\nabla_2 s(i,j) = s(i,j+1) - s(i,j)$, where i and j run over the pixel indices in the cross-range and down-range image axis. In this work, we have used the Nesterov algorithm from [32], which uses a regularization scheme together with a smoothed version of the l_1 norm to efficiently solve Equation 3.13.

Figure 3.5 presents the application of TVM to the SAR imaging of through-the-wall targets. The full data are collected over 1–3 GHz using 81 frequency bins. As seen in Figure 3.5b, the TVM, using only three frequency points, results in a less cluttered image and reconstructs the target shape better that the standard l_1 minimization in Figure 3.5c.

Figure 3.6 shows the TVM application to the MIMO imaging of a GPR target using the SPM evaluation of Green's function [33]. The full data set in the standard back-projection imaging uses 17 transmitters equally spaced from -0.96 to 0.96 m and 16 receivers equally spaced from -0.9 to 0.9 m at a height of 0.2 m above the ground. The dielectric constant

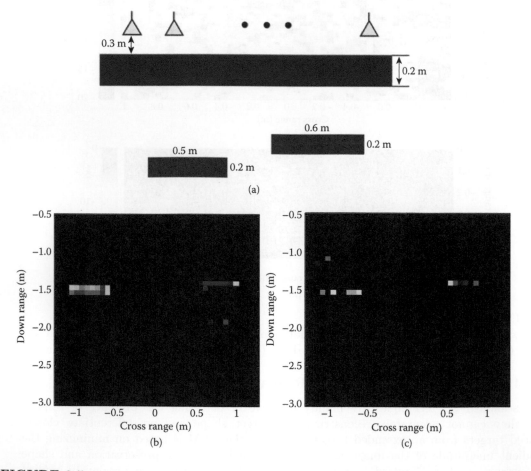

FIGURE 3.5
Imaging of targets behind wall using total variation minimization (TVM): (a) simulation geometry; (b) beamforming result with full data; (c) CS imaging result with limited data.   (*Continued*)

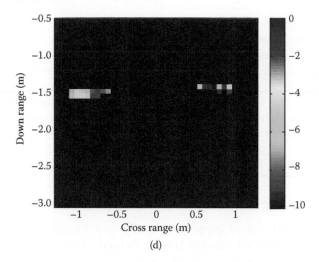

(d)

FIGURE 3.5 (Continued)
Imaging of targets behind wall using total variation minimization (TVM): (d) TVM imaging result with limited data.

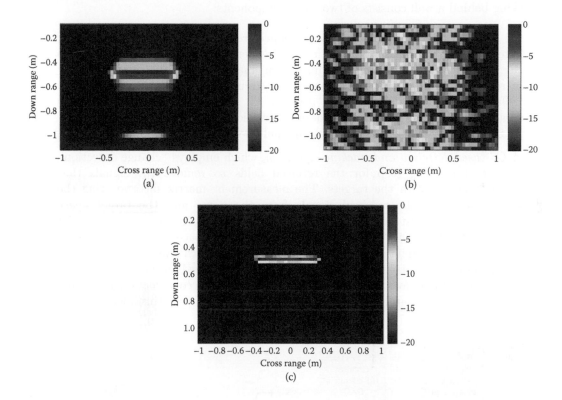

FIGURE 3.6
Ground-penetrating radar imaging of subsurface target using TVM: (a) imaging result with full data; (b) imaging result with limited data; (c) TVM imaging result with limited data.

and conductivity of the ground are $\varepsilon_b = 6$, $\sigma_b = 0.01\,\text{S/m}$, respectively. The operating frequency covers the range from 0.8 to 2 GHz with a total number of $M = 49$ frequency bins. The target is relatively large, a $0.8 \times 0.2\,\text{m}$ rectangular cylinder centered at $(0, -0.5\,\text{m})$. The sparse reconstructed image using TVM utilizes six randomly selected transmitters and six randomly selected receivers together with eight randomly measured frequencies at each receiver location. As seen, the TVM-based MIMO GPR accurately locates the target at its true position and provides high-resolution and a clean image, which closely resembles that obtained using MIMO imaging of the full data set.

3.3   CS-based tracking of multiple moving targets behind walls

Stepped frequency wideband radars supported by CS can be effectively applied in the detection and tracking of multiple moving targets behind clutter. In such applications, the scattered fields from the targets can be of the same order or even weaker than those from the clutter. Many techniques have been developed to mitigate such effects, including Doppler processing [34], time-reversal techniques [35], and change detection [36].

In addition to target tracking, one of our motivations is the human vital sign detection in cluttered environments, as depicted in Figure 3.7. In general, the system to achieve detection and tracking behind a wall consists of two main components:

1. *An antenna system*, which includes a stepped frequency wideband transmitting antenna with a proper bandwidth and frequency resolution step related to the desired range resolution and maximum unambiguous range and an array of receiving antennas backed by the radar circuitry. The antenna system also includes a steerable beamforming network for scanning to multiple locations simultaneously and expanding the desired directions of arrival (DoAs) steering or an SAR operation using a switch for the multielement array.

2. *A CS-based detection and tracking algorithm*, which employs "change detection" as a preprocessing step for the received fields to remove the signals that do not interact with the targets. The measurement matrix incorporating the relation between the received signals from the scene and the target space for tracking is constructed based on the spatial steering vectors and Fourier transform matrix. Due to the sparsity of the target space, the signals can be measured at a small number of random frequencies for each antenna element [37]. Accordingly, the sparse target space can be recovered based on the corresponding randomly reduced measurement matrix by a proper reconstruction algorithm (e.g., orthogonal matching pursuit (OMP) [38]). However, this measurement

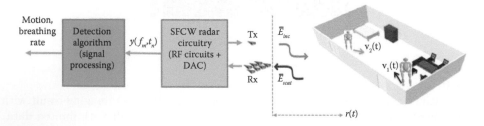

FIGURE 3.7
CS-based tracking.

matrix is highly coherent, which cannot always guarantee stable and accurate solutions. An improved reconstruction method needs to be sought to deal efficiently with the highly coherent columns. The resulting target locations can then be determined by merely selecting the highest peak values in the reconstructed target space for time instants. Because most CS reconstruction algorithms are associated with computationally intensive matrix operations (e.g., Cholesky decomposition), the parallelization of the CS algorithms may be necessary in order to apply this framework to real-time applications. As analyzed in [38], the OMP computational cost depends on the sparsity K of the recovered signal, the number measurements $M^{CS}N$, and the signal length MP for N antennas, M frequencies and P DoAs. Hence, the total computational cost of the CS reconstruction of the DoA–range space for a particular time instant is $O(KM^{CS}NMP)$. Based on the computational complexity of the conventional method; that is, $O(M^2NP)$ [34], and because $M^{CS} < M$ and K is practically small due to the sparse target space, the computational complexity of the proposed approach is less than that of the conventional method.

3.3.1 CS-based joint 3D target tracking

The procedure for a combination of DoA, range, and Doppler processing, in the reverse order compared to [34], can be represented as below in Equations 3.15 and 3.16. The bandwidth of the stepped frequency radar, which consists of N antenna elements with an inter-element spacing of d, is sampled by M frequencies (center frequency f_c).

The frequency bins f_m are uniformly distributed with a step size Δf. At each time instant, the received signals $y(n, f_m, t)$ from antenna elements for each frequency are spatially beamformed to resolve the targets along the DoA (θ). It is followed by the inverse Fourier transform of the frequency f_m to obtain the range r. Finally, the short-time Fourier transform is applied to capture the Doppler frequency f_D of the targets using the time window $h(t)$.

$$s(r,\theta,t) = \frac{1}{N} \sum_{m=1}^{M} \sum_{n=1}^{N} y(n, f_m, t) e^{-j(n-1)2\pi d \frac{f_c}{c} \sin\theta} e^{j2\pi f_m \frac{2r}{c}} \tag{3.15}$$

$$\xi(f_D, r, \theta, t) = \int_{t'} s(r, \theta, t') h(t' - t) e^{-j2\pi f_D t'} dt' \tag{3.16}$$

Employing CS for the joint DoA–range space and using Equation 3.15, the received signals and the joint DoA–range space can be related approximately as follows [37]:

$$y(n, f_m, t) = \sum_{l=1}^{M} \sum_{i=1}^{P} s(r_l, \theta_i, t) e^{j(n-1)2\pi d \frac{f_c}{c} \sin\theta_i} e^{-j2\pi f_m \frac{2r_l}{c}}, \tag{3.17}$$

where the angle of arrival is sampled by P directions within the desired DoA range $[\theta_1, \ldots, \theta_P]$. The range r_l is from 0 to a maximum unambiguous range r_m. The column vectors $\mathbf{y}$ and $\mathbf{x}$ are obtained by stacking $y(n, f_m, t)$ and $s(r_l, \theta_i, t)$. From Equation 3.16, the full measurement can be rewritten in matrix form as follows:

$$y = \Phi\Psi S, \tag{3.18}$$

where Ψ is the identity matrix, and $\Phi = [\phi_1, \phi_2, \ldots, \phi_{MP}]$ is the measurement matrix with MN rows and MP columns. In particular, the $((l-1)P+i)$th column is defined as follows:

$$\phi_{(l-1)P+i} = \varphi_r(r_l) \otimes \varphi_\theta(\theta_i), \tag{3.19}$$

where $\otimes$ represents the Kronecker product of two vectors. The effect of the wall can be incorporated in Φ using the far-field representation of Green's function, with the $((m-1)N+n, (l-1)P+i)$th element of Φ as follows:

$$\Phi_{(m-1)N+n,(l-1)P+1} = T_t(\boldsymbol{r}_n, \boldsymbol{r}, f_m)T_r(\boldsymbol{r}, \boldsymbol{r}_l, f_m)e^{j(n-1)2\pi d\frac{f_m}{c}\sin\theta_i}e^{-j2\pi f_m\frac{2r_l}{c}} \qquad (3.20)$$

Due to the K-sparsity of the joint DoA–range space, the CS-based approach signals are measured at a small number $M^{\mathrm{CS}}(< M)$ of random frequencies for each antenna element, reducing the data acquisition time by a factor of M/M^{CS}. Consequently, the reduced Φ^{CS} measurement matrix is constructed by randomly selecting $M^{\mathrm{CS}}N$ rows ($O(K\log(MP))$) of Φ which corresponds to the selected frequencies. The new measurements can be expressed as randomized projections. The reconstruction of the DoA–range space is formulated as $\|s\|_1$, s.t., $y^{CS} = \Phi^{CS}\Psi_S$, which can be solved by the OMP algorithm.

3.3.2　Results for multiple target tracking

Here we present simulated results for two scenarios. In the first scenario, we consider a stepped frequency radar for 3D tracking of two moving targets in free space. The radar consists of a transmitter operating between 2.22 and 2.58 GHz with 5 MHz frequency steps ($f_c = 2.4$ GHz). Hence, the number of frequency bins is $M = 73$. Given the bandwidth and frequency step, targets can be resolved with a range resolution of 0.417 m and a maximum unambiguous range of 30 m. The receiving array is positioned along the y-axis, centered at the origin. The array length is 0.75 m with 0.075 m displacement between elements, which corresponds to $N = 11$ elements. The range of the DoA angle is $[-45°, 45°]$ with a resolution of 1°. This scenario is shown in Figure 3.8a and consists of two moving spherical targets. Both move with the relative radial velocity of 6 m/s along straight paths. The first sphere approaches the radar, starting at a DoA angle of 15° and a distance of 18.125 m from the radar. The second moves away from the radar, starting at a DoA angle of $-18°$ and a distance of 15 m from the radar. For Doppler processing, a Gaussian window of 0.14 s is used.

In order to verify the accuracy, the simulation result of the proposed method is compared with the result using the conventional method with the full data set. The simulation is

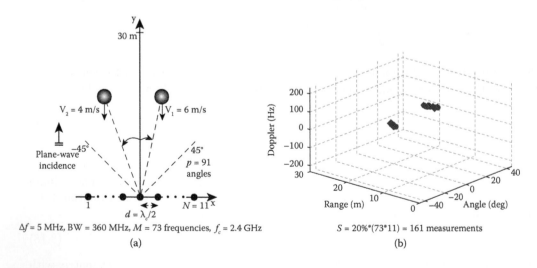

$\Delta f = 5$ MHz, BW = 360 MHz, $M = 73$ frequencies, $f_c = 2.4$ GHz

(a)

$S = 20\%^*(73^*11) = 161$ measurements

(b)

FIGURE 3.8

DoA–range–Doppler scene of two targets in free space. (a) The scene geometry and (b) Results. The blue trace is for CS and the red trace is for the conventional method.

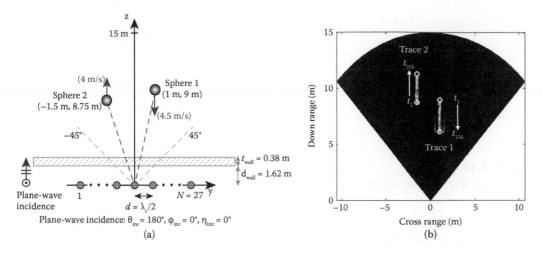

FIGURE 3.9
CS-based 2D tracking results for two targets behind a wall. (a) The scene geometry and
(b) Results.

performed for $M^{CS} = 15$ (20% of M). It is important to note that the number of randomly
selected frequencies M^{CS} is governed by the sparseness of the joint DoA–range space. The
resultant target locations are determined by selecting the highest peak values in the joint
DoA–range–Doppler space for a particular time instant. It is demonstrated in Figure 3.8b
that the proposed method shows good agreement with the conventional method.

In the second example, we consider 2D tracking of two moving spheres behind a wall as
depicted in Figure 3.9a. The radius of each sphere is 10 cm and the number of frequency
bins in this case is 81. The antenna array consists of 27 elements with inter-element spacing
of 6.25 cm. The wall thickness is 38 cm and it has a relative permittivity of 6.5 and a
loss tangent of 0.011. The white circles in the images of Figure 3.9b represent the exact
locations of the targets. As seen, the CS-based results, which use only 10% of the full data
measurements using the conventional method, accurately track the movement of the targets.

3.4 CS-based detection of human vital signs behind walls

There have been many approaches to extracting the vital signs of subjects, which mainly
focus on continuous wave (CW) Doppler radar [39], UWB radar [40], or stepped-frequency
continuous wave (SFCW) radar [41]. UWB and SFCW radar systems are both superior to
CW radar systems due to their capability of localization and multiple subject monitoring.
Nevertheless, SFCW radar possesses several advantages over the UWB radar system such
as high reliability, stability, and relatively easy implementation. Stepped-frequency radar,
however, suffers from long data acquisition time [15], which leads to aliasing while capturing
scattered signals. The application of CS for SFCW radar [42] can compress both the
measurement frequencies and the slow time samples, significantly decreasing the data
acquisition time and the amount of data to be processed.

We recently demonstrated the viability of CS-based SFCW radar for breathing rate
detection [43,44]. A PEC elliptical cylinder based on the set of data for thoracic dimensions
as presented in [45] is used to model a full respiration cycle as depicted in Figure 3.10 for an

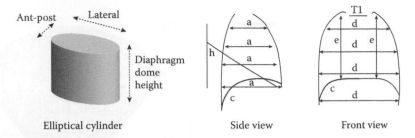

FIGURE 3.10
Rib-cage modeling for breathing.

average male with a height of 1.68 m. According to [46], during respiration the chest wall expands and shrinks periodically in both the anterior–posterior and lateral directions, so the travelled distance varies periodically around the nominal distance between the subject and the radar. Hence, in our simulation, the changes in the chest's anterior–posterior and lateral diameters are described as sinusoidal models around their nominal diameters. Hence the distance between a subject and the radar receiver can be expressed as follows:

$$d(t) = d_0 + m_b \sin(2\pi f_b t), \tag{3.21}$$

where d_0 is the distance between the receiver and thorax vibration center; m_b is the amplitude of the respiratory signal; f_b is the respiratory rate.

In conventional SFCW radar, we receive signals in m uniformly sampled frequencies, at n slow time indices. The inverse Fourier transform is applied to obtain range profiles for each time index; next, the range profile is used to locate the target. Finally, the forward Fourier transform is conducted along the slow time samples for the range bin corresponding to the subject location to show the respiratory rate in the frequency domain (see steps in Figure 3.11).

Two steps in this process render to CS implementation: the Fourier transform step, which relates the range profile and received baseband signals, and the inverse Fourier transform step, which relates the range to breathing rate. Considering a stepped-frequency radar centered at 3 GHz with a bandwidth of 2 GHz and a frequency step of 20 MHz (i.e., total number of frequency bins $M = 101$), targets can be resolved with a range resolution of 7.5 cm and a maximum unambiguous range of 0.5 m. In order to verify the accuracy, the simulation results of the CS-based method are compared with the results of the conventional method over 160 slow time samples for a human torso with a respiratory rate of 20 breaths/min. Figure 3.12 illustrates that the CS-based method shows a good match with the conventional method.

The proposed method can indeed detect the respiratory rate using fewer randomly selected frequencies and slow time samples. We have also reported on using micro-Doppler signatures for human motion detection in cluttered media such as forests [47–51]. The above CS method can be also extended to vital sign detection of humans behind a wall by incorporating the wall's Green's function as was discussed in Section 3.1. To demonstrate through-wall applicability, we consider another simulation as depicted in Figure 3.13a [52]. The scene consists of a 1.79 m stationary breathing human located behind a wall, which is assumed to be homogeneous ($\varepsilon_r = 2 + j0.1$), infinite, and composed of a single layer with a thickness of 2 cm. We consider a stepped-frequency radar centered at 3 GHz with a bandwidth of 2 GHz and a frequency step of 20 MHz. Hence, the total number of frequency bins is $M = 101$. The breathing rate is chosen to be 18 times/minute. The distance from the radar to the center of the human is 1.1 m. The radar is placed to face the wall with a standoff distance

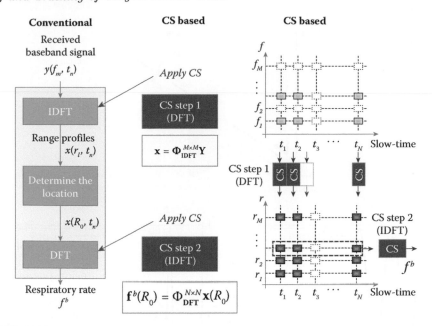

FIGURE 3.11
Conventional versus CS-based system for breathing.

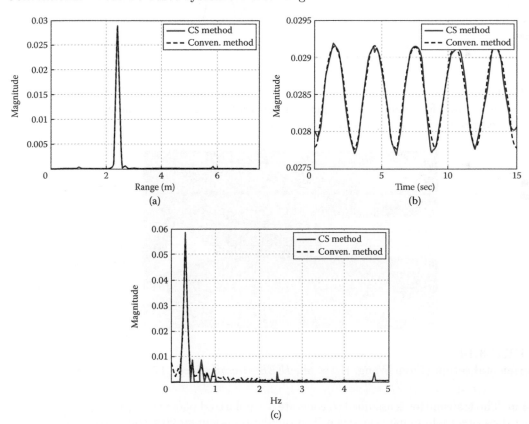

FIGURE 3.12
Simulation results for breathing rate in free-space CS versus conventional: (a) synthesized range profiles; (b) reconstructed respiratory pattern; (c) reconstructed respiratory rate.

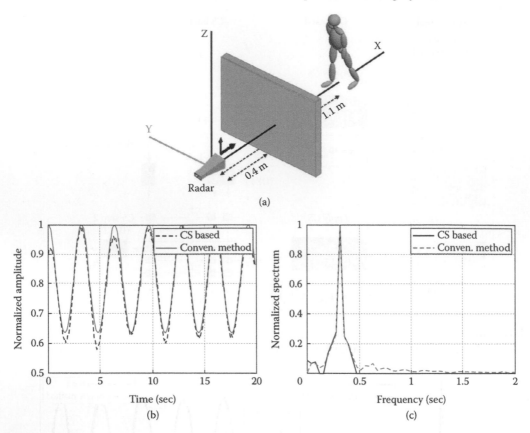

FIGURE 3.13
Simulation results for human breathing behind wall using CS: (a) simulation setup;
(b) detected respiration signal; (c) vital sign spectrum.

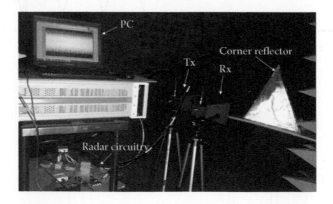

FIGURE 3.14
Experimental setup. (From Wang, H., et al., *IEEE Microw. Mag.*, 17(7), 53–63, 2016.)

of 0.4 m. The transmitter is assumed to generate a z-polarized plane wave propagating along
the $+x$ direction to illuminate the scene. The simulation is run for 20 seconds, corresponding
to 400 slow time samples. The results are shown in Figure 3.13b and c.

The CS experiment is performed for $M^{CS} = 30\%$ (which corresponds to 30% of 101)
and $N^{CS} = 160$ (which corresponds to 40% of 400). In order to verify the accuracy, the

simulation results of the CS-based method are compared with the results of the conventional method. Figure 3.14b and c demonstrates that the CS-based method can detect accurately the breathing rate of a human and shows a good match with the conventional method. It indicates that the human has a breathing rate of 18 times/min, corresponding to a frequency component of 0.33 Hz.

Experimental results on human respiration behind the wall are presented after a brief description of the hardware development in the following section.

3.5 CS-based receiver hardware design and experiments

3.5.1 CS-based stepped-frequency signal processing

Let us assume that the entire bandwidth of the system ranges from f_0 to f_{N-1}, with a center frequency f_c, and is sampled by N frequencies, with a step size Δf. To speed up data acquisition, the SFCW radar operates using a P-channel operating in parallel—thus reducing the acquisition time by a factor of N/P. Meanwhile, the target range space is divided into multiple range bins r_l from 0 to r_{N-1}, where r_{N-1} is the maximum unambiguous range. The relationship between the received baseband signal $y(f_n)$ from all the channels and the range profile $s(r_l)$ can be expressed as a Fourier transform as follows:

$$y(f_n) = \sum_{l=0}^{N-1} s(r_l)e^{-j2\pi f_n \frac{2r_l}{c}} \tag{3.22}$$

Equation 3.22 can be rewritten in a matrix form as $\boldsymbol{y} = \boldsymbol{\Theta s} = \boldsymbol{\Phi \Psi s}$, where $\boldsymbol{y}$ and $\boldsymbol{s}$ are the column vectors obtained by stacking $y(f_n)$ and $s(r_l)$, $e^{-j2\pi f_m 2r_c/c}$ and the measurement matrix $\boldsymbol{\Phi}$ is chosen as the $n \times N$ identity matrix when we use all frequencies in the conventional approach.

In the CS-based approach, however, we do not use all the frequency steps due to the sparseness of the target space. Hence, the baseband signals, $\mathbf{y}^{CS}$, are measured at a random subset $N^{CS}(<N/P)$ of frequencies for each channel. In this case, the data acquisition time is decreased by a factor of $(N/P)/N^{CS}$. Then, we follow the same procedure to retrieve the signal as explained in the previous sections.

3.5.2 Two-channel implementation

As an example, we illustrate the construction of a two-channel SFCW radar here, and the whole setup is shown in Figure 3.14. The radar has a center frequency of 3 GHz and bandwidth of 2 GHz, with a frequency step of 20 MHz, which corresponds to 101 frequency samples. In this case, the targets can be resolved with a range resolution of 7.5 cm and a maximum unambiguous range of 7.5 m based on radar parameters.

In general, a multiple channel SFCW radar system can transmit a set of frequencies simultaneously via one UWB antenna over multiple channels operating in parallel using a multiplexer, where a random subset of frequencies is transmitted and the target space is reconstructed by using a CS-based algorithm with sub-Nyquist sampling. These two strategies can reduce the data acquisition time by an order of magnitude.

3.5.3 Overview of the SFCW radar system

The two direct digital synthesizer (DDS) channels are synchronized using one master 1.2 GHz reference clock and integrated on one board to cover a 2 GHz bandwidth. Each

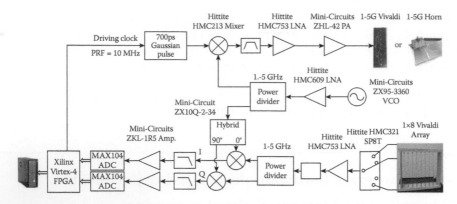

FIGURE 3.15

Block diagram of the multichannel stepped-frequency continuous wave radar system.

DDS channel synthesizes an IF signal with a bandwidth of 20 MHz, followed by a 50 times phase locked loop (PLL) to acquire the RF stepped-frequency signal. The center frequency of the second DDS channel is shifted by 1 GHz, so the total bandwidth of the stepped-frequency signal is 2 GHz. In our implementation, the RF stepped-frequency signal on each channel is first divided into two halves through a Mini-Circuits ZAPD-4+ power splitter. One half from each channel is combined using a multiplexer and then fed into a Mini-Circuits ZVE-8G power amplifier and transmitted through the transmitting antenna. Meanwhile, the other two halves are split again with a quadrature coupler like the Mini-Circuits ZAPDQ-4+90° power splitter to serve as the in-phase and quadrature-phase local oscillator. On the receiver side, the received signal is passed through a low noise amplifier (LNA) like that of the Analog Devices HMC753 wideband LNA. The received signal is then split into its four constituent components by mixing each with its corresponding local oscillator then filtering the unwanted components to acquire the baseband signals. Each baseband signal is then digitized and converted to a 14-bit digital signal and stored in a PC for further processing using a low-speed data acquisition card like the NI USB-6009. A block diagram is shown in Figure 3.15.

In the two-channel SFCW implementation, we use one reference clock to drive the two DDS chips to avoid any synchronization problems. This 1.2 GHz reference clock is divided by 24 in the first DDS chip to work as the control clock for the complex programmable logic device (CPLD), which is used to generate the digital signal to control the two DDS chips and is considered the master clock [53]. Synchronization of the two DDS chips is extremely important and is achieved by utilizing the synchronization pins on the two chips. Such a system can be extended to multiple channel operations upon increasing the number of DDS circuits as well as the number of multiplexer channels. Figure 3.16 shows a picture of the developed two-channel multiplexer.

3.5.4 UWB antenna

Vivaldi antennas have been widely used in UWB applications due to their simple structure, wide bandwidth, and high gain. A Vivaldi subarray consisting of eight antipodal Vivaldi elements [54,55] was utilized in this SFCW radar implementation. It has a gain of 11.5 ± 1.5 dB from 2 to 4 GHz and also has a constant radiation pattern over the entire frequency range.

The full Vivaldi array (shown in Figure 3.17) is comprised of eight subarrays and has an 0.8 wavelength at the highest operating frequency (4 GHz) to prevent grating lobes

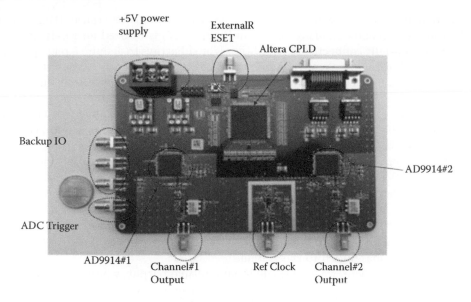

FIGURE 3.16
Photograph of the two-channel DDS board. (From Wang, H., et al., *IEEE Microw. Mag.*, 17(7), 53–63, 2016. With permission.)

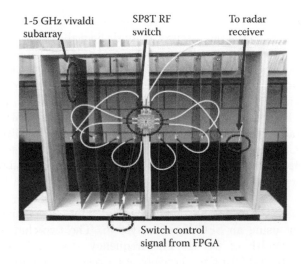

FIGURE 3.17
Eight-element Vivaldi ultra-wideband antenna.

within the limited scanning range. More details about the antenna design can be found in [55].

Based on adequate measured decoupling between the different subarrays, we can assume that all array elements are the same and independent of their location along the linear array; otherwise the calculation of the received signals will be impractical. To automate the full array; one single-pole eight-throw (SP8T) switch was designed and is shown in Figure 3.16.

In the SAR operation, one wideband Vivaldi element is used for transmitting, while an eight-element Vivaldi antenna array (shown in Figure 3.17) is utilized for receiving. The SP8T switch is sequentially connected to one of the eight subarrays to acquire a real aperture SAR image in the horizontal plane. For full imaging capability, both transmitting and receiving antennas can be mechanically steered in the elevation direction. Both simulation results and actual measurements indicate that the CS-based signal processing method allows for a reduction in the number of transmitted frequency points while still attaining performance comparable to that of the traditional inverse discrete Fourier transform (IDFT) method, which needs to process the full data set.

In MIMO operation, multiple transmitters and receivers are used to transfer more data at the same time. MIMO utilizes spatial diversity, as receiving multiple spatial streams simultaneously (if coherently combined) would significantly enhance performance. Upon combining these data streams arriving from different directions (angles) and at different times, the signal power is coherently added, thus increasing the signal power. By increasing the number of receiving and transmitting antennas, the signal power could increase linearly with every pair of antennas added to the system.

In our implementation here, we used single-input multiple-output, where the transmitter has a single antenna and the receiver has multiple independent antennas. The same setup used above for SAR can be utilized as well for MIMO operation, where one antenna is used as a transmitter and more than one antenna can be used as independent receivers.

3.5.5   Experimental results

An experiment was carried out [56] to identify the respiration of two stationary persons, standing at different distances to the radar sensor behind the wall, as shown in Figure 3.18a. The radar sensor has accurately separated two targets by the down range (i.e., distance to radar sensor), as shown in Figure 3.18b, which shows that the first person is at a down range of 3.2 m and a cross range of 2.1 m, whereas the second person is at a down range of 2 m and a cross range of 2.9 m. The range profile is demonstrated in Figure 3.18c, which introduces a respiratory rate of 18 breaths/min from the first person and 21 breaths/min from the second person, because radar return signals from the two persons are at distinct range bins. Both conventional and CS were applied on the slow-time sampled pulse at related range bins to acquire the Doppler frequency spectrum, as shown in Figure 3.18d and e, which show a Doppler frequency of 0.29 Hz from the first person and 0.34 Hz from the second person.

To validate the proposed CS technique, one non-line-of-sight experiment was conducted in an indoor environment using an SFCW radar system. The two-channel SFCW radar system had a total bandwidth of 2 GHz. The frequency step and duration for each frequency step were 20 MHz and 50 μs, respectively. As a result, a maximum unambiguous distance of 7.5 m was achieved. To be consistent with the simulation setup, the frequency steps were programmed to be 101 and a total of 400 frames were collected during the experiments.

In this experiment, one sedentary subject with a height of 1.79 m was breathing normally and standing 1.1 m away from the radar system. A wooden wall was located 0.4 m away from the radar. When $M^{CS} = 30$ (30% of 101) was performed on frequency samples, it was shown (Figure 3.19a) that the respiration signal detected with the CS-based method matched well with the one recovered using the conventional method. The experiment was performed for $M^{CS} = 40$ and $N^{CS} = 160$ (40% of 400) as well, and a respiration rate of 0.33 Hz was accurately detected using the CS-based method, as shown in Figure 3.19b.

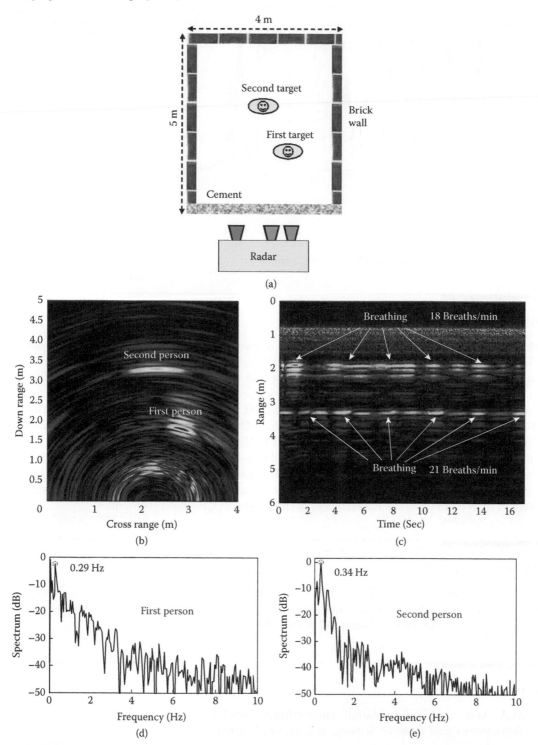

FIGURE 3.18
Localization and respiration detection of two persons through a cement wall: (a) experimental setup; (b) localization of the two human targets; (c) range profile versus time; (d) Doppler frequency spectrum due to the breathing of the first person; (e) Doppler frequency spectrum due to the breathing of the second person.

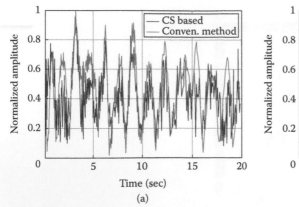 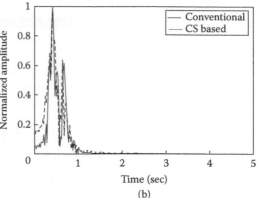

FIGURE 3.19
(a) Detected respiration signal; (b) vital sign spectrum.

3.6   Conclusion

In this chapter, we presented highlights of some of our work on imaging and tracking of targets behind or inside building walls using CS and discussed a CS-based method for detection of human vital signs. For through-the-wall imaging and tracking, our approach used the far-field approximation of multilayered Green's functions to fully account for all the wall effects, whereas for the intrawall and GPR imaging, we used a saddle-point approximation of the half-space Green's function. In addition to standard CS, we also presented the use of TVM in sparse reconstruction of through-the-wall and GPR targets. Finally, in this chapter, we outlined the hardware design of a multiple channel step-frequency radar system that included a low-complexity receiver system for efficient implementation of the proposed CS-based target imaging, tracking, and human vital sign detection. Simulated and experimental results were presented to demonstrate the effectiveness of the proposed CS-based techniques.

Acknowledgments

The authors thank Dr. Wenji Zhang, Dr. Vinh Dang, Dr. Haofei Wang, Dr. Yazhou Wang, Mr. Lingyun Ren, Ms. Sabikun Nahar, and Mr. Nghia Tran for their contributions in generating some of the results presented in this chapter.

References

[1] M.A. Amin and K. Sarrabandi, guest editors, Special issue of *IEEE Transactions on Geosciences and Remote Sensing*, vol. 47, no. 5, 2009.

[2] M. G. Amin, editor, *Through-the-Wall Radar Imaging*, CRC Press, 2010.

[3] F. Ahmad, M. G. Amin and S. A. Kassam, Synthetic aperture beamformer for imaging through a dielectric wall, *IEEE Trans. Aerospace Electronic Systems*, vol. 41, pp. 271–283, 2005.

[4] M. Dehmollaian and K. Sarabandi, Refocusing through building walls using synthetic aperture radar, *IEEE Trans. Geosci. Remote Sens.*, vol. 46, pp. 1589–1599, 2008.

[5] F. Ahmad, Y. Zhang, and M. Amin, Three-dimensional wideband beamforming for imaging through a single wall, *IEEE Geosci. Remote Sens. Lett.*, vol. 5, pp. 176–179, 2008.

[6] C. Le, T. Dogaru, L. Nguyen, and M. A. Ressler, Ultrawideband (UWB) radar imaging of building interior: Measurements and predictions, *IEEE Trans. Geosci. Remote Sens.*, vol. 47, no. 5, pp. 1409–1420, 2009.

[7] Y. Wang and A. Fathy, Advanced system level simulation platform for three-dimensional UWB through-wall imaging SAR using time-domain approach, *IEEE Trans. Geosci. Remote Sens.*, vol. 50, pp. 1986–2000, 2012.

[8] W. Zhang and A. Hoorfar, Three-dimensional real-time through-the-wall radar imaging with diffraction tomographic algorithm, *IEEE Trans. Geosci. Remote Sens.*, vol. 51, pp. 4155–4163, 2013.

[9] W. Zhang, A. Hoorfar, C. Thajudeen, and F. Ahmad, Full polarimetric beam-forming algorithm for through-the-wall radar imaging, *Radio Sci.*, 46, RS0E16, 2011, doi:10.1029/2010RS004631.

[10] W. Zhang and A. Hoorfar, Three-dimensional synthetic aperture radar imaging through multilayered walls, *IEEE Trans. Antennas Propag.*, vol. 62, pp. 459–462, 2014.

[11] Y. Yoon and M. G. Amin, Compressed sensing technique for high-resolution radar imaging, in *Proceedings of the SPIE, Signal Processing, Sensor Fusion, and Target Recognition XVII*, 2008.

[12] J. Ender, On compressive sensing applied to radar, *Signal Processing*, vol. 90, pp. 1402–1414, 2010.

[13] R. Baraniuk and P. Steeghs, Compressive radar imaging, *Proc. IEEE Radar Conf.*, pp. 128–133, 2007.

[14] M. Herman and T. Strohmer, High-resolution radar via compressive sensing, *IEEE Trans. Signal Process*, vol. 57, pp. 2275–2284, 2009.

[15] A. C. Gurbuz, J. H. McClellan, and W. R. Scott, A compressive sensing data acquisition and imaging method for stepped frequency GPRs, *IEEE Trans. Signal Processing*, vol. 57, pp. 2640–2650, 2009.

[16] Q. Huang, L. Qu, B. Wu, and G. Fang, UWB through-wall imaging based on compressive sensing, *IEEE Trans. Geosci. Remote Sens.*, vol. 48, pp. 1408–1415, 2010.

[17] W. Zhang, M. G. Amin, F. Ahmad, A. Hoorfar, and G. E. Smith, Ultrawideband impulse radar through-the-wall imaging with compressive sensing, *Intl. J. Antennas Propag.*, vol. 2012, p. 11, 2012.

[18] K. E. Browne, R. J. Burkholder, and J. L. Volakis, Fast optimization of through-wall radar images via the method of Lagrange multipliers, *IEEE Trans. Antennas Propgation*, vol. 61, pp. 320–328, 2013.

[19] M. Amin and F. Ahmad, Compressive sensing for through-the-wall radar imaging, *J. Electron. Imaging*, 22(3), 030901, 2013, doi:10.1117/1.JEI.22.3.030901.

[20] W. Zhang and A. Hoorfar, A generalized approach for SAR and MIMO radar imaging of building interior targets with compressive sensing, *IEEE Antennas Wireless Propag. Lett.*, vol. 14, pp. 1052–1055, 2015.

[21] W. Zhang, A. Hoorfar and L. Li, Through-the-wall target localization with time reversal music method, *Progress Electromagnetic Res. (PIER)*, vol. 106, pp. 75–89, 2010.

[22] M. Aftanas, J. Sachs, M. Drutarovsky, and D. Kocur, Efficient and fast method of wall parameter estimation by using UWB radar system, *Frequenz. J.*, vol. 63, no. 11–12, pp. 231–235, 2009.

[23] P. Protiva, J. Mrkvica, and J. Machac, Estimation of wall parameters from time-delay-only through-wall radar measurements, *IEEE Trans. Antennas Propag.*, vol. 59, no. 11, pp. 4268–4278, 2011.

[24] C. Thajudeen and A. Hoorfar. Wall parameters estimation using a hybrid time-delay-only and reflected wave ratio technique, in *IEEE International Symposium on Antennas and Propagation*, Orlando, FL, July 2013.

[25] C. Thajudeen and A. Hoorfar, A modified time-domain reflectometry for the estimation of interior-wall parameters, in *Proceedings of the IEEE International Symposium on Antennas and Propagation*, pp. 567–568, 2013.

[26] C. Thajudden, W. Zhang, and A. Hoorfar, Theory and experiments on imaging of walls' interior structures using diffraction tomography, in *IEEE International Symposium on Antennas and Propagation*, Chicago, IL, 2012.

[27] W. C. Chew, *Waves and Fields in Inhomogeneous Media*, Piscataway, NJ: IEEE Press, 1997.

[28] S. Chen, D. Donoho, and M. Saunders, Atomic decomposition by basis pursuit, *SIAM J. Sci. Comp.*, vol. 20, no. 1, pp. 33–61, 1998.

[29] P. Boufounos, M. Duarte, and R. Baraniuk, Sparse signal reconstruction from noisy compressive measurements using cross validation, in *Proceedings of the IEEE Workshop on Statistical Signal Processing*, pp. 299–303, 2007.

[30] Romberg, J., l1-magic, http://www.acm.caltech.edu/l1magic.

[31] C. Vogel and M. Oman, Fast, robust total variation-based reconstruction of noisy, blurred images, *IEEE Trans. Image Processing*, vol. 7, no. 6, pp. 813–824, 1998.

[32] S. Becker, J. Bobin, and E. J. Candes, NESTA: A fast and accurate first-order method for sparse recovery, *SIAM J. Imaging Sci.*, vol. 4, no. 1, pp. 1–39, 2011.

[33] A. Hoorfar and W. Zhang, MIMO ground penetrating radar imaging using total variation minimization, in *IEEE International Symposium on Antennas and Propagation*, Puerto Rico, June 2016.

[34] S. S. Ram, Y. Li, A. Lin, and H. Ling, Doppler-based detection and tracking of humans in indoor environments, *J. Franklin Inst.*, vol. 345, no. 6, pp. 679–699, 2008.

[35] A. E. Fouda and F. L. Teixeira, Imaging and tracking of targets in clutter using differential time-reversal techniques, *Waves Random Complex Media*, vol. 22, no. 1, pp. 66–108, 2012.

[36] F. Ahmad and M. Amin, Through-the-wall human motion indication using sparsity-driven change detection, *IEEE Trans. Geosci. Remote Sens.*, vol. 51, no. 2, pp. 881–890, 2013.

[37] V. Dang and O. Kilic, Joint DoA-range-Doppler tracking of moving targets based on compressive sensing, in *2014 IEEE Antennas and Propagation Society International Symposium (APSURSI)*, 6–11 July 2014.

[38] J. Tropp and A. Gilbert, Signal recovery from random measurements via orthogonal matching pursuit, *IEEE Trans. Info. Theory*, vol. 53, no. 12, pp. 4655–4666, 2007.

[39] C. Li, J. Ling, J. Li, and J. Lin, Accurate Doppler radar noncontact vital sign detection using the RELAX algorithm, *IEEE Trans. Instrum. Meas.*, vol. 59, no. 3, pp. 687–695, 2010.

[40] Y. Wang, Q. Liu, and A. E. Fathy, CW and pulse-Doppler radar processing based on FPGA for human sensing applications, *IEEE Trans. Geosci. Remote Sens.*, vol. 51, no. 5, pp. 3097–3107, 2013.

[41] L. Liu and S. Liu, Remote detection of human vital sign with stepped-frequency continuous wave radar, *IEEE J. Sel. Topics Appl. Earth Observ.*, vol. 7, no. 3, pp. 775–782, 2014.

[42] D. Donoho, Compressed sensing, *IEEE Trans. Info. Theory*, vol. 52, no. 4, pp. 1289–1306, 2006.

[43] V. Dang, T. Phan, and O. Kilic, Compressive sensing based approach for detection of human respiratory rate, in *APS-URSI 2015*, Vancouver, BC, Canada, July 2015.

[44] L. Ren, N. Tran, H. Wang, A. Fathy, and O. Kilic, Analysis of micro-Doppler signatures for vital sign detection using UWB impulse Doppler radar, in *IEEE Radio and Wireless Week*, Austin, Texas, 24–27 January 2016.

[45] F. Bellemare, A. Jeaneret, and J. Couture, Sex differences in thoracic dimensions and configuration, *Am. J. Respir Crit. Care Med.*, vol. 168, pp. 305–312, 2003.

[46] A. Lazaro, D. Girbau, and R. Villarino, Analysis of vital signs monitoring using and IR-UWB radar, *PIER*, vol. 100, pp. 265–284, 2010.

[47] J. M. Garcia-Rubia, O. Kilic, N. Tran, V. Dang, and Q. Nguyen, Analysis of moving human micro–Doppler signature in forest environments, *Progress Electromagnetics Res.*, vol. 148, pp. 1–14, 2014.

[48] O. Kilic, J. M. Garcia-Rubia, N. Tran, V. Dang, and Q. Nguyen, Tracking of moving human micro-Doppler signature in forest environments with swaying tree components by wind, *Radio Sci.*, 2015, vol. 50(3), pp. 238–248, doi: 10.1002/2014RS005555.

[49] N. Tran, O. Kilic, J. M. Garcia-Rubia, V. Dang, and Q. Nguyen, Micro-Doppler signature of human walking in forest environment, in *U.S. National Committee on International Union of Radio Science (USNC-URSI) Meeting*, Boulder, CO, 8–11 January 2014.

[50] J. M. Garcia-Rubia, O. Kilic, and N. Tran, Propagation effects on micro-Doppler radar signature for human discrimination in forests, in *URSI Commission F Triennial Open Symposium on Radiowave Propagation & Remote Sensing*, Ottawa, Canada, April 2013.

[51] N. Tran, V. Dang, and O. Kilic, Near-field interactions for micro-Doppler signature of human motion in forest using FMM on hybrid platforms, ACES, Williamsburg, VA, USA, 22–26 March 2015.

[52] H. Wang, V. Dang, L. Ren, Q. Liu, L. Ren, E. Mao, O. Kilic, and A. E. Fathy, An elegant solution: An alternative ultra-wideband transceiver based on stepped-frequency continuous-wave operation and compressive sensing, *IEEE Microw. Mag.*, vol. 17, no. 7, pp. 53–63, 2016.

[53] W. Hongyu, W. Haofei, L. Ren, and E. Mao, Low spurious noise frequency synthesis based on a DDS-driven wideband PLL architecture, *J. Beijing Institute Technol.*, vol. 22, pp. 514–518, 2013.

[54] Y. Wang, Y. Yang, and A.E. Fathy, Ultra-wideband Vivaldi arrays for see- through-wall imaging radar applications, in *2009 IEEE AP-S. International Symposium, Antennas and Propagation*, Charleston, SC, June 2009.

[55] Y. Yang, Y. Wang, and A.E. Fathy, Design of compact Vivaldi antenna arrays for UWB see through wall applications, *Progr. Electromagnetics Res.*, vol. 82, pp. 401–418, 2008.

[56] Y. Wang, UWB pulse radar for human imaging and Doppler detection applications, PhD dissertation, University of Tennessee, 2012.

4

Recovery Guarantees for High-Resolution Radar Sensing with Compressive Illumination

Nithin Sugavanam and Emre Ertin

CONTENTS

4.1   Introduction

Radar imaging systems acquire information about the scene of interest by transmitting pulsed waveforms and processing the received backscatter energy to form an estimate of the range, angle of arrival, Doppler velocity, and amplitude of the reflectors in the scene. These range profiles from multiple pulses and/or multiple antenna elements can be processed jointly to solve a multitude of inference tasks including detection, tracking, and classification [1]. In this chapter, we focus on coherent multiple-input and multiple-output (MIMO) radar systems with closely separated antennas, such that the angle of arrival of each scatterer in the illuminated scene is approximately the same for all phase centers. The main advantage of coherent MIMO radar is its ability to synthesize a large virtual array with fewer antenna elements for improved spatial processing. Additionally, MIMO radar systems with multiple transmit and receive elements employing independent waveforms on transmit can provide spatial processing gains by exploiting the diversity of channels between targets and radar [2,3].

This chapter focuses on the problem of estimation of range, angle of arrival, and amplitude of reflectors in the scene using a modulated wideband pulse $\varphi_i(t)$ for the ith transmitter of bandwidth B and duration τ using N_T transmitters and N_R receivers. Assuming the support of the observed delays is known to lie on an interval T_u (termed *range swath* in the radar literature), the received signal at receiver l can be expressed as follows:

$$y_l(t) = \sum_{\substack{k=1 \\ i=1}}^{\substack{k=K \\ i=N_T}} \alpha_{l,i}(\theta_k)\phi_i(t - \Delta_k)x_k + w_l(t), \qquad (4.1)$$

where $w_l(t)$ is the additive receiver noise and $\alpha_{l,i}(\Theta)$ is the array factor for the lth receiver and the ith transmitter, which is a function of the angle of arrival Θ_k, round-trip delay time Δ_k, and the complex scattering coefficients x_k of the kth target, as shown in Figure 4.1. For the case of a single transmitter and receiver setup, the problem simplifies to delay estimation and the received signal $y(t)$ is as follows:

$$y(t) = \sum_{k=1}^{k=K} \phi(t - \Delta_k)x_k + w(t), \qquad (4.2)$$

where $w(t)$ is the receiver noise.

Conventionally, matched filtering is implemented in the analog domain where the number of samples is reduced to $N = BT_u$ to cover the delay support. However, implementation of the matched filter requires Nyquist rate sampling, which is proportional to the bandwidth. This, in turn, limits the resolution and dynamic range of the analog-to-digital converter (ADC) needed for direct digital implementation of the radar, because the resolution of the ADC is roughly inversely proportional to the maximum sampling rate [4].

If a linear frequency modulated (LFM) waveform $\varphi(t) = e^{j\beta t^2}$ is used on transmit, the matched filtering can be approximately implemented by mixing the received signal with a reference LFM waveform and low-pass filtering the mixer output.

At the receiver output, the waveform delayed by Δ appears as a sinusoidal tone whose frequency is given by $\beta\Delta$, as shown in Figure 4.2. This preprocessing step is termed *stretch processing* [1,5] and can result in a substantially reduced sampling rate for the ADC used in the receiver if the delay support T_u is smaller than the pulse length τ. Specifically, the received signal at the stretch processor's output can be written as $y(t) = \sum_{n=1}^{N} x_n e^{j(n\beta\Delta)t}$. If the scene consists of K point targets, where K is much less than the number of delay bins N, well-known results [6,7] from compressive sensing (CS) show that successful reconstruction with sub-Nyquist samples is possible with the number of measurements M

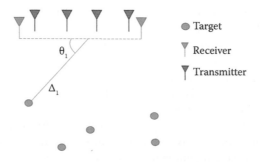

FIGURE 4.1
Illustration of the multiple-input and multiple-output (MIMO) system with sparse targets.

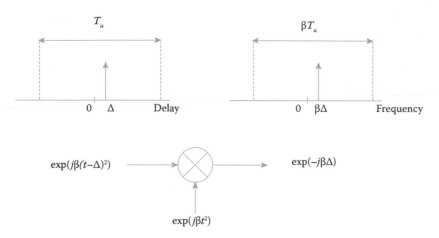

FIGURE 4.2
Effect of stretch processing.

scaling with $K \log N$. Furthermore, there are numerous practical algorithms, with provable performance, that are either based on convex relaxation [8–11] or greedy methods [12,13] for solving the reconstruction problem from compressed samples.

Motivated by these advances, compressive sensing techniques have been applied to a variety of problems in radar [14]: range profile estimation [15], waveform design using frequency hopping codes for estimation in range, Doppler velocity and angle domain [16], waveform design using multi-objective optimization of a combination of mutual coherence and signal-to-interference ratio [17], single-pulse systems for range and Doppler velocity estimation [18], single-pulse multiple transmit and receive system for range, Doppler velocity and azimuth estimation and target detection [19–21], remote sensing [22], and estimation of range, angle of arrival, and Doppler velocity using stepped frequency multipulse MIMO radar employing stochastic waveforms in each transmitter [23,24]. CS radar sensors based on pure random waveforms [25] and the Xampling framework [26,27] have also been implemented. A parallel research thrust [28] provided an average case recovery guarantee for the problem of angle of arrival estimation with randomly located antenna elements, under the idealized assumption of orthogonality between received waveforms from different range bins. The problem of subsampling in array elements was also posed as a matrix completion problem in a gridless estimation setting [29,30] and a condition established on the number of antenna elements that need to be observed in order to recover the entire low-rank data matrix.

Here, we pursue an alternative strategy based on the observation that LFM waveforms and analog stretch processing converts the range estimation problem into an equivalent sparse frequency spectrum estimation problem. It is well known that uniform subsampling in this setting has poor performance [31], and therefore uniform subsampling of a classical stretch processor cannot be used. While nonuniform random subsampling possesses good theoretical guarantees [11,31,32], its implementation with commercially available ADCs still requires them to be rated at the Nyquist rate to accommodate close samples. In this chapter, we pursue an alternative strategy and push randomization of the transmit signal structure to obtain compressive measurements at the stretch processor output. The proposed scheme can be readily implemented by utilizing a small number of random parameters in waveform generation and uniform sampling of ADCs with high analog bandwidth on the receiver. ADCs with analog bandwidth that exceeds their maximum sampling rate by several factors are readily available commercially and used routinely in

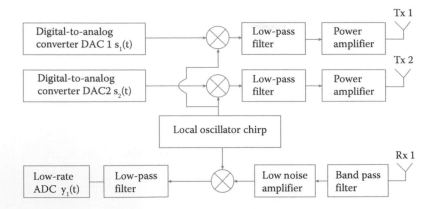

FIGURE 4.3
Block diagram of the transmitter and receiver.

passband sampling. This compressive radar structure, termed *compressive illumination*, was first proposed by Ertin [33] and utilized a linear combination of sinusoids to modulate an LFM waveform at the transmitter with randomly selected center frequencies, while maintaining the simple standard stretch processing receiver structure as shown in Figure 4.3. The output of the stretch processor receiver is given by $y(t) = \sum_{n=1}^{N} x_n \sum_{k=1}^{N_c} e^{j\varphi_{n,k}} e^{j(n\beta\Delta + \omega_k)^t}$, where $\varphi_{n,k}$ is a predetermined known complex phase and N_c is the number of tones modulating the LFM waveform. We observe that under the proposed compressive sensor design, each delayed copy of the transmitted waveform is mapped to a multitone spectra with a known structure.

The rest of the chapter is organized as follows. Section 4.2 states the model for the case of a single transmitter and receiver setup and presents the recovery guarantees. Section 4.3 presents the setup for a multiple transmitter and receiver along with the associated guarantees, and Section 4.4 provides simulation results that verify our theoretical results.

4.2 Single transmitter and receiver

4.2.1 Signal and target model

We now consider the compressive illumination framework proposed in [33,34] for the problem of range estimation. The chirp rate of the transmitted LFM waveform is fixed at $\frac{\beta}{\tau}$, where β is the bandwidth of the chirp waveform, τ is the pulse duration, and $B = g\beta$ is the system bandwidth for $g \geq 1$ with carrier frequency f_c. We denote the unambiguous time interval as $t_u = t_{\max} - t_{\min}$, where $t_{\max} = \frac{2R_{\max}}{c}$ and $t_{\min} = \frac{2R_{\min}}{c}$; $R_{\max}$, $R_{\min}$ are the maximum and minimum ranges in the area of interest, respectively, while c is the velocity of light in a vacuum. The whole space of the range is discretized into grids based on the radar's resolution; therefore we get $N = Bt_u$ grids. The frequency interval from $[0, B]$ is divided into N grids such that $f(t) = \frac{iR}{N}$, $i = 0, \ldots, N - 1$, which are used as frequencies for the multitone waveform that further modulate the chirp waveform. From the possible N modulated waveforms, a subset of size N_c is chosen at random for transmission. We simplify this selection model by considering the independent Bernoulli random variables as indicator variables to select LFM waveforms such that N_c waveforms are selected on average. Let $\gamma_i \in \{0, 1\}$ be the random variable indicating that $f(i)$ is part of the subset of size N_c. It is evident that $Pr(\gamma_i = 0) = 1 - \frac{N_c}{N}$ and $Pr(\gamma_i = 1) = \frac{N_c}{N}$. The chosen LFM waveforms are then scaled by independent random variables given by a sequence of independent and identical complex phases with probability density function as $f_\Phi(\varphi_i) = \frac{1}{2\pi}$, $\varphi_i \in [0, 2\pi]$. We define the sequence of random variables that model the selection and phase:

$$c_i = \gamma_i \exp(j\Phi_i). \tag{4.3}$$

The transmitted signal as a result of this procedure is given by the following:

$$s(t) = \sum_{i=0}^{N-1} \frac{c_i}{\sqrt{N_c}} \exp\left(j 2\pi (\hat{f}(i)t + \frac{\beta}{2\tau}t^2)\right) r\left(\frac{t - \frac{\tau}{2}}{\tau}\right),$$

where $r(\frac{t-\frac{\tau}{2}}{\tau}) = 1$ if $t \in [0, \tau]$ and 0 otherwise and $\hat{f}(i) = f_c + f(i)$. Under the assumption that $B \gg \beta$, the bandwidth of $s(t) \approx B$. The receiver utilizes stretch processing at the same chirp rate as the transmitter and a fixed reference frequency $f_d = f_c$ to demodulate the carrier frequency. The sampling rate employed at the receiver is $F_s = \frac{\beta t_u}{\tau}$. The total number of samples in the pulse duration τ is $M = \beta t_u$. The normalized output samples of the stretch processor due to the target at different delay bins $\Delta_m = \frac{m}{g\beta}$ are as follows:

$$y_k = \frac{1}{\sqrt{MN_c}} \sum_{i=0}^{N-1} \sum_{m=0}^{N-1} c_i \exp(-j 2\pi f(i)\Delta_m) \left(2\pi j \left(f(i) - \frac{\beta\Delta m}{\tau}\right) \frac{k}{F_s}\right) x_m + w_k,$$

where $k = 0, \ldots, M - 1$, w_k is the measurement noise process with a mean of 0 and variance of σ^2, $p = \frac{\tau}{t_u} \in \mathbf{Z}$, and x_m is the complex scattering coefficient due to a target at the delay bin Δ_m. This can be compactly written as follows:

$$\mathbf{y} = \mathbf{A}\mathbf{x} + \mathbf{w}, \tag{4.4}$$

where $\mathbf{y}, \mathbf{w} \in \mathbf{C}^M$, $\mathbf{A} \in \mathbf{C}^{M \times N}$, and $\mathbf{x} \in \mathbf{C}^N$. The sensing matrix $\mathbf{A}$ can be represented as a series of deterministic matrices corresponding to the response to each of the chirp waveforms scaled by zero mean random coefficients as shown:

$$\mathbf{A} = \sum_{i=0}^{N-1} c_i \mathbf{H}_i \overline{\mathbf{A}} \mathbf{D}_i. \tag{4.5}$$

The individual components are as follows:

$$\overline{\mathbf{A}} = \frac{1}{\sqrt{MN_c}} \left[\overline{\mathbf{A}}(0) \cdots \overline{\mathbf{A}}(N-1)\right]$$

$$\overline{\mathbf{A}}(r) = \left[1 \quad \exp\left(-2\pi j \frac{r}{N}\right) \quad \cdots \quad \exp\left(-2\pi j \frac{r(M-1)}{N}\right)\right]^T,$$

$$\mathbf{D}_i = \text{diag}\left[1 \quad \exp\left(-j 2\pi \frac{i}{N}\right) \quad \cdots \quad \exp\left(-j 2\pi \frac{i(N-1)}{N}\right)\right],$$

$$\mathbf{H}_i = \text{diag}\left[1 \quad \exp\left(j 2\pi \frac{ip}{M}\right) \quad \cdots \quad \exp\left(j 2\pi \frac{ip(M-1)}{M}\right)\right], \tag{4.6}$$

where $i = 0, \ldots, N - 1$ and $r = 0, \ldots, N - 1$; $\overline{\mathbf{A}} \in \mathbf{C}^{M \times N}$ are the samples from tones that correspond to each delay bin generated as a result of the dechirping process in case of a single-chirp system, $\mathbf{H}_i \in \mathbf{C}^{M \times M}$ is the shift in frequency due to the ith chirp waveform, and $\mathbf{D}_i \in \mathbf{C}^{N \times N}$ is the phase term associated with different delay bins due to the ith chirp.

Alternatively, we can express each column of the sensing matrix, which corresponds to the signal return from a point scatterer in each range bin, as follows:

$$\mathbf{A} = \begin{bmatrix} \mathbf{A}(0) & \cdots & \mathbf{A}(N-1) \end{bmatrix}$$

$$\mathbf{A}(m) = \mathbf{E}_m \mathbf{F} \mathbf{G}_m \mathbf{c}, \quad \text{where}$$

$$\mathbf{E}_m = \text{diag} \begin{bmatrix} 1 & \exp\left(-j2\pi\tfrac{m}{N}\right) & \cdots & \exp\left(-j2\pi\tfrac{m(M-1)}{N}\right) \end{bmatrix}$$

$$\mathbf{F} = \frac{1}{\sqrt{MN_c}} \begin{bmatrix} \mathbf{F}(0) & \cdots & \mathbf{F}(N-1) \end{bmatrix}$$

$$\mathbf{F}(r) = \begin{bmatrix} 1 & \exp\left(2\pi j\tfrac{rp}{M}\right) & \cdots & \exp\left(2\pi j\tfrac{rp(M-1)}{M}\right) \end{bmatrix}^T,$$

$$\mathbf{G}_m = \text{diag} \begin{bmatrix} 1 & \exp\left(-j2\pi\tfrac{m}{N}\right) & \cdots & \exp\left(-j2\pi\tfrac{m(N-1)}{N}\right) \end{bmatrix},$$

$$\mathbf{c} = \begin{bmatrix} c_0 & \cdots & c_{N-1} \end{bmatrix}^T, \tag{4.7}$$

where $r, m = 0, \ldots, N-1$; $\mathbf{E}_m \in \mathbf{C}^{M \times M}$ represents the tone generated due to the target present at the mth delay bin; $\mathbf{F} \in \mathbf{C}^{M \times N}$ are the different chirp center frequencies; $\mathbf{G}_m \in \mathbf{C}^{M \times M}$ is the phase term due to different chirp frequencies for a particular delay bin m; and $\mathbf{c} \in \mathbf{C}^N$ is the random vector that selects the chirp waveforms and scales them. A closer inspection of the matrix $\mathbf{F}$ reveals that the center frequencies used to modulate the chirp waveform are being aliased into lower frequency tones. We assume that $p \in \mathbf{Z}$ in order to simplify the analysis as we get subsampled discrete Fourier transform matrices. We impose an additional condition that p should be co-prime with M in order for N frequency tones to be uniformly mapped onto M frequency bins, where $M \leq N$. A simple example of $p = M + 1$, which makes p co-prime with M, circularly maps the N possible frequencies into M bins.

We consider a statistical model similar to that in [19] for the sparse range profile of targets. We assume that the targets are located at N discrete locations corresponding to the different delay bins. The support of the K-sparse range profile is chosen uniformly from all possible subsets of size K. The complex amplitude of nonzero components is assumed to have an arbitrary magnitude and uniformly distributed phase in $[0, 2\pi]$. We also empirically study the performance of the proposed illumination system for targets not located on the grid, which is solved by overdiscretizing the search space.

4.2.2   Problem statement

4.2.2.1   Compressed sensing terminology

For $\mathbf{x} \in \mathbf{C}^M \|\mathbf{x}\|_0$ is called the ℓ_0 *semi-norm*, which is given as the number of nonzero elements in a vector. This semi-norm plays a key role in the formulation of the fundamental problem in compressed sensing. We denote $\|\mathbf{x}\|_1 = \sum_i |x_i|$ as the ℓ_i norm. The spectral or operator norm of the matrix is given as $\|\mathbf{A}\|_{op} = \sigma_{max}(\mathbf{A})$, the largest singular value of the matrix. Mutual coherence $\mu(\mathbf{A})$ is a measure of the correlation between the columns of matrix $\mathbf{A}$, which plays an important role in characterizing the sensing matrix. The mutual coherence is given as follows:

$$\mu(\mathbf{A}) = \max_{i \neq j} \frac{|\langle \mathbf{A}_i, \mathbf{A}_j \rangle|}{\|\mathbf{A}_i\|_2 \|\mathbf{A}_j\|_2},$$

where $\mathbf{A}_i$, $\mathbf{A}_j$ are the columns of matrix $\mathbf{A}$.

Another fundamental property for the measurement matrices is called the *restricted isometry property* (RIP). A measurement matrix $\mathbf{A} \in \mathbf{C}^{M \times N}$ is said to satisfy the RIP of order K, if for any K-sparse vector $\mathbf{x} \in \mathbf{C}^N$:

$$(1 - \delta)\|\mathbf{x}\|_2^2 \leq \|\mathbf{A}x\|_2^2 \leq (1 + \delta)\|\mathbf{x}\|_2^2,$$
$$\text{equivalently, } \delta_K(\mathbf{A}) = \max_{\substack{\Gamma \\ card(\Gamma) \leq K}} \|\mathbf{A}_\Gamma^* \mathbf{A}_\Gamma - \mathbf{I}\| \leq \delta, \tag{4.8}$$

where Γ is an index set that selects the columns of $\mathbf{A}$, $card(\Gamma)$ refers to the number of elements in the set $\delta \leq 1$, and $\mathbf{A}_\Gamma$ is the restriction of $\mathbf{A}$ having columns indexed by Γ. Table 4.1 summarizes the characteristics of some well-studied random sensing schemes as well as our proposed scheme.

Baraniuk et al. [35] showed that random matrices with i.i.d. entries from either Gaussian or sub-Gaussian probability distribution satisfy the RIP condition, such that for any $\delta \in [0, 1]$ $\delta_K \leq \delta$ if the number of measurements $M \sim O(K \log(N/K))$. Although these unstructured random matrices have remarkable recovery guarantees, they do not represent any practical measurement scheme, which leads us to consider classical linear time-invariant systems.

This leads to a structured measurement matrix that is either a partial or subsampled Toeplitz or circulant matrix. The RIP condition of order K for partial Toeplitz matrices in the context of channel estimation was established by Haupt et al. [36]. They showed that if the number of measurements $M \sim O(K^2 \log N)$, then $\delta_K \leq \delta$. This quadratic scaling of the number of measurements with respect to sparsity was improved in [37–39]. Romberg [37] considered an active imaging system that used waveforms with a random symmetric frequency spectrum and acquired compressed measurements using a random subsampler or random demodulator at the receiver to estimate the sparse scene. The resultant system was a randomly subsampled circulant matrix representing the convolution and compression process. It was shown that for a given sparsity level K, the condition that $\delta_{2K} \leq \delta$ is satisfied if the number of measurements $M \geq \alpha_6 \delta^{-2} \min(K(\log N)^6, (K \log N)^2)$, where $\alpha_6 > 0$ is a universal constant independent of the size of the problem and δ.

Rauhut et al. [38] improved the guarantees for the random circulant matrix case. They consider a deterministically sampled random waveform in the time domain with samples following the Rademacher distribution, which is modeled as a subsampled Toeplitz or circulant matrix with entries sampled from the Rademacher distribution. It was shown that

TABLE 4.1
Measures that Characterize Sensing Matrices

Matrix Type of Size $M \times N$	Mutual Coherence	Spectral Norm	Reference
Random matrix with (NM) independent random entries	$2\sqrt{\frac{\log N}{M}}$	$\sqrt{\frac{N}{M}} + 1$	[49,50]
Toeplitz block matrix with $(N + M)$ random entries	$\mathcal{O}\left(\sqrt{\frac{\log N}{M}}\right)$	$\mathcal{O}\left(\sqrt{\frac{N}{M}}\right)$	[44]
LFM waveform modulated with $N_c \ll N$ randomly selected tones for single transmitter and receiver	$\mathcal{O}\left(\sqrt{\frac{\log N}{M}}\right)$	$\mathcal{O}\left(2\sqrt{\frac{N \log(N+M)}{M}}\right)$	[45,46]

for a given sparsity level K, $\delta_K \leq \delta$ with high probability if the number of measurements $M \geq \alpha_7 \max(\delta^{-1}(K \log N)^{3/2}, \delta^{-2}K(\log N \log K)^2)$, where α_7 is a universal constant. In the subsequent work by Krahmer et al. [39], the relation between sparsity level and number of measurements was improved and more general random variables were considered, such as vectors following sub-Gaussian distribution to generate a Toeplitz or circulant matrix. It was shown that, for a given sparsity level K, the condition $\delta_K \leq \delta$ is satisfied if the number of measurements $M \geq \alpha_8\delta^{-2}K(\log K \log N)^2$, where the constant α_8 is a function of only the sub-Gaussian norm of the random variables generating the matrix. We adopt a method similar to that in [36], establish the RIP condition of order K, and obtain a similar result stating that $\delta_K \leq \delta$ if $M \geq a\delta^{-2}K^2 \log N$, where $a > 0$ is independent of δ.

4.2.2.2 Main problem and related results

Given a sparse scene with targets following the statistical model discussed in the previous section, and the measurement scheme in Equation 4.4 with $M \ll N$ and sparsity level $K \ll N$, the goal of compressed sensing [6] is to recover the sparse or compressible vector $\mathbf{x}$ using the minimum number of measurements in $\mathbf{y}$ constructed using random linear projections $\mathbf{A}$. The search for the sparsest solution can be formulated as an optimization problem given below:

$$\min_{\mathbf{x}} ||\mathbf{x}||_0, \text{ subject to } ||\mathbf{Ax} - \mathbf{y}||_2 \leq \eta, \tag{4.9}$$

where η^2 is the noise variance. This problem is nondeterministic, polynomial time and hence intractable as shown by Foucart and Rauhut [40], and many approximate solutions have been found. One particular solution is to use the convex relaxation technique to modify the objective as an ℓ_1 norm minimization instead of the nonconvex ℓ_0 norm, which is given by the following:

$$\min_{\mathbf{x}} ||\mathbf{x}||_1, \text{ subject to } ||\mathbf{Ax} - \mathbf{y}||_2 \leq \eta. \tag{4.10}$$

This approach has been shown to successfully recover sparse or compressible vectors [41,42], given that the submatrices formed by columns of the sensing matrix are well conditioned. Our analysis is based on least absolute shrinkage and selection operator (LASSO) [9], which is a related method that solves the optimization problem in Equation 4.10. It has been shown [43] that, for an appropriate choice of λ and with measurement matrix conditions being satisfied, the support of the solution of the below-mentioned optimization problem coincides with the support of the solution of the intractable problem in Section 4.2.2.2:

$$\min_{\mathbf{x}} \lambda||\mathbf{x}||_1, +\frac{1}{2}||\mathbf{Ax} - \mathbf{y}||_2^2, \tag{4.11}$$

The goal of our analysis is to show that the measurement model satisfies the conditions on mutual coherence given by Candes and Plan [43] and to find a bound on the sparsity level of the range profile that guarantees successful support recovery of almost all sparse signals using LASSO with high probability from noisy measurements. Table 4.2 shows the support recovery with different sensing schemes.

Although the upper bound on the sparsity level that guarantees successful support recovery for our scheme has an additional $\log(N+M)$ penalty compared to an unstructured Gaussian matrix as shown by Candes and Plan [43] and block Toeplitz matrices with entries sampled from sub-Gaussian distribution as shown by Bajwa [19,44], our sensing scheme has a simple implementation and far lower storage requirements as compared to systems employing stochastic waveforms. In addition, we also provide an estimate of the number of measurements required for the sensing matrix representing our scheme to satisfy the RIP condition or order K with high probability. If the sensing matrix satisfies the

TABLE 4.2

Support Recovery Guarantees for Different Sensing Matrices

Recovery Guarantees from Noisy Measurements with Component-Wise Noise Variance σ^2

Matrix Type of Size $M \times N$	Sparsity for Successful Recovery	Minimum Signal Strength	Reference
Random matrix with (NM) independent random entries	$\mathcal{O}\left(\frac{M}{N}\right)$	$\mathcal{O}(\sigma\sqrt{2\log N})$	[43]
Toeplitz block matrix with $(N+M)$ random entries	$\mathcal{O}\left(\frac{M}{\log N}\right)$	$\mathcal{O}\left(\sigma\sqrt{2\log N}\right)$	[44]
LFM waveform modulated with $N_c \ll N$ randomly selected tones for single transmitter and receiver	$\mathcal{O}\left(\frac{M}{\log N \log(N+M)}\right)$	$\mathcal{O}\left(\sigma\sqrt{2\log N}\right)$	[46]

RIP property or order $2K$, given by $\delta_{2K} \leq \delta \approx \sqrt{2} - 1$ with high probability, then all K-sparse vectors are successfully recovered with a reconstruction error of an oracle estimator that knows the support of the sparse vector or the support of the K largest elements [42], which is termed *uniform recovery guarantees*. The next section presents our main results in detail.

4.2.3 Recovery guarantees for single transmitter and receiver

In order to obtain the nonasymptotic recovery guarantee for our system, we find estimates of the tail bounds for mutual coherence and spectral or operator norms of the measurement matrix. Using these estimates, we deduce the conditions required for successful support recovery as well as RIP conditions for order K to hold. For proofs of these results, we refer the readers to [45,46].

Lemma 4.2.1 *The operator norm of the sensing matrix in Equation 4.5 is bounded with high probability by the following:*

$$||A||_{op} \leq 2\sqrt{\frac{N}{M} \log(M + N)}. \tag{4.12}$$

The following result on mutual coherence is obtained using concentration inequalities of quadratic forms of random vectors having the sub-Gaussian distribution given in [47], which is extended to the complex domain in [46].

Lemma 4.2.2 *The mutual coherence of the sensing matrix $\boldsymbol{A}$ scales as follows:*

$$\mu(\mathbf{A}) \sim \mathcal{O}\left(\sqrt{\frac{\log(N)}{M}}\right), \tag{4.13}$$

with high probability.

We first state the support recovery guarantee for our proposed scheme.

Theorem 4.2.3 *Consider a compressive radar system with the measurement model* $y = Ax + w$, *where* $A \in C^{M \times N}$ *is defined in Equation 4.5 such that the target scene* x *is drawn from a* $K-$*sparse model with support* S, *complex unknown amplitudes and observed in the i.i.d. noise process* $\mathbf{w} \sim \mathcal{CN}(0, \sigma^2 \mathbf{I})$. *The support of the targets in the scene can be recovered using a LASSO estimator with high probability for a system using* M *samples and* N_C *tones with* $M \sim O(\log^3 N)$ *and* $N_C \sim O(N)$, *if the number of targets in the scene is* $K \sim O(\frac{M}{\log^2(N)})$, *with minimum amplitude* $\min_{k \in S} |x_k| > \frac{8}{\sqrt{1-\varepsilon}} \sigma \sqrt{2 \log(N)}$.

Theorem 4.2.4 *For the measurement matrix* A *given in Equation 4.5 and any* $\delta \in [0, 1]$, *the RIP in Section 4.2.2.1* $\delta_K(A) \leq \delta$ *is satisfied with high probability if the number of measurements* M *scales with* $M \sim O(K^2 \log(N))$.

4.3 Multiple transmitters and receivers

We consider N_T transmitters and N_R collocated receivers that function as an MIMO radar system. The system employs the compressive illumination framework analyzed by Ertin et al. [45] for estimating the target range and angle of arrival. The transmit antenna elements are placed with a spacing of $d_T = 0.5$. and the receive antenna elements are placed at $d_R = 0.5 N_T$ relative to the wavelength λ_C of the carrier in order to obtain the virtual array with aperture length $N_T N_R$. The angle of arrival characterized by $\cos \Theta \in [-1, 1]$ is partitioned into $N_\Theta = N_T N_R$ grids. The other parameters associated with the chirp waveform and the modulating tones are defined in Section 4.2. From the possible N modulated waveforms, a subset of size $N_C N_T$ is chosen at random for transmission, where N_C is the number of modulating tones per transmitter. We simplify this selection model by considering N independent indicator random variables following a Bernoulli distribution to select the tones that modulate LFM waveforms such that $N_C N_T$ waveforms are selected on average. Let $\hat{\gamma}_i \in \{0, 1\}$ be the indicator random variable such that $P(\hat{\gamma}_i = 1) = \frac{N_C N_T}{N}$ and $P(\hat{\gamma}_i = 0) = 1 - \frac{N_C N_T}{N}$. The chosen LFM waveform is then scaled by a sequence of independent and identical complex exponentials with a uniformly distributed phase and probability density function $f_\Phi(\varphi_i) = \frac{1}{2\pi}, \varphi_i \in [0, 2\pi]$. We define the sequence of random variables $\hat{c}_i$ that model this selection and scaling process as follows:

$$\hat{c}_i = \hat{\gamma}_i \exp(j\Phi_i). \tag{4.14}$$

The selected waveforms are then assigned to one of the N_T transmitters. Let $\xi(i)$ be the discrete random variable that indicates the assignment of waveform i to transmitter l given by the uniform probability $P(\xi(i) = l) = \frac{1}{N_T}, l = 0, \dots, N_{T-1}, i = 0, \dots, N - 1$. The transmitted signal from all the N_T transmitters can be written as follows:

$$s(t) = \sum_{i=1}^{N} \hat{c}_i \frac{\exp \left(j 2\pi \left[\hat{f}(i)t + \frac{\beta}{2\tau} t^2 \right] \right)}{\sqrt{N_c N_T}} r \left(\frac{t - \frac{\tau}{2}}{\tau} \right), \tag{4.15}$$

where $r \left(\frac{t - \frac{\tau}{2}}{\tau} \right) = 1$ if $t \in (0, \tau)$ and 0 otherwise. The instantaneous frequency of the transmitted and received modulated LFM signal as a function of time is illustrated in Figure 4.4 for the case of $N_T = 2$, $N_R = 1$.

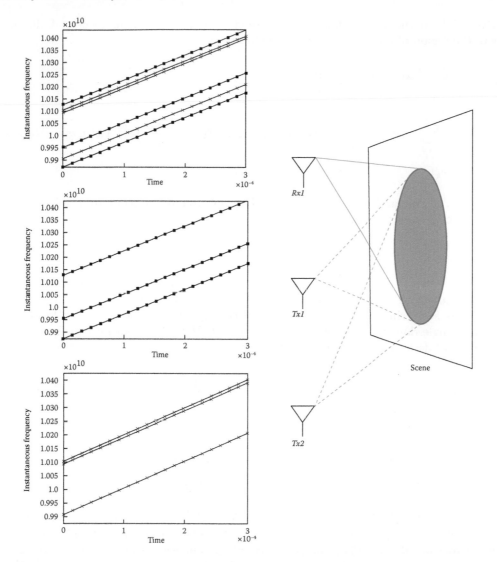

FIGURE 4.4
The time–frequency representation of the waveform employed by two transmitters Tx_1 and Tx_2 and received signal at one of the collocated receivers Rx_1, which is a modulation of a linear frequency modulated waveform by a train of sinusoids with random frequencies.

The sampling rate at the receiver after stretch processing is $F_S = \frac{\beta t_u}{\tau}$, which leads to $M = \beta t_u$ samples of stretch processor output at each receiver. Each delay bin is denoted by $\Delta_m = \frac{m}{B}$, $m = 0, 1, \ldots, N - 1$. Each angle bin is denoted by $\Theta_v \in \left\{ \frac{2i}{N_T N_R} \middle| i = \frac{-N_T N_R}{2}, \ldots, \frac{N_T N_R - 1}{2} \right\}$. The receiver and transmitter steering vectors as a function of the angle of arrival Θ are defined as follows:

$$\alpha_{\mathbf{R}}(\theta_v) = \begin{bmatrix} 1 & \cdots & \exp(j2\pi d_R (N_R - 1)\theta_v) \end{bmatrix}^T, \text{ and}$$
$$\alpha_{\mathbf{T}}(\theta_v) = \begin{bmatrix} 1 & \cdots & \exp(j2\pi d_T (N_T - 1)\theta_v) \end{bmatrix}^T, \tag{4.16}$$

respectively. The normalized samples at the stretch processor output at receiver k due to targets in the region of interest is given by the following:

$$
y_k(n) = \sum_{v=0}^{N_T N_R - 1} \frac{\alpha_R(\theta_v; k)\alpha_T(\theta_v; \xi(i))}{\sqrt{N_T N_R N_c M}}
$$

$$
\times \sum_{\substack{m=0 \\ i=1}}^{\substack{m=N-1 \\ i=N}} \hat{c}_i \exp(-j2\pi f(i)\Delta_m) \exp\left(-j2\pi\left(f(i) - \frac{\beta\Delta_m}{\tau}\right)\frac{k}{F_s}\right) x(v, m) \quad (4.17)
$$

$$
+ w_k(n),
$$

where $k = 1, \ldots, N_R$, and $n = 0, \ldots, M - 1$. The concatenated output from all the N_R receivers can be compactly written as follows:

$$
\mathbf{y} = \mathcal{A}\mathbf{x} + \mathbf{w}, \quad (4.18)
$$

where $\mathbf{y} = [\mathbf{y}_1 \cdots \mathbf{y}_{N_R}]^T$, $\mathbf{y}_k = [y_k(0) \cdots y_k(M-1)]^T \in \mathbf{C}^M$, $\mathbf{w} \in \mathbf{C}^{N_R M}$ is the zero mean additive noise following a complex Gaussian distribution with variance σ^2, and $\mathbf{x} \in \mathbf{C}^{N N_T N_R}$ is the complex scattering amplitude associated with targets at all possible grid locations in the range-angle domain. The sensing matrix $\mathcal{A} \in \mathbf{C}^{N_R M \times N_\Theta N}$ can be expressed as a random series of deterministic matrices as follows:

$$
\mathcal{A} = \sum_{i=1}^{N} \hat{c}_i (\bar{\alpha}_\mathbf{R} \bar{\alpha}_\mathbf{T}(\xi(\mathbf{i}))) \otimes (\mathbf{H}_i \bar{\mathbf{A}} \mathbf{D}_i), \quad (4.19)
$$

where, $\bar{\alpha}_\mathbf{R} = \sqrt{\frac{1}{N_R N_T}} \left[\alpha_\mathbf{R}(\theta_0) \quad \cdots \quad \alpha_\mathbf{R}(\theta_{N_\theta - 1})\right]$

$$
\bar{\alpha}_\mathbf{T}(\xi(\mathbf{i})) = diag(\exp(2\pi j d_T \xi(i)\theta_0 \cdots \exp(2\pi j d_T \xi(i)\theta_{N_\theta - 1}))), \quad (4.20)
$$

where $i = 0, \ldots, N - 1$ and $r = 0, \ldots, N - 1$, $\bar{\mathbf{A}} \in \mathbf{C}^{M \times N}$, $\mathbf{H}_i \in \mathbf{C}^{M \times M}$, and $\mathbf{D}_i \in \mathbf{C}^{N \times N}$ are defined in Equation 4.6, $\bar{\alpha}_\mathbf{R} \in \mathbf{C}^{N_R M \times N_\Theta N}$ is the matrix consisting of receiver steering vectors for all the bins of angle of arrival, $\otimes$ represents the Kronecker product, and $\bar{\alpha}_\mathbf{T}(\xi(\mathbf{i})) \in \mathbf{C}^{N_\Theta N \times N_\Theta N}$ is the random diagonal matrix with diagonal elements as the randomly chosen transmitter's component of the steering vector for all the angle bins.

Each column of the sensing matrix $\mathcal{A}$ can be written as follows:

$$
\mathcal{A}(m, v) = (\alpha_\mathbf{R}(\theta_\mathbf{v}) \otimes (\mathbf{E_m F G_m}))\hat{\mathbf{c}}(v)
$$

$$
\hat{c}_r(v) = \hat{c}_r \alpha_T(\theta_v; \xi(r)), \quad (4.21)
$$

where $m = 0, \ldots, N - 1$, $v = 0, \ldots, N_\Theta - 1$, $r = 0, \ldots, N - 1$, $\mathbf{E}_m \in \mathbf{C}^{M \times M}$, $\mathbf{F} \in \mathbf{C}^{M \times N}$, and $\mathbf{G}_m \in \mathbf{C}^{M \times M}$ are defined in Equation 4.7, and $\hat{\mathbf{c}}(s) \in \mathbf{C}^N$ is the random vector with independent components expressed as a product of two random variables: a variable that selects the chirp waveform and an independent variable that allocates the waveform to a particular transmitter.

We assume that the targets are located at the $NN_\Theta = NN_R N_T$ discrete locations corresponding to different delay bins and angle bins. The support of the K-sparse range profile is chosen uniformly from all possible subsets of size K. The complex amplitude of nonzero components is assumed to have an arbitrary magnitude and uniformly distributed phase in $[0, 2\pi]$.

4.3.1 Recovery guarantees for the MIMO system

Similar to the single transmitter and receiver case, we state the bounds on the spectral norm and mutual coherence of the structured sensing matrix representing the proposed MIMO system. Using these results, we state the conditions for successful support recovery and RIP condition of order K to hold. For proofs of these results, we refer the readers to [45,46].

Lemma 4.3.1 *The operator norm of the sensing matrix in Equation 4.19 is bounded with high probability by the following:*

$$||\mathcal{A}||_{op} \leq 2\sqrt{\frac{N_T N}{M} \log(N_R M + N_R N_T N)}. \tag{4.22}$$

Lemma 4.3.2 *The mutual coherence of the sensing matrix $\mathcal{A}$ scales as follows:*

$$\mu(\mathcal{A}) = \mathcal{O}\left(\sqrt{\frac{\log(NN_R N_T)}{M}}\right) \tag{4.23}$$

with high probability.

Theorem 4.3.3 *Consider a compressive MIMO radar system with the measurement model $\boldsymbol{y} = \mathcal{A}\boldsymbol{x} + \boldsymbol{w}$, where $\mathcal{A} \in \boldsymbol{C}^{N_R M \times N_R N_T N}$ is de ned in Equation 4.19 such that the target scene $\boldsymbol{x}$ is drawn from a K-sparse model with complex unknown amplitudes and observed in the i.i.d. noise process $\boldsymbol{w} \sim \mathcal{CN}(0, \sigma^2 I)$. The support of the targets in the scene can be recovered using a LASSO estimator with high probability for a system using M samples at each receiver and N_c tones at each transmitter with $M \sim O(\log^3(NN_R N_T))$ and $N_C \sim O(N/N_T)$, if the target scene consists of K targets with $K \sim O(N_R M/\log^2(2NN_R N_T))$ of minimum amplitude:*

$$\min_{k \in \boldsymbol{S}} |x_k| > \frac{8}{\sqrt{1-\epsilon}} \sigma \sqrt{2 \log(NN_R N_T)}, \tag{4.24}$$

Theorem 4.3.4 *For the measurement matrix $\mathcal{A}$ given in Equation 4.19 and any $\delta \in [0, 1]$, the RIP condition in Section 4.2.2.1 as $\delta_K(\mathcal{A}) \leq \delta$ is satis ed with high probability if the number of measurements M scale with $M \sim O(K^2 \log(N_R N_T N))$.*

4.4 Simulation results

In this section, we conduct simulation studies to study the performance of the proposed compressive radar sensor as a function of system parameters. The fixed parameters of the simulations are given in Table 4.3.

4.4.1 Effect of multiple tones on mutual coherence

We first study the effect of increasing the number of tones in a single transmitter and receiver setting. We compare the proposed illumination scheme with a Toeplitz matrix with independent and identical elements sampled from a complex standard normal distribution. The Toeplitz sensing matrix represents the impulse response of the linear time-invariant system with a randomly distributed waveform with independent entries as input. From Figure 4.5, the coherence of a system employing a single tone is high for lower sampling rates. Increasing the number of tones improves the mutual coherence and as the number of

TABLE 4.3

Parameters for Simulation Results

Parameter	Value
Bandwidth B	500×10^6 Hz
Range interval	$[0, 100]$m
Number of range bins N	334
Unambiguous time interval t_u	6.6×10^{-7} s
pulse duration τ	6.86×10^{-5} s
Number of samples per receiver M	111

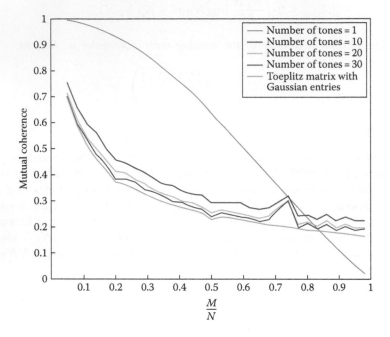

FIGURE 4.5

Mutual coherence of a single transmitter system with a single receiver as a function of the undersampling ratio $\frac{M}{N}$ as the number of chirps N_C is increased along with the mutual coherence of the random Toeplitz matrix.

modulating tones increase, the mutual coherence of the system converges in mean to the mutual coherence of structured random Toeplitz matrix.

Next, we compare the coherence of the proposed system as the number of chirps increases with a multiple-input system employing samples from a Gaussian distribution, which leads to a partial block Toeplitz measurement matrix with random Gaussian entries.

From Figure 4.6, it is evident that as the number of transmitters and modulating tones increase, the randomness in the waveform increases and hence the mutual coherence of the system approaches that of a system employing random waveforms with independent samples.

4.4.2 On grid

First, we consider the noiseless case and consider the reconstruction error as a performance criterion. In Figure 4.7a through d, the probability of successful recovery (defined as reconstruction error $< 10^{-5}$) is shown as a function of the sparsity ratio (the ratio of

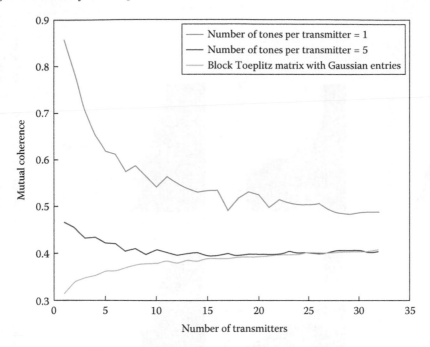

FIGURE 4.6

Mutual coherence of a multiple transmit system with a single receiver as a function of the number of transmitters N_T as the number of chirps N_C is increased along with the mutual coherence of the random block Toeplitz matrix.

the number of targets in the scene to the number of measurements $\frac{K}{M}$) and undersampling ratio $\frac{\beta}{B} = \frac{M}{N}$. We observe that for a sufficiently high number of modulating tones, the performance characterized by the phase transition plot is similar to that of a system employing stochastic waveforms on transmit. Next, we consider noisy measurements and adjust the undersampling ratio to 0.5. We characterize the values of the signal-to-noise ratio (SNR) and sparsity ratio where the performance criterion for support recovery (here defined by the area under the curve (AUC) of receiver operating characteristic (ROC), [AUC] exceeding a threshold of 0.99) is satisfied with high probability. The results are given in Figure 4.7e through l. Next, we fix the SNR at 20 dB and study the criterion for support recovery (defined by the AUC exceeding a threshold of 0.99) as the undersampling ratio and the sparsity level are varied. It is evident that as the number of tones increases, the performance of the system approaches the performance of the system employing random waveforms.

Figure 4.7a through l illustrates the probability that the reconstruction error is below 10^{-5} in the noiseless setting as a function of undersampling and sparsity ratio. Figure 4.7a through l shows the probability that $AUC > 0.99$ as a function of SNR at a fixed undersampling ratio $\frac{\beta}{B} = 0.5$. Figure 4.7a through l shows the probability that $AUC > 0.99$ as a function of the undersampling ratio $\frac{\beta}{B} = \frac{M}{N}$ for a fixed signal to ratio of 20 dB.

Next, we characterize the performance of the MIMO system for support recovery using the receiver operating characteristics for successful support recovery in Figure 4.8. The noise is generated from a complex Gaussian distribution with a variance that guarantees an SNR of 12 dB. The target support is generated uniformly from the grid and the amplitudes are sampled from a complex Gaussian distribution. We fix the number of targets $K = 120$, consider an MIMO system with $N_T = 16$ transmitters and $N_R = 8$ receivers, and vary the

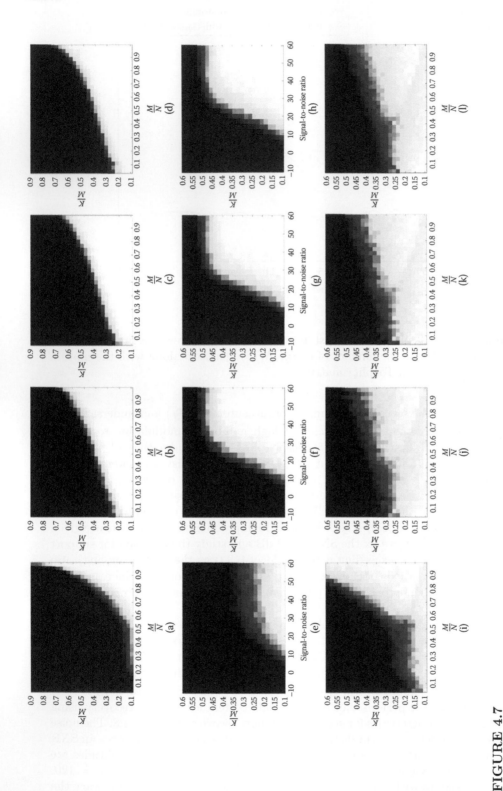

FIGURE 4.7

(a) to (d) illustrate the probability of reconstruction error is below 10^{-5} in the noiseless setting as a function of undersampling and sparsity ratio; (e) to (h) shows the probability that $AUC > 0.99$ as a function of signal-to-noise ratio at a fixed undersampling ratio $\frac{\beta}{B} = 0.5$; (i) to (l) shows the probability that $AUC > 0.99$ as a function of undersampling ratio $\frac{\beta}{B} = \frac{M}{N} = 0.5$ for a fixed signal to ratio of 20 dB.

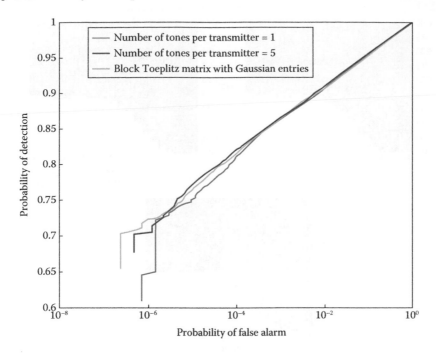

FIGURE 4.8
Receiver operating characteristics with number of targets $K = 120$ and signal-to-noise ratio (SNR) = 12 dB. The number of transmitters $N_T = 16$, and the number of receivers $N_R = 8$. The number of chirps per transmitter is increased and it is compared to the ROC obtained for a system employing random waveforms.

number of modulating tones per transmitter. We also compare the ROC curve for a system employing random waveforms with samples from a Gaussian distribution.

It is evident that the performance of the proposed reduced complexity compressive system approaches the performance of the random code system for a small number of modulating tones at each transmitter.

We further show the effect of noise variance on the support recovery guarantee in the form of a phase transition diagram in Figure 4.9a through c, where the criterion used is that even the area under the ROC curve exceeds AUC > 0.95. We observe a region in noise variance and number of targets where the support recovery is guaranteed with high probability.

4.4.3 Off grid

In this section, we consider targets that are uniformly sampled from the interval $[0, t_u]$ without any restriction to lie on the range grid considered in the derivation of our theoretical result for the setup with a single transmitter and receiver. For recovery, we use a fine grid at a resolution four times that of a Nyquist grid and use the iteratively reweighted ℓ_1 method [48] that minimizes a nonconvex objective function that approximates an ℓ_0 norm. We declare a successful detection if $|\hat{r} - r| \leq \frac{1}{2B}$, where $\hat{r}$ is the estimated range and r is the true range of the target, and B is the system bandwidth. We declare only a single detection if more than one estimated target falls in that region. The detection rate is defined as the ratio between the number of successful detections and the number of targets. All other detections are declared false alarms. The false alarm rate is defined as the ratio between

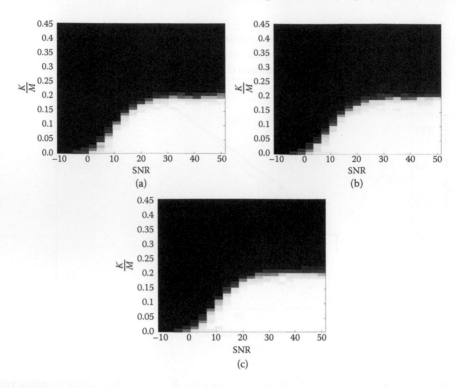

FIGURE 4.9

(a) $\mathbf{A}$ with $N_C = 1$ chirps; (b) $\mathbf{A}$ with $N_C = 5$ chirps; (c) Teoplitz sensing matrix. The intensity values of the images is 1 if the area under the curve for receiver operating characteristics curve exceeds 0.95 and 0 otherwise as the SNR and number of targets in the scene is varied. The number of transmitters and receivers in the system is $N_T = 16$ and $N_R = 8$, respectively.

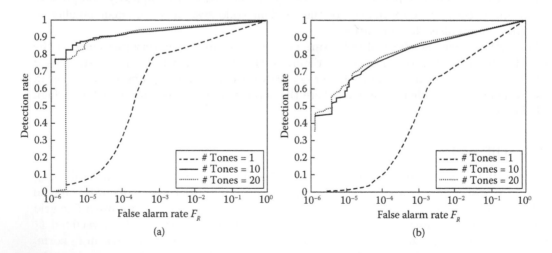

FIGURE 4.10

The detection rate as a function of the false alarm rate for recovering off-grid targets using iterative reweighted ℓ_1 on an over-discretized grid with an undersampling ratio $\frac{\beta}{B} = \frac{1}{3}$ and SNR = 12 dB. (a) $K = 10$ targets; (b) $K = 20$ targets.

the number of false detections to the total number of grid points. Figure 4.10 illustrates the performance improvement achieved as the number of modulating tones is increased for an *SNR* of 12 dB.

4.5   Conclusion

In this chapter, a new compressive acquisition scheme, termed *compressive illumination*, was examined for high-resolution radar sensing. We showed that a system comprising multitone LFM transmit waveforms and a uniformly subsampled stretch processor results in a structured random sensing matrix with provable recovery guarantees for delay and angle of arrival estimation in sparse scenes. The recovery guarantees for the proposed compressive illumination scheme are comparable to those of random Toeplitz matrices with a much larger number of random elements. The proposed scheme is well matched to practical implementation utilizing a small number of random parameters and uniform sampling ADCs on receive. Our simulation shows that targets both on and off the grid can be detected using sparsity regularized recovery algorithms.

Acknowledgments

This research was partially supported by Army Research Office Grant W911NF-11-1-0391 and National Science Foundation (NSF) Grant IIS-1231577.

References

[1] M. A. Richards, *Fundamentals of Radar Signal Processing*, New York: McGraw-Hill, 2014.

[2] P. Stoica and J. Li, *MIMO Radar Signal Processing*, Hoboken, NJ: Wiley, 2008.

[3] S. Baskar and E. Ertin, A software defined radar platform for waveform adaptive MIMO radar research, in *2015 IEEE Radar Conference (RadarCon)*, May 2015, pp. 1590–1594.

[4] R. H. Walden, Analog-to-digital converter survey and analysis, *IEEE Journal on Selected Areas in Communications*, vol. 17, no. 4, pp. 539–550, 1999.

[5] R. Middleton, Dechirp-on-receive linearly frequency modulated radar as a matched-filter detector, *IEEE Transactions on Aerospace and Electronic Systems*, vol. 48, no. 3, pp. 2716–2718, 2012.

[6] D. L. Donoho, Compressed sensing, *IEEE Transactions on Information Theory*, vol. 52, no. 4, pp. 1289–1306, 2006.

[7] E. Candes, J. Romberg, and T. Tao, Robust uncertainty principles: Exact signal reconstruction from highly incomplete frequency information, *IEEE Transactions on Information Theory*, vol. 52, no. 2, pp. 489–509, 2006.

[8] E. Candes and T. Tao, The dantzig selector: Statistical estimation when p is much larger than n, *The Annals of Statistics*, vol. 35, no. 6, pp. 2313–2351, 2007.

[9] R. Tibshirani, Regression shrinkage and selection via the LASSO, *Journal of the Royal Statistical Society: Series B*, vol. 73, no. 3, pp. 273–282, 1996.

[10] Z. Yang and L. Xie, On gridless sparse methods for line spectral estimation from complete and incomplete data, *IEEE Transactions on Signal Processing*, vol. 63, no. 12, pp. 3139–3153, 2015.

[11] G. Tang, B. N. Bhaskar, P. Shah, and B. Recht, Compressed sensing off the grid, *IEEE Transactions on Information Theory*, vol. 59, no. 11, pp. 7465–7490, 2013.

[12] J. Tropp and A. Gilbert, Signal recovery from random measurements via orthogonal matching pursuit, *IEEE Transactions on Information Theory*, vol. 53, no. 12, pp. 4655–4666, 2007.

[13] D. Needell and J. Tropp, Cosamp: Iterative signal recovery from incomplete and inaccurate samples, *Applied and Computational Harmonic Analysis*, vol. 26, no. 3, pp. 301–321, 2009.

[14] L. Potter, E. Ertin, J. Parker, and M. Cetin, Sparsity and compressed sensing in radar imaging, *Proceedings of the IEEE*, vol. 98, no. 6, pp. 1006–1020, 2010.

[15] K. Gedalyahu and Y. Eldar, Time-delay estimation from low-rate samples: A union of subspaces approach, *IEEE Transactions on Signal Processing*, vol. 58, no. 6, pp. 3017–3031, 2010.

[16] C.-Y. Chen and P. Vaidyanathan, Compressed sensing in MIMO radar, in *42nd Asilomar Conference on Signals, Systems and Computers, 2008*, October 2008, pp. 41–44.

[17] Y. Yu, A. Petropulu, and H. Poor, Measurement matrix design for compressive sensing based MIMO radar, *IEEE Transactions on Signal Processing*, vol. 59, no. 11, pp. 5338–5352, 2011.

[18] M. Herman and T. Strohmer, High-resolution radar via compressed sensing, *IEEE Transactions on Signal Processing*, vol. 57, no. 6, pp. 2275–2284, 2009.

[19] T. Strohmer and B. Friedlander, Analysis of sparse MIMO radar, *Applied and Computational Harmonic Analysis*, vol. 37, no. 3, pp. 361–388, 2014.

[20] T. Strohmer and H. Wang, Accurate imaging of moving targets via random sensor arrays and kerdock codes, *Inverse Problems*, vol. 29, no. 8, p. 085001, 2013.

[21] D. Dorsch and H. Rauhut, Refined analysis of sparse MIMO radar, *Journal of Fourier Analysis and Applications*, pp. 1–45, 2016. doi:10.1007/s00041-016-9477-7.

[22] M. Hügel, H. Rauhut, and T. Strohmer, Remote sensing via ℓ_1-minimization, *Foundations of Computational Mathematics*, vol. 14, no. 1, pp. 115–150, 2014.

[23] Y. Yu, A. Petropulu, and H. Poor, CSSF MIMO radar: Compressive-sensing and step-frequency based MIMO radar, *IEEE Transactions on Aerospace and Electronic Systems*, vol. 48, no. 2, pp. 1490–1504, 2012.

[24] B. Li and A. Petropulu, RIP analysis of the measurement matrix for compressive sensing-based MIMO radars, in *IEEE 8th Sensor Array and Multichannel Signal Processing Workshop (SAM), 2014*, June 2014, pp. 497–500.

[25] M. Shastry, R. Narayanan, and M. Rangaswamy, Sparsity-based signal processing for noise radar imaging, *IEEE Transactions on Aerospace and Electronic Systems*, vol. 51, no. 1, pp. 314–325, 2015.

[26] E. Baransky, G. Itzhak, N. Wagner, I. Shmuel, E. Shoshan, and Y. Eldar, Sub-nyquist radar prototype: Hardware and algorithm, *IEEE Transactions on Aerospace and Electronic Systems*, vol. 50, no. 2, pp. 809–822, 2014.

[27] O. Bar-Ilan and Y. C. Eldar, Sub-nyquist radar via Doppler focusing, *IEEE Transactions on Signal Processing*, vol. 62, no. 7, pp. 1796–1811, 2014.

[28] M. Rossi, A. Haimovich, and Y. Eldar, Spatial compressive sensing in MIMO radar with random arrays, in *46th Annual Conference on Information Sciences and Systems (CISS), 2012*, March 2012, pp. 1–6.

[29] Y. Chi, Sparse MIMO radar via structured matrix completion, in *Global Conference on Signal and Information Processing (GlobalSIP)*, 2013 IEEE, December 2013, pp. 321–324.

[30] S. Sun, W. Bajwa, and A. Petropulu, MIMO-MC radar: A MIMO radar approach based on matrix completion, *IEEE Transactions on Aerospace and Electronic Systems*, vol. 51, no. 3, pp. 1839–1852, 2015.

[31] M. Duarte and R. Baraniuk, Spectral compressive sensing, *Applied and Computational Harmonic Analysis*, vol. 35, no. 1, pp. 111–129, 2013.

[32] A. Gilbert, M. Strauss, and J. Tropp, A tutorial on fast fourier sampling, *IEEE Signal Processing Magazine*, vol. 25, no. 2, pp. 57–66, 2008.

[33] E. Ertin, Frequency diverse waveforms for compressive radar sensing, in *2010 International Waveform Diversity and Design Conference (WDD)*, August 2010, pp. 000216–000219.

[34] E. Ertin, L. Potter, and R. Moses, Sparse target recovery performance of multi-frequency chirp waveforms, in *19th European Signal Processing Conference, 2011*, August 2011, pp. 446–450.

[35] R. Baraniuk, M. Davenport, R. A. DeVore, and M. B. Wakin, A simple proof of the restricted isometry property for random matrices, *Constructive Approximation*, vol. 28, no. 3, pp. 253–263, 2008.

[36] J. Haupt, W. Bajwa, G. Raz, and R. Nowak, Toeplitz compressed sensing matrices with applications to sparse channel estimation, *IEEE Transactions on Information Theory*, vol. 56, no. 11, pp. 5862–5875, 2010.

[37] J. Romberg, Compressive sensing by random convolution, *SIAM Journal on Imaging Sciences*, vol. 2, no. 4, pp. 1098–1128, 2009.

[38] H. Rauhut, J. Romberg, and J. A. Tropp, Restricted isometries for partial random circulant matrices, *Applied and Computational Harmonic Analysis*, vol. 32, no. 2, pp. 242–254, 2012.

[39] F. Krahmer, S. Mendelson, and H. Rauhut, Suprema of chaos processes and the restricted isometry property, *Communications on Pure and Applied Mathematics*, vol. 67, no. 11, pp. 1877–1904, 2014.

[40] S. Foucart and H. Rauhut, *A Mathematical Introduction to Compressive Sensing*, Basel: Birkhauser, 2013.

[41] S. S. Chen, D. L. Donoho, and M. A. Saunders, Atomic decomposition by basis pursuit, *SIAM Review*, vol. 43, no. 1, pp. 129–159, 2001.

[42] E. Candes, The restricted isometry property and its implications for compressed sensing, *Comptes Rendus Mathematique*, vol. 346, no. 9–10, pp. 589–592, 2008.

[43] E. Candes and Y. Plan, Near-ideal model selection by ℓ_1 minimization, *The Annals of Statistics*, vol. 37, no. 5A, pp. 2145–2177, 2009.

[44] W. Bajwa, Geometry of random Toeplitz-block sensing matrices: Bounds and implications for sparse signal processing, *Proceedings of SPIE*, vol. 8365, p. 836505, June 2012.

[45] N. Sugavanam and E. Ertin, Recovery guarantees for MIMO radar using multi-frequency LFM waveform, in *2016 IEEE Radar Conference (RadarConf)*, May 2016, pp. 1–6.

[46] N. Sugavanam and E. Ertin, Recovery guarantees for multifrequency chirp waveforms in compressed radar sensing, *CoRR*, vol. abs/1508.07969, 2015.

[47] M. Rudelson and R. Vershynin, Hanson-wright inequality and sub-Gaussian concentration, *Electronic Communications in Probability*, vol. 18, no. 82, pp. 1–9, 2013.

[48] E. J. Candès, M. B. Wakin, and S. P. Boyd, Enhancing sparsity by reweighted l1 minimization, *Journal of Fourier Analysis and Applications*, vol. 14, no. 5, pp. 877–905, 2008.

[49] T. T. Cai and T. Jiang, Limiting laws of coherence of random matrices with applications to testing covariance structure and construction of compressed sensing matrices, *The Annals of Statistics*, vol. 39, no. 3, pp. 1496–1525, 2011.

[50] K. Davidson and S. Szarek, Local operator theory, random matrices and banach spaces, in *Handbook on the Geometry of Banach Spaces*, vol. 1, pp. 317–366, 2001.

5

Compressive Sensing for Inverse Synthetic Aperture Radar Imaging

Alessio Bacci, Elisa Giusti, Sonia Tomei, Davide Cataldo, Marco Martorella, and Fabrizio Berizzi

CONTENTS

5.1 Introduction

In the last few decades, the increasing demand for high-resolution images in applications such as automatic target recognition (ATR) and automatic target classification (ATC) for surveillance and homeland security has attracted the attention of the research community worldwide [35]. In this scenario, inverse synthetic aperture radar (ISAR) imaging techniques have been widely studied since they allow to obtain high-resolution images of moving targets by means of Fourier-based imaging methods [11,53]. In particular, the transmission of large bandwidth signals and the coherent integration of the received echoes from different aspect angles are the key aspects to obtain images with fine resolution both in range and cross-range. By the way, in some cases, the transmitted bandwidth and/or the coherent processing time (CPI) might be not enough for the desired resolution. In addition, system malfunctioning, transmission disruption, or data compression may be the causes of missing samples, both in the frequency and/or in the slow-time domain. In such cases, conventional Fourier-based imaging methods, such as the range–Doppler (RD) method, suffer from coarse resolution and distortions [11]. To enhance the results obtained via RD methods, different super-resolution (SR) techniques have been studied in the literature, such as apodization techniques [14,15] and bandwidth extrapolation (BWE) techniques [56]. The main disadvantages of such SR methods are that their performance is very sensitive to noise and clutter and that these methods generally computationally expensive.

However, recent results in signal processing have demonstrated the ability of compressive sensing (CS) to reconstruct a sparse or compressible signal from a limited number of measurements with a high probability by solving an optimization problem [13]. Such a technique has been successfully applied for data storage reduction in medical imaging [30] and radar imaging applications [3], among others. The successful application of CS relies on the *sparsity* of the signal, that is, the characteristic of the signal of being represented by few nonzero entries in a suitable orthonormal basis. Different basis can be identified which support different sparsity index such as that proposed in [18]. In the case of radar imaging, the applicability of CS has been justified by observing that at high frequencies a radar signal is sparse in the image domain, that is, when represented in the 2D Fourier basis, since it can be approximated by the superimposition of few prominent scatterer responses with respect to the pixels in the image in the radar image plane [20,22]. Another property which must be satisfied by the radar signal for the application of CS is the restricted isometry property (RIP). The RIP denotes the characteristic of the matrices which define the basis on which the signal is sparse of being orthonormal, at least when dealing with sparse vectors. It has been demonstrated that the Fourier matrices which define the domain in which the ISAR signal is sparse satisfy the RIP property [7,16], hence demonstrating the applicability of CS reconstruction to the ISAR imaging problem. CS reconstruction capabilities have been tested for a number of radar applications, such as synthetic aperture radar (SAR) imaging and ISAR [44], ground penetrating radars [25], multiple-input multiple-output (MIMO) radar [1,57] 3D ISAR imaging, [43] and interferometric ISAR (InISAR) [2]. Specifically, three different applications of CS to the ISAR imaging problem have been described in the literature [48]. First of all, CS can be applied for the reconstruction of ISAR images from data with random missing samples. For example, in [51] and [50], CS has been suggested as an effective ISAR image reconstruction tool in the case of data with missing samples in the slow-time and in the frequency domain, respectively. In addition, given the CS capability of reconstructing images from undersampled data, the sampling constraints related to Nyquist's sampling theorem can be overcome, as demonstrated in [4]. CS for data storage reduction has been suggested also in [12,29,61]. In [24], CS has been suggested as an effective reconstruction tool to recover a signal from its principal component extracted via principal component analysis (PCA) to reduce the clutter effects and enhance the target return extraction. In this framework, the advantages of CS can be mainly associated with the capability of obtaining good quality images from a limited number of measurements, thus reducing the amount of data to be stored and processed.

On the contrary, CS processing can be successfully applied to achieve resolution enhancement in both the delay-time/range and the Doppler/cross-range domain, as suggested in [60]. Specifically, CS can be applied to reconstruct the image from data considering a larger data support in the frequency/slow-time domain, thus enhancing the resolution in range and cross-range [23].

Finally, the CS can be used to process data that are not complete in one domain, presenting large portions of missing samples. Data can be incomplete, or *gapped*, in the slow-time domain such as data acquired by a multiple target tracking radar which collects data from different targets in different and not adjacent slow-time intervals, as shown in [59,62]. On the contrary, data can be gapped in the frequency domain such as the signal obtained by transmitting over different nonadjacent channels within a large bandwidth [58].

This latter case coincides with the case of passive radars which exploit the DVB-T signals transmitted for broadcast TV services.

The aim of this chapter is to provide a common framework for the application of CS to ISAR imaging. Specifically, an ISAR signal model in the framework of CS-based reconstruction is provided in Section 5.2. A review of the application in which CS-based reconstruction allows an enhancement of performance is given in Section 5.3 together with some results. In particular, the effectiveness of such a method and its performances against different SNR values are proven by using the image contrast as image quality criteria on real data. In addition, a comparison between CS and other SR techniques will be provided. Conclusion and remarks are given in Section 5.4.

5.2 Signal model

5.2.1 ISAR signal model

Let us consider the geometry in Figure 5.1, where $T_z(z_1, z_2, z_3)$ is a Cartesian reference system embedded on the target; $T_\xi(\xi_1, \xi_2, \xi_3)$ is a Cartesian reference system centered on the transmitter; R_{TxTg}, R_{RxTg}, and R_{TxRx} represent the transmitter–target, receiver–target, and transmitter–receiver distances, β is the bistatic angle, and $\mathbf{i}_{LoSBi}$ is the unit vector identifying the bisector of the bistatic angle [23].

It is worth pointing out that the bistatic geometry in Figure 5.1 can be used to model the received signal in either the active or the passive radar system, and for both monostatic and bistatic configurations.

It is well known that under the *stop and go* assumption, the target motion can be split into two components: a translational motion denoted by $R_0(t)$, and a rotational component

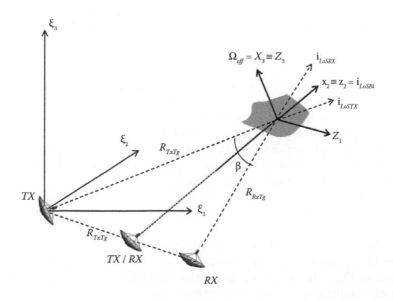

FIGURE 5.1
Bistatic geometry representation.

which can be described by means of the total angular rotation vector, $\mathbf{\Omega}_T(n)$, while the motion component which contributes to the synthetic aperture formation is denoted by the effective rotation angle, $\mathbf{\Omega}_{eff}(n)$, where n denotes the slow time. It is worth pointing out that the translational component is assumed to be perfectly compensated by means of autofocusing techniques [21,33] before the application of the CS-based imaging algorithm. However, it is worth remarking that different signal models have been proposed in the literature which include the motion compensation step while solving the optimization problem related to CS, such as the one proposed in [45]. In addition, it is assumed that the target is stationary during the transmission and reception of a sweep or a pulse; thus, the slow time can be considered a discrete variable, which will be denoted by n. The received signal can be defined as

$$s_R(t,n) = \int_V \gamma'(\mathbf{z}) s_T(t - \tau(\mathbf{z},n), n) h(n)\, d\mathbf{z}, \qquad (5.1)$$

where $s_T(t,n)$ is the transmitted signal within the nth sweep, t denotes the fast time, $n = 1,\ldots,N$, N is the number of transmitted sweeps, $\tau(\mathbf{z},n) = \frac{R_{T_x T_g}(\mathbf{z},n) + R_{R_x T_g}(\mathbf{z},n)}{c}$ represents the delay time of a point on the target with coordinates $\mathbf{z}$ in the Cartesian reference system T_z, at the nth sweep, c is the speed of the light in a vacuum, $h(n)$ is the signal support in the slow-time domain, and $\gamma'(\mathbf{z})$ is the target reflectivity function defined within the volume V occupied by the target. At the output of the matched filter, after Fourier transforming along the time domain, the baseband signal in the frequency/slow-time domain can be expressed as

$$S_R(f,n) = W(f,n) \int_V \gamma'(\mathbf{z}) e^{-j2\pi f \tau(\mathbf{z},n)}\, d\mathbf{z}, \qquad (5.2)$$

where f denotes the frequency which is the Fourier transform variable of the fast time t. A special attention should be given to the signal support $W(f,n)$ which defines the region occupied by the signal $S_R(f,n)$ in the frequency/slow-time domain. At the output of the matched filter, $W(f,n)$ is given by

$$W(f,n) = |S_T(f,n)|^2 h(n), \qquad (5.3)$$

where $h(n) = u(n) - u(n - N)$ and $u(n)$ is the unit step discrete function. When the transmitted signal is unchanged over adjacent sweeps, which is usually the case for active radar systems, Equation 5.3 can be rewritten as $W(f,n) = |S_T(f)|^2 h(n)$.

In the case of passive radars, according to the cross ambiguity function (CAF) batches algorithm proposed in [40], to form the RD image, both the transmitted and the received signals are split into N batches of fixed temporal length, so that each batch is associated with a different sweep. Since the broadcast signal is content dependent, the transmitted signal can change from batch to batch; therefore, Equation 5.3 is appropriate for the passive case as well.

Assume that the effective rotation angle of the target is constant within the observation time, $\mathbf{\Omega}_{eff}(n) \approx \mathbf{\Omega}_{eff}$, and assume that the Cartesian reference system T_z is oriented so that z_3 is parallel to $\mathbf{\Omega}_{eff}$. In this case, under the straight iso-range approximation, the signal after motion compensation can be rewritten as [36]

$$S(f,n) = W(f,n) \int_{z_1} \int_{z_2} \gamma(z_1, z_2) e^{-j\frac{4\pi f}{c} \cos\left(\frac{\beta(n)}{2}\right)(z_1 \cos(\Omega_{eff}n) + z_2 \sin(\Omega_{eff}n))}\, dz_1 dz_2, \qquad (5.4)$$

where $\gamma(z_1, z_2) = \int_{z_3} \gamma'(\mathbf{z}) dz_3$ is the target reflectivity function projection onto the image plane.

Suppose that the target consists of K point-like scatterers and that the interactions among the scatterers can be neglected, its reflectivity function becomes

$$\gamma(z_1, z_2) = \sum_{k=1}^{K} \sigma_k \delta\left(z_1 - z_1^{(k)}, z_2 - z_2^{(k)}\right), \tag{5.5}$$

where σ_k and $\left(z_1^{(k)}, z_2^{(k)}\right)$ denote complex amplitude and the range/cross-range coordinates of the kth scatterer, respectively, and $\delta(.,.)$ denotes the 2D Dirac delta function.

Under this hypothesis and assuming that the bistatic angle variation is relatively small within the observation time, i.e. $\beta(n) \approx \beta(0) = \beta_0$, Equation 5.4 can be rewritten as

$$S(f, n) = W(f, n) \sum_{k=1}^{K} \sigma_k e^{-j\frac{4\pi f}{c} k_0 (z_1^{(k)} \cos(\Omega_{eff} n) + z_2^{(k)} \sin(\Omega_{eff} n))}, \tag{5.6}$$

where $k_0 = \cos\left(\frac{\beta_0}{2}\right)$.

Changes in the bistatic angle affect the PSF of the system and can cause defocusing effects as demonstrated in [36].

By defyining the spatial frequencies as

$$\begin{aligned} Z_1(f, n) &= \frac{2 f k_0 \cos(\Omega_{eff} n)}{c} \\ Z_2(f, n) &= \frac{2 f k_0 \sin(\Omega_{eff} n)}{c} \end{aligned} \tag{5.7}$$

Equation 5.6 can be rewritten as

$$\begin{aligned} S(Z_1, Z_2) &= W(Z_1, Z_2) \sum_{k=1}^{K} \sigma_k e^{-j2\pi\left(z_1^{(k)} Z_1 + z_2^{(k)} Z_2\right)} \\ &= W(Z_1, Z_2) \Gamma(Z_1, Z_2), \end{aligned} \tag{5.8}$$

where $\Gamma(Z_1, Z_2) = \sum_{k=1}^{K} \sigma_k e^{-j2\pi\left(z_1^{(k)} Z_1 + z_2^{(k)} Z_2\right)}$ is the Fourier transform of the target reflectivity function $\gamma(z_1, z_2)$.

Equation 5.8 clearly shows that the motion-compensated signal is related to the Fourier transform of the target reflectivity function. In particular, the inverse Fourier transform of Equation 5.8 is given by

$$\begin{aligned} I(z_1, z_2) &= w(z_1, z_2) \otimes \otimes \gamma(z_1, z_2) \\ &= \sum_{k=1}^{K} \sigma_k w(z_1 - z_1^{(k)}, z_2 - z_2^{(k)}), \end{aligned} \tag{5.9}$$

where the symbol $\otimes\otimes$ denotes the 2D convolution operator, and $w(z_1, z_2)$ is the inverse Fourier transform of $W(Z_1, Z_2)$ and coincides with the PSF of the system in the range/cross-range domain.

It is worth pointing out that the knowledge of the modulus of the effective target rotation vector, Ω_{eff}, is needed to calculate the inverse Fourier transform. It is usually known in SAR systems where the motion of the platform is known *a priori*. Conversely, in ISAR systems, the target is non-cooperative and, hence, Ω_{eff} is unknown. As a consequence, Ω_{eff} should be estimated before forming the ISAR image.

Under the hypothesis that the aspect angle variation within the observation time is sufficiently small ($\Delta\theta = \Omega_{eff}T_{obs} < 10°$) [11], the spatial frequencies can be approximated as follows:

$$Z_1(f, n) \approx \frac{2fk_0}{c}$$
$$Z_2(f, n) \approx \frac{2f_0k_0\Omega_{eff}n}{c}. \tag{5.10}$$

This means that the region in the Fourier domain where the signal is defined can be approximated by a rectangular domain. Therefore, the two spatial frequencies can be assumed to be independent of each other and a 2D fast fourier transform can be applied to form the ISAR image.

In particular, under the assumption of small aspect angle variation, the received signal after motion compensation can be rewritten as

$$S(f, n) = W(f, n) \sum_{k=1}^{K} \sigma_k e^{-j\frac{4\pi k_0}{c}\left(fz_1^{(k)} + f_0z_2^{(k)}\Omega_{eff}n\right)}. \tag{5.11}$$

Let the variables (τ, υ) represent the delay-time and Doppler frequency, respectively, which are defined as

$$\tau = \frac{2z_1k_0}{c}$$
$$\upsilon = \frac{2f_0k_0\Omega_{eff}z_2}{c}. \tag{5.12}$$

By substituting Equation 5.12 into Equation 5.11, the motion-compensated signal becomes

$$S(f, n) = W(f, n) \sum_{k=1}^{K} \sigma_k e^{-j2\pi(f\tau_k + n\upsilon_k)} \tag{5.13}$$

It is worth noting that the summation in Equation 5.13 represents the 2D Fourier transform of the target reflectivity function in the delay-time/Doppler domain. Equation 5.12 suggests how to scale the ISAR image from the delay-time/Doppler domain to the range/cross-range domain and vice versa. The use of Equations 5.12 and 5.13 allows for an ISAR image to be formed when Ω_{eff} remains unknown. Nevertheless, to scale the ISAR image using spatial coordinates along the cross-range, Ω_{eff} must be estimated [32].

In a real scenario, both the variables in the signal domain (f, n) and in the image domain (τ, υ) are discrete variables. It is therefore appropriate to write the signal in Equation 5.13 as a function of discrete variables, defined as follows:

$$\begin{aligned} f &= f_0 + m\Delta f & m &= 1, ..., M \\ t &= nT_R & n &= 1, ..., N \\ \upsilon &= p\Delta\upsilon & p &= 1, ..., P \\ \tau &= q\Delta\tau & q &= 1, ..., Q, \end{aligned} \tag{5.14}$$

where Δf is the frequency step, T_R is the pulse repetition interval, $\Delta\upsilon = \frac{1}{NT_R} = \frac{1}{T_{obs}}$ is the Doppler frequency pixel spacing, and $\Delta\tau = \frac{1}{Q\Delta f}$ is the delay-time pixel spacing.

It is worth remarking that usually $Q = M$ and $P = N$ and when this holds, Δv and $\Delta \tau$ coincide with the Doppler and the delay-time resolutions, respectively.

By exploiting Equation 5.14, the discrete version of the received signal after motion compensation can be written as

$$S(m,n) = C \cdot W(n,m) \cdot \sum_{k=1}^{K} \sigma_k e^{-j2\pi \frac{mq_k}{Q}} e^{-j2\pi \frac{np_k}{P}}, \tag{5.15}$$

where $W(m,n) = (u(n) - u(n-N)) \cdot (u(m) - u(m-M))$ is the signal support in the discrete slow-time/frequency domain in the case of flat spectrum and $u(.)$ is the unit step discrete function.

Equation 5.15 highlights that the received signal after motion compensation is linked to the 2D Fourier transform of the target reflectivity function. Therefore, by using the inverse Fourier transform of Equation 5.15, the ISAR image which represents an approximation of the target reflectivity function can be expressed as

$$I(q,p) = \sum_{k=1}^{K} \sigma_k \cdot w(q - q_k, p - p_k). \tag{5.16}$$

5.2.2 ISAR signal model in a CS framework

For the sake of simplicity, it is possible to write the relation between $S(m,n)$ and $I(q,dp)$ in a matricial form. To use a notation coherent with all the applications described in this document, the ISAR data after motion compensation, $S(m,n)$, will be identified by the matrix $\overline{\mathbf{S}}_c$, while the ISAR image, $I(q,p)$, will be identified by the matrix $\mathbf{I}$, as follows:

$$\overline{\mathbf{S}}_c = \mathbf{\Psi}_x \mathbf{I} \mathbf{\Psi}_y^T, \tag{5.17}$$

where $\mathbf{\Psi}_x \in \mathbb{C}^{M \times Q}$ and $\mathbf{\Psi}_y \in \mathbb{C}^{N \times P}$ are the Fourier dictionaries that perform the range compression and the cross-range compression, respectively.

The generic element of the Fourier dictionaries is defined as follows:

$$\begin{aligned}
[\mathbf{\Psi}_y]_{n,p} &= e^{-j2\pi \frac{np}{P}} \\
[\mathbf{\Psi}_x]_{m,q} &= e^{-j2\pi \frac{mq}{Q}},
\end{aligned} \tag{5.18}$$

where $\mathbf{\Psi}_x$ and $\mathbf{\Psi}_y$ are defined as *complete Fourier matrices*, in which $P = N$ and $Q = M$.

The signal in Equation 5.17 represents the *complete* data from which the ISAR image can be obtained by applying Fourier-based methods with a resolution given by the theoretical bounds [5]. In a real scenario, the signal acquired by the system can suffer from data loss due to hardware malfunctioning or compression requirements. In these cases, the acquired signal is assumed to be obtained from the signal in Equation 5.17 after a sensing process

$$\begin{aligned}
\mathbf{S} &= \mathbf{\Phi}_x \mathbf{\Psi}_x \mathbf{I} \mathbf{\Psi}_y^T \mathbf{\Phi}_y^T \\
&= \mathbf{\Theta}_x \mathbf{I} \mathbf{\Theta}_y^T,
\end{aligned} \tag{5.19}$$

where $\mathbf{S} \in \mathbb{C}^{M' \times N'}$ and $\mathbf{I} \in \mathbb{C}^{M \times N}$. The matrices $\mathbf{\Phi}_x \in \mathbb{C}^{M' \times M}$ and $\mathbf{\Phi}_y \in \mathbb{C}^{N' \times N}$ represent the sensing matrices which performs the random selection of samples in the frequency/slow-time domain, respectively, and $M' < M$ and $N' < N$ are the data dimensions after the sensing operation.

$\mathbf{\Theta}_x$ and $\mathbf{\Theta}_y$ are the undercomplete Fourier matrices which satisfy the RIP property and provide stronger noncoherence than the Gaussian matrix [60]. These matrices are defined by taking into account the indexes of the pulses and frequency bins selected by the sensing process. In particular, the generic element of these matrices are defined as follows:

$$[\mathbf{\Theta}_y]_{i,p} = e^{-j2\pi\frac{pn_i}{P}}$$
$$[\mathbf{\Theta}_x]_{l,q} = e^{-j2\pi\frac{qm_l}{Q}}, \tag{5.20}$$

where $\{n_i\}_{i=1}^{N'}$ are the pulse indexes, $\{m_l\}_{l=1}^{M'}$ are the frequency bin indexes, and p and q are defined in Equation 5.14.

According to the CS formulation, the image reconstruction process in case of noise is based on a minimization problem, given by [13]

$$\hat{\mathbf{I}} = \min_{\mathbf{I}} \|\mathbf{I}\|_{\ell_0} \quad s.t. \quad \|\mathbf{S} - \mathbf{\Theta_x I \Theta_y}^T\|_F^2 \leq \epsilon, \tag{5.21}$$

where $\|\cdot\|_F$ denotes the Frobenius norm and $\|\cdot\|_{\ell_0}$ denotes the ℓ_0 norm. The minimization algorithm in Equation 5.21 can be solved via 2D-SL0 algorithm [19]. It exploits a continuous Gaussian function to approximate the l_0 norm of a signal. In this way, the l_0 norm of $\mathbf{I}$ in Equation 5.21 can be approximated as follows:

$$\| \mathbf{I} \|_0 \approx PQ - \sum_{p=1}^{P}\sum_{q=1}^{Q} \exp\left\{-\frac{|\mathbf{I}_{p,q}|^2}{2\delta^2}\right\} \quad \text{when } \delta \to 0 \tag{5.22}$$

Then, a *projective steepest ascent* optimal approach can be introduced to find the minimum value of $\|\mathbf{I}\|$, that is, the sparest solution of Equation 5.21, which makes this algorithm work well and fast.

It is worth pointing out that the image at the output of the CS reconstruction algorithm cannot be directly compared with the image at the output of the RD algorithm because there is no relationship between the pixel size and the spatial resolution of the image in the Rayleigh sense [11]. For this reason, raw data are reconstructed from the CS image, as shown in Figure 5.2, by means of the Fourier dictionaries as

$$\hat{\mathbf{S}}_c = \mathbf{\Psi}_x \hat{\mathbf{I}} \mathbf{\Psi}_y^T, \tag{5.23}$$

At this point, an ISAR image can be obtained by means of the conventional RD algorithm applied to the reconstructed data and can be fairly compared with the image obtained via RD algorithm applied to the original data.

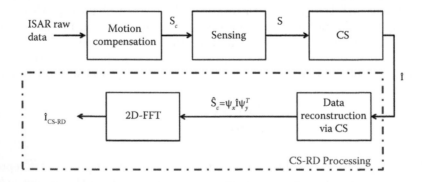

FIGURE 5.2
CS image reconstruction block diagram.

5.3 Application of CS in ISAR imaging

5.3.1 Data compression

ISAR signals usually consist of huge matrices of data which determine the computational burden of the image processing. To reduce the computational load, a data compression can be performed. By the way, as the data compression increases, the quality of ISAR images obtained via conventional Fourier-based methods decreases. In particular, when the sample structure of the data does not meet the Nyquist requirement, the ISAR image suffers from high sidelobes and distortion. In this scenario, the capabilities of CS of reconstructing the ISAR image from a compressed signal, in which the number of samples is lower than the Nyquist bound, is fundamental [4]. CS can be applied in the delay-time/frequency dimension, Doppler/slow-time dimension, or both frequency/slow-time. Without loss of generality, we will focus our attention on the 2D formulation [41]. The advantage of the 2D fomulation in Equation 5.19 with respect to 1D methods or staking operations that can be found in the literature is that this formulation deals with smaller matrices, thus reducing computational burden [44]. The key concept of the data compression in the CS framework is illustrated in Figure 5.3, in which the data before and after the sensing operation and the ISAR image are schematically represented. As can be easily noticed, the data after the sensing operation are obtained by randomically discarding some samples in the frequency/slow-time domain.

The sensing matrices, that is, $\mathbf{\Phi}_x$ and $\mathbf{\Phi}_y$, are then obtained from the identity matrix by putting element equal to zero corresponding to the row/column to be eliminated.

An application of CS for data compression and reconstruction is presented in [2], where an InISAR system is used to obtain a 3D reconstruction of non-cooperative moving targets. Since three ISAR images are needed for InISAR processing, the amount of data to be acquired and stored is huge and data compression may help to overcome hardware issues. It is quite obvious that data reconstruction must recover image information relative to both amplitude and phase. An analysis of the reconstruction capabilities of the CS-ISAR processing is here reported in terms of imaging results, correctly reconstructed scatterers, and RMSE on the amplitude and phase value for different values of compression rate (CR). More specifically, two performance evaluation approaches are proposed:

- "Truth-based" performance analysis, which is based on the comparison of the original ISAR image (which represents the "truth") with the one reconstructed using the CS-based approach after data compression.

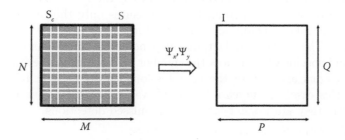

FIGURE 5.3
Data compression concept.

- "Image-quality-based" performance analysis, in which standard image quality indexes are computed on the CS-reconstructed ISAR images.

The truth-based performance analysis consists of a comparison between the "true image" and the "estimated image." The comparison approach is based on the main scatterer positions and amplitudes in the two images and aims at measuring the distortions after the application of data compression and CS reconstruction method. Specifically, such distortions are evaluated in terms of

1. False alarms, defined as peaks which are generated in the reconstructed image at a position where no scatterers are present in the "true" image.

2. Missed detections, when no peak is generated in the reconstructed ISAR image where a scatterer is instead detected in the "true" image.

3. Correctness of the amplitude estimation. For all the "correct detections," the error in the reconstruction of the scatterers complex amplitude is also considered. In this case, the relative root mean square error (RRMSE) is computed on the amplitude and phase values as follows:

$$\text{RRMSE}_{amp} = \sqrt{\frac{1}{N_{\text{CD}}} \sum_{k=1}^{N_{\text{CD}}} \left(\frac{|\mathbf{A}_T(k)| - |\mathbf{A}_{SR}(k)|}{|\mathbf{A}_T(k)|} \right)^2} \qquad (5.24)$$

$$\text{RRMSE}_{ph} = \sqrt{\frac{1}{N_{\text{CD}}} \sum_{k=1}^{N_{\text{CD}}} \left(\frac{\angle\mathbf{A}_T(k) - \angle\mathbf{A}_{SR}(k)}{\angle\mathbf{A}_T(k)} \right)^2}, \qquad (5.25)$$

where N_{CD} is the number of "correct detections." $\mathbf{A}_T$ and $\mathbf{A}_{SR}$ are the vectors containing the complex amplitude of the scatterers in the "true" image and the CS-reconstructed image, respectively.

The scatterers are extracted from the image by using the CLEAN technique. Scatterers that represent the 90% of the image energy are extracted.

On the contrary, "Image-quality-based" performance analysis methods consist of directly calculating image quality indexes. Two different quality indexes are used here:

1. Image contrast (IC), defined as the ratio between the standard deviation and the mean value of the image intensity:

$$\text{IC} = \frac{\sqrt{\mathbb{A}\{[I^2(q,p) - \mathbb{A}\{I^2(q,p)\}]^2\}}}{\mathbb{A}\{I^2(q,p)\}}, \qquad (5.26)$$

where $\mathbb{A}$ denotes the spatial mean operator and I is the image amplitude. IC is a popular indicator of image quality, but it measures only the image sharpness, and it does not take account for the introduction of artifacts in the super-resolved image. The IC value, in fact, may depend on the number of reconstructed scatterers, and specifically, the less the number of scatterers, the higher the IC value. If scatterers are lost in the compression processing, the IC value may increase. In this case, an increase in the IC value denotes better performance when, in truth, they are worse because of the missed detections.

2. Signal-to-noise ratio (SNR), computed as follows:

$$\text{SNR} = 20 \log_{10} \left(\frac{\frac{1}{N_T} \sum_{(q,p) \in \mathcal{T}} |I(q,p)|^2}{\frac{1}{N_B} \sum_{(q,p) \in \mathcal{B}} |I(q,p)|^2} \right), \qquad (5.27)$$

where N_T denotes the number of pixels associated with the target under test and N_B is the number of pixel associated with the background in the ISAR image. The ISAR image obtained by applying the RD algorithm to the full data is shown in Figure 5.4. An X-band interferometric ground-based linear frequency modulated continuous wave (LFMCW) radar, named PIRAD, has been used for collecting data in a real scenario. PIRAD is a multichannel ground-based radar prototype designed and built within the Department of Information Engineering, University of Pisa, with the collaboration of the CNIT Radar and Surveillance System (RaSS) national laboratory [26,27]. The acquisition parameters are listed in Table 5.1.

Results obtained by applying the RD to compressed data for $CR = 32\%$ and $CR = 64\%$ are shown in Figure 5.5. Only the data acquired by one channel are considered for data compression and image reconstruction.

The reconstructed image obtained by means of the proposed CS-RD processing is shown in Figure 5.6 for $CR = 32\%$ and $CR = 64\%$. The numerical performance parameters are summarized in Table 5.2. As is evident from results of both visual inspection and numerical performance evaluation, CS-RD is able to restore ISAR images even with high CR.

5.3.2 Resolution enhancement

It is well known that the spatial resolutions in ISAR images are strictly related to the transmitted bandwidth and the observation time. In some cases, it may happen that these

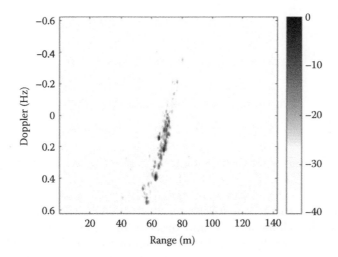

FIGURE 5.4
Full data ISAR image.

TABLE 5.1
PIRAD's Acquisition Parameter

TX power	33 dBm
Bandwidth	300 MHz
PRF	1000 Hz
T_{obs}	1 s
Range resol.	0.5 m

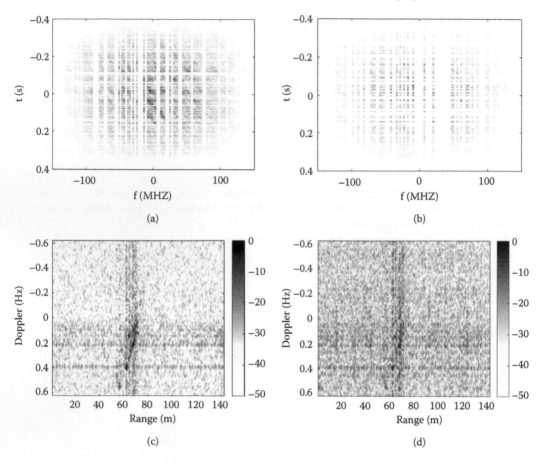

FIGURE 5.5
Compression results: (a) Compressed data $CR = 32\%$, (b) $CR = 64\%$, (c) RD image $CR = 32\%$, and (d) $CR = 64\%$.

two parameters are not large enough to obtain the desired range/cross-range resolution. Different techniques have been proposed in the literature to enhance the resolution of ISAR images, such as spatially variant apodization techniques and BWE-based methods [39,47,55,63]. By the way, these techniques are efficient when the gaps to be filled are small with respect to the whole spectral and observation time occupancy of the signal. Conversely, CS can be successfully applied to a version of the received signal in which the dimensions in the frequency and the slow-time domain have been increased, as schematically represented in Figure 5.2. In this case, the available samples are assumed to be the result of a sensing operation performed over a larger domain and the signal is reconstructed over a larger bandwidth and a longer observation time, which leads to a finer resolution. (see Figure 5.7)

A COSMO-SkyMed (CSK) SAR image is used to test the SR algorithms. Parameters of interest about the dataset under test are summarized in Table 5.3.

The SAR image used to obtain the results in this chapter was acquired in April 2008 by satellite-1 of the CSK constellation, and it covers the area of Istanbul. Once a moving target is detected, an SAR subimage containing that target is cropped from the whole SAR image. The ISAR image is then obtained by applying the inverse Omega-Key to recover the SAR data of the moving target, and the ISAR processing to get a well-focused image

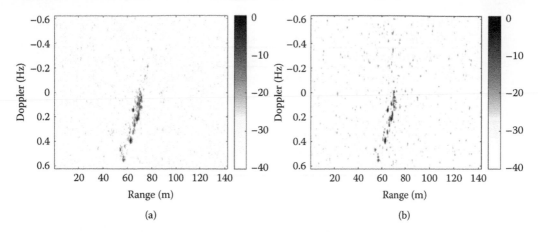

FIGURE 5.6
Reconstructed image via CS-RD: (a) $CR = 32\%$ and (b) $CR = 64\%$.

TABLE 5.2
Data Compression and Reconstruction: Numerical Performance Results

CR %	0	10	32	50	64
SNR [dB]	42.11	40.81	38.89	39.86	42.30
I.C.	1.49	1.57	1.78	2.1	2.54
N.o.S.	20	19	18	15	14
C.D.	–	18	16	14	11
F.A.	–	1	2	1	3
M.D.	–	2	4	6	9
$RMSE_{amp}\%$	–	1.95	4.21	6.65	10.71
$RMSE_{ph}\%$	–	6.43	10.38	13.79	14.14

C.D.: Correct detections, F.A.: False Alarm, I.C.: image contrast, M.D.: Missing Detection, N.o.S.: Number of scatterers, RMSE_amp: root mean square error on the amplitude, RMSE_ph: root mean square error on the phase.

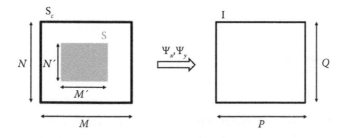

FIGURE 5.7
Resolution enhancement concept.

of the moving target. The ISAR image is shown in Figure 5.8a and the frequency/slow-time signal is shown in Figure 5.8b. The complex signal is halved along both the frequency and the slow-time domain, as shown in Figure 5.8d, to obtain a low-resolution ISAR image of the same target, which is shown in Figure 5.8c.

A comparison between CS-based resolution enhancement techniques and state-of-art SR techniques is provided here. In particular, BWE [6,37,49], Capon's minimum variance

TABLE 5.3

CSK Dataset Information of Interest

CSK Information	Values
Transmitted signal waveform	Linear Chirp
Image processing algorithm	OMEGA-KEY
Radar frequency	9.6 GHz
Frequency bandwidth	277.515 MHz
Coherent integration time	1.6340 s
Theoretical range resolution	0.5405 m
Theoretical Doppler resolution	0.6120 Hz

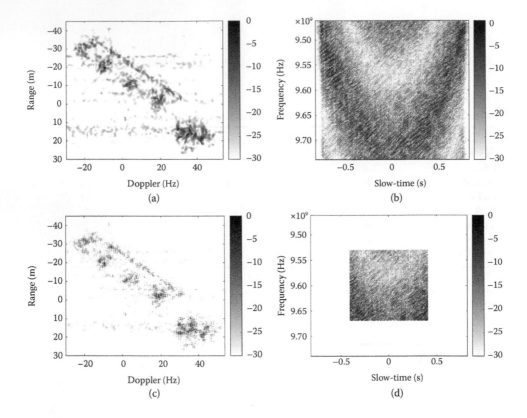

FIGURE 5.8

(a) High-resolution CSK refocused ISAR image, (b) relative frequency/slow-time complex signals of high-resolution ISAR image, and (c) low-resolution ISAR image, and (d) relative frequency/slow-time complex signals of low-resolution ISAR image.

method (MVM) [8,17,31,52], amplitude and phase estimation of a sinusoid (APES) [28], super spatially variant apodization (SSVA) [46,54], and CS algorithm are applied to the low-resolution image in Figure 5.8c to restore the original ISAR image in Figure 5.8a. SR results are shown in Figure 5.9 for BWE, MVM, APES, and SSVA and Figure 5.10 for CS.

From Figure 5.9, it appears that BWE Figure 5.9a is not able to successfully reconstruct the original image. MVM and APES results are almost identical, and SSVA and CS return similar results as well. MVM and APES (Figure 5.9b and c, respectively) are not able to

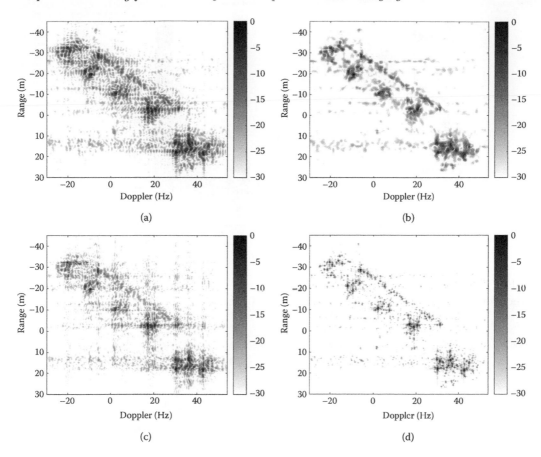

FIGURE 5.9
CSK dataset: Results of (a) BWE, (b) Capon's MVM, (c) APES, and (d) SSVA applied to the low-resolution ISAR image.

restore the original image resolution, but the target shape is better defined with respect to the low-resolution image in Figure 5.8c and to the BWE result in Figure 5.9a. On the contrary, SSVA and CS (Figures 5.9d and 5.10, respectively) allow to enhance the image resolution with respect to the original image even though the nominal resolution is the same. This is because the CPI reduction allows to mitigate the range–Doppler migration effect. When the resolution is restored by means of any SR technique, the range migration is strongly attenuated, and the super-resolved image appears better focused than the original one [10].

Results of the truth-based performance analysis method for the CSK dataset are shown in Figure 5.11 for BWE, Capon's MVM, APES, and SSVA and in Figure 5.12 for CS. The CLEAN is applied on both the high-resolution image in Figure 5.8a and the super-resolved in Figures 5.9 and 5.10. The CLEAN iteratively detects the main scatterers in the SAR/ISAR image under test until a certain constraint is satisfied. Such a constraint can be arbitrarily set and may depend on the residual energy, the scatterers amplitude, or a fixed desired number of extracted scatterers. When the CLEAN is applied on the original high-resolution image, the scatterers detection is stopped when the residual energy is less than the 15% of the total image energy. One-hundred thirty scatterers are then detected in the "true" image. Such a threshold is empirically set. On the contrary, the constraint for the peak

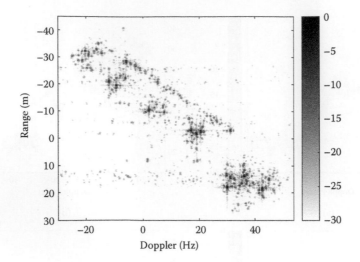

FIGURE 5.10
CSK dataset: Result of compressed sensing applied to the low-resolution ISAR image.

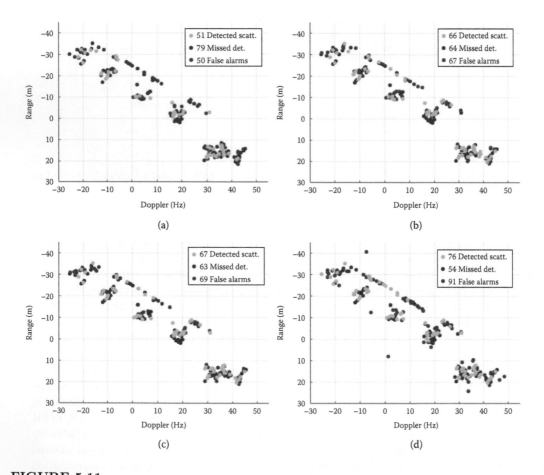

FIGURE 5.11
CSK dataset: Truth-based performance analysis results for (a) BWE, (b) Capon's MVM, (c) APES, and (d) SSVA.

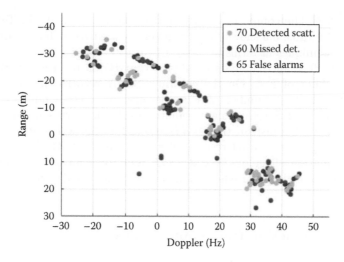

FIGURE 5.12
CSK dataset: Truth-based performance analysis result for compressed sensing.

detection in the super-resolved images depends on the measured amplitude of the detected peaks. The CLEAN detects the peaks in the super-resolved images until the measured amplitude is greater than the 80% of the weakest detected scatterer in the "truth" image. As a result, in different images, different number of peaks are detected, but a fair peak detection strategy is guaranteed. It is worth pointing out that the original high-resolution image and all the super-resolved images are normalized with respect to their own energy. Results in terms of the number of detected scatterers (N.o.S.), correct detections (C.D.), missed detections (M.D.), and false alarms (F.A.) are shown in Figure 5.11 and summarized in Table 5.4. Performance analysis results in terms of IC, SNR, measured resolution, and scatterers amplitude RRMSE are reported in Table 5.4, where $\hat{\delta}_{rng}$ and $\hat{\delta}_\nu$ denote the estimated range and Doppler resolution, respectively.

Results in Table 5.4 confirm that SSVA and CS applied after the CPI/bandwidth reduction allow to mitigate the range migration effect and increase the IC with respect to the original high-resolution image. CS allows for the best SNR, whereas BWE and Capon's MVM give the worst results. The measured resolution values return the best performance for CS and the worst for APES. It is worth noting that MVM returns slightly better results with respect to APES in terms of resolution enhancement, but APES has better performance

TABLE 5.4
Resolution Enhancement on CSK Dataset: Numerical Performance Results

	High-res.	BWE	MVM	APES	SSVA	CS
IC	1.759	1.444	1.483	1.656	1.818	2.181
SNR [dB]	41.749	36.069	35.907	38.666	40.260	43.146
$\hat{\delta}_{rng}$ [m]	0.492	0.743	0.836	0.908	0.577	0.554
$\hat{\delta}_\nu$ [Hz]	0.638	0.786	0.988	1.085	0.713	0.588
N.o.S.	130	101	133	136	167	144
C.D.	−	51	65	67	76	70
M.D.	−	79	65	63	54	60
F.A.	−	50	68	69	91	74
RRMSE	−	0.410	0.451	0.471	0.533	0.700

in terms of IC and SNR. The comparison between the resolution results in Table 5.4 with the theoretical values in Table 5.3 shows that CS generally returns better performance with respect to the other techniques in terms of image quality.

The truth-based performance analysis highlights that BWE returns a high number of M.D. and a small number of F.A. On the contrary, SSVA returns the smallest number of M.D., but also the highest number of F.A. because of a large number of detected peaks. CS, MVM, and APES return intermediate results in terms of M.D. and F.A. The RRMSE index highlights instead that BWE allows for the best scatterers amplitude estimation, whereas CS gives the worst result. BWE gives the best RRMSE because of the range migration attenuation effect of CS and SSVA. On the contrary, CS allows for the best scatterers amplitude estimation when compared with an ideal "true" image without migration effects [10].

5.3.3 Image reconstruction from gapped data

A gapped data is defined as an incomplete set of data, in which a large amount of samples are missing in the frequency domain, the slow-time domain, or both. Gaps in the frequency domain can be associated with the transmission in nonadjacent frequency ranges, for example, in the case in which some bandwidths are dedicated to other services, or in the case of passive radars which exploit the DVB-T signals. A data with gaps in the slow-time domain can be associated with the signal acquired by a multi-target radar system, in which different targets are illuminated for different time intervals. A schematic representation of the concept of gapped data is shown in Figure 5.13. The difference between a compressed data and a gapped data is that in a compressed data the missing samples in a domain are usually associated with very small gaps, which can be the results of a system malfunctioning. In a gapped data, the missing samples correspond to large gaps that can be associated with transmission disruption in the frequency or the slow-time domain.

As illustrated in [38], for passive ISAR with DVB-T signals, a higher range resolution can be achieved by processing multiple adjacent channels. Nevertheless, such channels are separated by gaps where no signal is transmitted. When such a multichannel signal is used for ISAR imaging, grating lobes will appear in the final ISAR images if a Fourier transform (FT)-based imaging algorithm is applied. Moreover, as already stated in the introduction, the VHF/UHF band is not filled completely with broadcast channels. Therefore, even if hardware was developed which is able to acquire a multichannel DVB-T signal to obtain a wider bandwidth, the available channels may be nonadjacent. In this case, the gaps become larger than those that are present between adjacent channels, leading to more evident grating lobes in the final ISAR images. Such grating lobes in an ISAR image may be interpreted as false scatterers (artifacts), which may mislead the

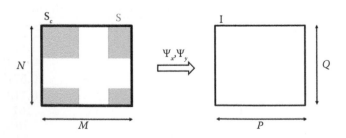

FIGURE 5.13
Gapped data concept.

image interpretation and degrade the performance in target classification or recognition applications. Furthermore, due to the large spectral gaps, the grating lobes cancellation algorithm presented in [38] does not work in this case. The application of the CS-based image reconstruction technique has been proposed in [42]. Here some new results are shown. These data are acquired by exploiting a technological demonstrator developed by the RaSS Lab of the Italian National Interuniversity Consortium for Telecommunications (CNIT) named software-defined Multiband Array Passive Radar (SMARP) installed in Livorno within the Naval Academy area [9]. It is a multiband passive radar demonstrator based on a software-defined solution and oriented to coastal surveillance applications. To propose an innovative solution, the SMARP demonstrator architecture shows advances especially in multiband receiving array antenna (UHF and S band) with dual polarization reception, software-defined multiband flexible receiver based on commercially available solutions, digital array processing techniques, and advanced radar signal processing algorithms implemented on COTS (commercial off-the-shelf) processing architectures (multicore Central Progessing Units [CPU] and Graphical Processing Units [GPU]).

Results obtained with passive radar data are shown in Figure 5.14. Figure 5.14a and b shows the full data and the ISAR image obatined by applying the conventional RD technique to such a data. Gapped data is obtained by discarding the central channel (Figure 5.14b) leading to a distorted version of the ISAR image obtained with RD technique (Figure 5.14d). The data reconstructed by exploiting the CS-based technique and the ISAR image obtained with CS-RD are shown in Figure 5.14e and f, respectively. The effectiveness of the proposed CS-based method to overcome incompleteness in the available data is quite evident.

Results obtained with active radar data are shown in Figure 5.15. A target extracted from a CSK SAR image is used [34].

5.4 Conclusion and remarks

In this chapter, the capabilities of CS for ISAR image reconstruction and resolution enhancement have been demonstrated. In particular, a CS-based ISAR method based on the application of the RD processing to the signal obtained from the CS-estimated image has been proposed. This method allows for a fair comparison between the ISAR image obtained via the conventional RD processing and the image obtained with the CS-based ISAR method. The present method has been applied to real datasets and an extended performance analysis performed by comparing the proposed method with conventional RD techniques and SR methods. In particular, to assess the capabilities of the proposed CS-ISAR reconstruction technique, two different criteria have been identified, based on the image contrast and the number of C.D. with respect to the number of artifacts. These two criteria allow to assess the superiority in terms of reconstruction ability of CS with respect to other conventional reconstruction methods, such as RD, or SR techniques.

Acknowledgment

Copyright ©-2012 - European Defence Agency. All rights reserved. The opinions expressed herein reflect the author's view only. Under no circumstances shall the European Defence Agency be held liable for any loss, damage, liability, or expense incurred or suffered that is claimed to have resulted from the use of any of the information included herein.

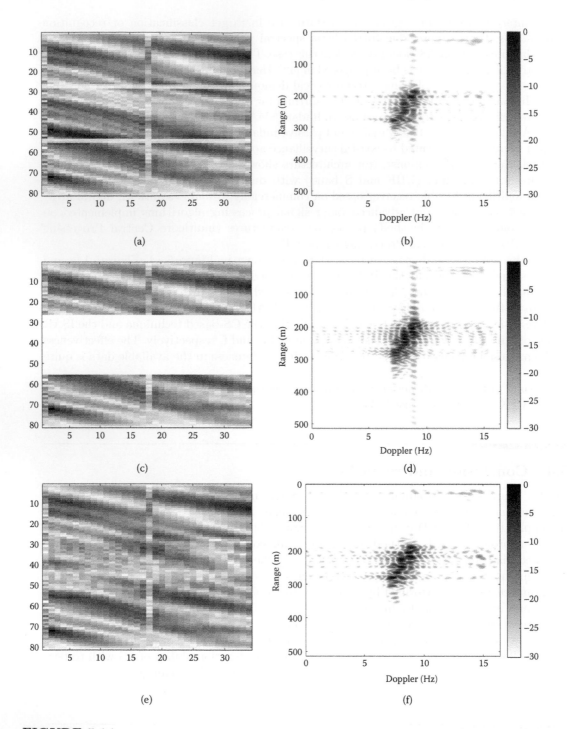

FIGURE 5.14
P-ISAR Results: (a) full data, (b) RD image with full data, (c) gapped data, and (d) RD image with gapped data (e) data reconstructed with CS and (f) CS reconstructed ISAR image.

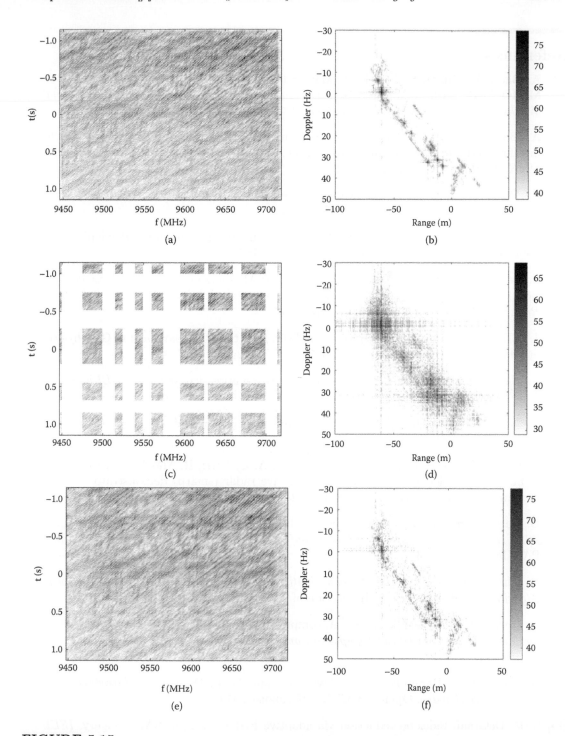

FIGURE 5.15
Active ISAR results: (a) full data, (b) RD image with full data, (c) gapped data, and (d) RD image with gapped data (e) data reconstructed with CS and (f) CS-reconstructed ISAR image.

References

[1] A. Bacci, E. Giusti, S. Tomei, M. Martorella, and F. Berizzi, Time-slotted FMCW MIMO ISAR with compressive sensing image reconstruction. In *3rd International Workshop on Compressive Sensing Applied to Radar (CoSeRa2015)* (Sept 2013).

[2] A. Bacci, D. Staglianò, E. Giusti, S. Tomei, F. Berizzi, and M. Martorella, 3d interferometric ISAR via compressive sensing. In *2014 11th European Radar Conference (EuRAD)* (Oct 2014), pp. 233–236.

[3] R. Baraniuk and P. Steeghs, Compressive radar imaging. In *2007 IEEE Radar Conference* (April 2007), pp. 128–133.

[4] B. C. Zhang, W. Hong, and Y. Wu, Sparse microwave imaging: Principles and applications. *Science China* 55(8) (2012), 1722–1754.

[5] F. Berizzi, E. Dalle Mese, and M. Martorella, ISAR imaging of oscillating targets by range-instantaneous-Doppler technique. In *The Record of the IEEE 2000 International Radar Conference, 2000* (2000), pp. 475–480.

[6] J. P. Burg, Maximum entropy spectral analysis. PhD thesis, Department of Geophysics, Stanford University, Stanford, CA, 1975.

[7] E. Candes and M. Wakin, An introduction to compressive sampling. *IEEE Signal Processing Magazine* 25(2) (2008), 21–30.

[8] J. Capon, High-resolution frequency-wavenumber spectrum analysis. *Proceedings of the IEEE* 57(8) (1969), 1408–1418.

[9] A. Capria, D. Petri, C. Moscardini, M. Conti, A. C. Forti, R. Massini, M. Cerretelli, et al., Software-defined multiband array passive radar (smarp) demonstrator: A test and evaluation perspective. In *OCEANS 2015—Genova* (May 2015), pp. 1–6.

[10] D. Cataldo and M. Martorella, Super-resolution for bistatic distortion mitigation. In *2016 IEEE Radar Conference (RadarConf) (2016 IEEE RadarConf)* (Philadelphia, PA, May 2016), pp. 387–392.

[11] V. Chen and M. Martorella, *Inverse Synthetic Aperture Radar Imaging; Principles, Algorithms and Applications*. Institution of Engineering and Technology, 2014.

[12] P. Cheng, H. Liu, and Z. Zhao, A compressive sensing-based technique for ISAR imaging. In *2012 5th Global Symposium on Millimeter Waves (GSMM)* (May 2012), pp. 466–469.

[13] H. Chinmay, R. Baraniuk, M. A. Davenport, and M. F. Duarte, *An Introduction to Compressive Sensing*. OpenStax CNX., 27 August 2014.

[14] S. R. DeGraaf, Sidelobe reduction via adaptive FIR filtering in SAR imagery. *IEEE Transactions on Image Processing* 3(3) (1994), 292–301.

[15] S. R. DeGraaf, SAR imaging via modern 2-D spectral estimation methods. *IEEE Transactions on Image Processing* 7(5) (1998), 729–761.

[16] D. Donoho, Compressed sensing. *IEEE Transactions on Information Theory* 52(4) (2006), 1289–1306.

[17] W. Featherstone, H. Strangeways, M. Zatman, and H. Mewes, A novel method to improve the performance of Capon's minimum variance estimator. In *Tenth International Conference on Antennas and Propagation (Conf. Publ. No. 436)* (Apr 1997), vol. 1, pp. 322–325.

[18] C. Feng, L. Xiao, and Z. Wei, Compressive sensing ISAR imaging with stepped frequency continuous wave via gini sparsity. In *2013 IEEE International Geoscience and Remote Sensing Symposium (IGARSS)* (July 2013), pp. 2063–2066.

[19] A. Ghaffari, M. Babaie-Zadeh, and C. Jutten, Sparse decomposition of two dimensional signals. In *IEEE International Conference on Acoustics, Speech and Signal Processing, 2009. ICASSP 2009* (2009), pp. 3157–3160.

[20] E. Giusti, A. Bacci, S. Tomei, and M. Martorella, Compressive sensing based ISAR: Performance evaluation. In *2015 16th International Radar Symposium (IRS)* (June 2015), pp. 398–403.

[21] E. Giusti, S. Tomei, A. Bacci, M. Martorella, and F. Berizzi, Autofocus for CS based ISAR imaging in the presence of gapped data. In *2nd International Workshop on Compressive Sensing Applied to Radar (CoSeRa2013)* (Sept 2013).

[22] E. Giusti, Q. Wei, A. Bacci, S. Tomei, and M. Martorella, Super resolution ISAR imaging via compressing sensing. In *EUSAR 2014; 10th European Conference on Synthetic Aperture Radar* (June 2014), pp. 1–4.

[23] E. Giusti, Q. Wei, A. Bacci, S. Tomei, and M. Martorella, Super resolution ISAR imaging via compressing sensing. In *EUSAR 2014; 10th European Conference on Synthetic Aperture Radar* (June 2014), pp. 1–4.

[24] S. Gunnala, L. Camacho, and S. Tjuatja, Target detection above rough surfaces in microwave imaging using compressive sampling. In *2010 IEEE International Geoscience and Remote Sensing Symposium (IGARSS)* (July 2010), pp. 3498–3501.

[25] A. C. Gurbuz, J. McClellan, and W. Scott, Compressive sensing for GPR imaging. In *Conference Record of the Forty-First Asilomar Conference on Signals, Systems and Computers, 2007. ACSSC 2007* (Nov 2007), pp. 2223–2227.

[26] S. Lischi, R. Massini, L. Musetti, D. Stagliano, F. Berizzi, B. Neri, and S. Saponara, Low cost FMCW radar design and implementation for harbour surveillance applications. In *International Conference on Electronic Applications, APPLEPIES 2014* (May 2014).

[27] S. Lischi, R. Massini, D. Stagliano, L. Musetti, F. Berizzi, B. Neri, and M. Martorella, X-band compact low cost multi-channel radar prototype for short range high resolution 3D-InISAR. In *2014 11th European Radar Conference (EuRAD)* (Oct 2014), pp. 157–160.

[28] P. Lopez-Dekker and J. Mallorqui, Capon- and APES-based SAR processing: Performance and practical considerations. *IEEE Transactions on Geoscience and Remote Sensing 48*(5) (2010), 2388–2402.

[29] H. Lu and Z. Qianqian, 2D ISAR imaging scheme in complex baseband echo domain based on compressive sensing. In *2014 International Conference on Information and Communications Technologies (ICT 2014)* (May 2014), pp. 1–5.

[30] M. Lustig, D. Donoho, J. Santos, and J. Pauly, Compressed sensing MRI. *IEEE Signal Processing Magazine 25*(2) (2008), 72–82.

[31] S.L. Marple, M. Adeli, and H. Liu, Super-fast algorithm for minimum variance (Capon) spectral estimation. In *2010 Conference Record of the Forty Fourth Asilomar Conference on Signals, Systems and Computers (ASILOMAR)* (Nov 2010), pp. 1832–1836.

[32] M. Martorella, Novel approach for ISAR image cross-range scaling. *IEEE Transactions on Aerospace and Electronic Systems 44*(1) (2008), 281–294.

[33] M. Martorella, F. Berizzi, and B. Haywood, Contrast maximisation based technique for 2-D ISAR autofocusing. *IEE Proceedings Radar, Sonar and Navigation 152*(4) (2005), 253–262.

[34] M. Martorella, E. Giusti, F. Berizzi, A. Bacci, and E. D. Mese, ISAR based techniques for refocusing non-cooperative targets in SAR images. *IET Radar, Sonar Navigation 6*(5) (2012), 332–340.

[35] M. Martorella, E. Giusti, L. Demi, Z. Zhou, A. Cacciamano, F. Berizzi, and B. Bates, Target recognition by means of polarimetric ISAR images. *IEEE Transactions on Aerospace and Electronic Systems 47*(1) (2011), 225–239.

[36] M. Martorella, J. Palmer, J. Homer, B. Littleton, and I. Longstaff, On bistatic inverse synthetic aperture radar. *IEEE Transactions on Aerospace and Electronic Systems 43*(3) (2007), 1125–1134.

[37] T. G. Moore, B. W. Zuerndorfer, and E. C. Burt, *Enhanced Imagery Using Spectral-Estimation-Based Techniques, Lincoln Laboratory Journal 10*, 2 (1997).

[38] D. Olivadese, E. Giusti, D. Petri, M. Martorella, A. Capria, and F. Berizzi, Passive ISAR with DVB-T signals. *IEEE Transactions on Geoscience and Remote Sensing 51*(8) (2013), 4508–4517.

[39] D. Olivadese, M. Martorella, and F. Berizzi, Multi-channel P-ISAR grating lobes cancellation. In *IET International Conference on Radar Systems (Radar 2012)* (Oct 2012), pp. 1–5.

[40] D. Petri, C. Moscardini, M. Martorella, M. Conti, A. Capria, and F. Berizzi, Performance analysis of the batches algorithm for range-Doppler map formation in passive bistatic radar. In *IET International Conference on Radar Systems (Radar 2012)* (Oct 2012), pp. 1–4.

[41] W. Qiu, E. Giusti, A. Bacci, M. Martorella, F. Berizzi, H. Zhao, and Q. Fu, Compressive sensing for passive ISAR with DVB-T signal. In *2013 14th International Radar Symposium (IRS)* (June 2013), vol. 1, pp. 113–118.

[42] W. Qiu, E. Giusti, A. Bacci, M. Martorella, F. Berizzi, H. Zhao, and Q. Fu, Compressive sensing-based algorithm for passive bistatic ISAR with DVB-T signals. *IEEE Transactions on Aerospace and Electronic Systems 51*(3) (2015), 2166–2180.

[43] W. Qiu, M. Martorella, J. Zhou, H. Zhao, and Q. Fu, Three-dimensional inverse synthetic aperture radar imaging based on compressive sensing. *IET Radar, Sonar Navigation 9*(4) (2015), 411–420.

[44] W. Qiu, H. Zhao, J. Zhou, and Q. Fu, High-resolution fully polarimetric ISAR imaging based on compressive sensing. *IEEE Transactions on Geoscience and Remote Sensing 52*(10) (2014), 6119–6131.

[45] R. Raj, R. Lipps, and A. Bottoms, Sparsity-based image reconstruction techniques for ISAR imaging. In *2014 IEEE Radar Conference* (May 2014), pp. 0974–0979.

[46] H. Stankwitz and M. Kosek, Sparse aperture fill for SAR using super-SVA. In *Proceedings of the 1996 IEEE National Radar Conference*, 1996 (1996), pp. 70–75.

[47] G. Thomas and N. Gadhok, Sidelobe apodization in fourier imaging. In *Conference Record of the Thirty-Fifth Asilomar Conference on Signals, Systems and Computers, 2001* (Nov 2001), vol. 2, pp. 1369–1373.

[48] S. Tomei, A. Bacci, E. Giusti, M. Martorella, and F. Berizzi, Compressive sensing-based inverse synthetic radar imaging from incomplete data. *IET Radar, Sonar Navigation* 10(2) (2016), 386–397.

[49] K. Vos, *A Fast Implementation of Burg's Method*. http://www.arxiv.org (accessed August 2013).

[50] H. Wang, Y. Quan, M. Xing, and S. Zhang, ISAR imaging via sparse probing frequencies. *IEEE Geoscience and Remote Sensing Letters* 8(3) (2011), 451–455.

[51] Y. Wang, J. Kang, and R. Zhang, ISAR imaging with random missing observations based on non-iterative signal reconstruction algorithm. In *2014 12th International Conference on Signal Processing (ICSP)* (Oct 2014), pp. 1876–1879.

[52] Y. Wang, J. Li, and P. Stoica, Rank-deficient robust Capon filter bank approach to complex spectral estimation. *IEEE Transactions on Signal Processing* 53(8) (2005), 2713–2726.

[53] D. Wehner,*High resolution radar*. Norwood, MA, Artech House, Inc., 484, 1 (1987).

[54] X. Xu and R. Narayanan, SAR image enhancement using noninteger Nyquist SVA technique. In *IEEE Antennas and Propagation Society International Symposium, 2002* (2002), vol. 4, pp. 298–301.

[55] X. Xu and R. M. Narayanan, Enhanced resolution in SAR/ISAR imaging using iterative sidelobe apodization. *IEEE Transactions on Image Processing* 14(4) (2005), 537–547.

[56] X. Xu and R. M. Narayanan, Enhanced resolution in SAR/ISAR imaging using iterative sidelobe apodization. *IEEE Transactions on Image Processing* 14(4) (2005), 537–547.

[57] Y. Yu, A. Petropulu, and H. Poor, MIMO radar using compressive sampling. *IEEE Journal of Selected Topics in Signal Processing* 4(1) (2010), 146–163.

[58] L. Zhang, Z.-J. Qiao, M. Xing, Y. Li, and Z. Bao, High-resolution ISAR imaging with sparse stepped-frequency waveforms. *IEEE Transactions on Geoscience and Remote Sensing* 49(11) (2011), 4630–4651.

[59] L. Zhang, Z. J. Qiao, Xing, M. D., J. L. Sheng, R. Guo, and Z. Bao, High-resolution ISAR imaging by exploiting sparse apertures. *IEEE Transactions on Antennas and Propagation* 60(2) (2012), 997–1008.

[60] L. Zhang, M. Xing, Qiu, C.-W., J. Li, J. Sheng, Y. Li, and Z. Bao, Resolution enhancement for Inversed Synthetic Aperture Radar imaging under low SNR via improved compressive sensing. *IEEE Transactions on Geoscience and Remote Sensing* 48(10) (2010), 3824–3838.

[61] F. Zhu, Q. Zhang, Y. Xiang, and Y. Feng, Compressive sensing in ISAR spectrogram data transmission. In *2nd Asian-Paci c Conference on Synthetic Aperture Radar, 2009. APSAR 2009* (Oct 2009), pp. 89–92.

[62] F. Zhu, Q. Zhang, J. Yan, F. Gu, and S. Liu, Compressed sensing in ISAR imaging with sparse sub-aperture. In *2011 IEEE CIE International Conference on Radar (Radar)* (Oct 2011), vol. 2, pp. 1463–1466.

[63] L. Zhuang, X. Liu, and Z. Zhou, Enhanced resolution for sparse aperture radar imaging using super-SVA. In *Asia-Paci c Microwave Conference, 2007. APMC 2007* (Dec 2007), pp. 1–4.

6

A Novel Compressed Sensing–Based Algorithm for Space–Time Signal Processing Using Airborne Radars

Jing Liu, Mahendra Mallick, Feng Lian, and Kaiyu Huang

CONTENTS

6.1 Introduction

A large number of compressed sensing–based algorithms have been applied to radar systems [1–9]. Such algorithms reconstruct the target scene from fewer measurements than traditional algorithms. Baraniuk and Steeghs [1] demonstrated that compressed sensing can eliminate the need for a matched filter at the receiver and that it has the potential to reduce the required sampling rate. Liu, Wei, and Li [2] presented an adaptive clutter suppression algorithm for airborne random pulse repetition interval radar by using prior knowledge of the clutter boundary in the Doppler spectrum. Yang and Zhang [3] focused on monostatic chaotic multiple-input multiple-output radar systems and analyzed, theoretically and numerically, the performance of sparsity-exploiting algorithms for the parameter estimation of targets at low signal-to-noise ratios (SNRs). In the context of synthetic aperture radar, compressed sensing–based data acquisition and imaging algorithms are given in [4–9].

Space–time adaptive processing (STAP) is a signal processing technique that was originally developed for detecting slowly moving targets using airborne radars [10–13]. It represents the simultaneous adaptive application of both Doppler filtering and spatial beamforming [14,15] and allows the suppression of clutter that neither technique can address individually. While much of the early work in STAP focuses on the simplest case of side-looking monostatic radars with uniform linear arrays (ULAs), STAP techniques have also been applied to bistatic radars, conformal arrays, space-based radars, and other applications [16]. However, the traditional STAP algorithm uses a large number of training cells to estimate the space–time covariance matrix, which requires a large computer memory and is time-consuming.

In recent years, a number of compressed sensing–based methods have been proposed to detect an unknown number of moving targets in strong clutter directly using space–time data, which significantly reduces the computational complexity [17–20]. Zhang, Li, and Meng [17] reconstructed the entire radar scene, the direction of arrival (DOA)–Doppler plane, using a compressed sensing–based approach. In [18], the problem of clutter was addressed by applying a mask to the signal in the DOA–Doppler plane before penalizing. However, this was based on the assumption of known clutter ridge location. Parker and Potter's work [19] is a combination of the traditional STAP algorithm and compressed sensing. Sun, Meng, and Wang [20] proposed a new direct data domain approach using sparse representation (D3SR) to estimate the high-resolution space–time spectrum with only a test cell. However, the method assumes that the location of the targets is known *a priori*.

The classical model of compressed sensing, $\mathbf{y} = \boldsymbol{\Phi}\mathbf{x} + \mathbf{e}$, is adopted in the above work, where $\mathbf{y} \in \mathbb{R}^M$, $\mathbf{e} \in \mathbb{R}^M$, $\boldsymbol{\Phi}$, and $\mathbf{x} \in \mathbb{R}^N$ represent the measurement vector, measurement noise vector, $M \times N$ sensing matrix, and a K-sparse vector (with large N), respectively. The measurement vector $\mathbf{y}$ represents the received echo signal snapshot from the fixed range cell, and $\mathbf{x}$ is the collection of the amplitude of the original transmitted signals (including targets, clutter, or both) from the whole DOA–Doppler plane. The sensing matrix $\boldsymbol{\Phi}$ contains Spatial-Doppler steering vectors in columns, which are deterministic in nature. All the above work assumes that the sparse vector $\mathbf{x}$ can be reconstructed based on the sensing matrix $\boldsymbol{\Phi}$ perfectly. However, the coherence of the sensing matrix is not low due to high resolution of the DOA–Doppler plane, which does not guarantee a good reconstruction of the sparse vector with large probability [21–26].

In this chapter, a novel algorithm called the *general similar sensing matrix pursuit* is proposed to reconstruct the K-sparse signal based on the original deterministic sensing matrix. Our proposed algorithm consists of two components: the offline and online components. The goal of the offline processing is to construct a similar sensing matrix with low coherence. The online processing begins when the measurements arrive. An orthogonal matching pursuit (OMP) algorithm [23] is then used to find an approximate estimate of the true support set, which contains the indices of the columns that contribute to the original sparse vector. Finally, the correct subspace is identified, and the original K-sparse signal is reconstructed perfectly.

In this chapter, we consider the application of detecting an unknown number of moving targets in high clutter using an airborne radar system. Because the airborne radar scenario has a high clutter-to-signal ratio (CSR) (e.g., CSR > 20 dB), the prominent elements of the spectral distribution focus along the clutter ridge in the DOA–Doppler plane. Therefore, it is reasonable to assume that the received data of the test cell are sparse in the DOA–Doppler plane [20]. Our proposed general similar sensing matrix pursuit (GSSMP) algorithm is then used to reconstruct the sparse signal representing the radar scene. Our simulation results show that both the strong elements (clutter) and the weak elements (targets) are estimated accurately in the reconstructed DOA–Doppler plane, and consequently the targets are successfully distinguished from the clutter. This is due to the fact that our proposed GSSMP algorithm can efficiently cope with the deterministic sensing matrix with high coherence.

Our chapter has the following three main contributions:

1. A novel algorithm, the GSSMP algorithm, is proposed to cope with the deterministic sensing matrix with high coherence.

2. A novel compressed sensing–based algorithm is used to detect multiple moving targets in high clutter directly on the space–time data, without the need to know the clutter ridge location or the target region.

3. Our proposed algorithm uses only the data from the test cell, which significantly reduces the computational complexity.

The chapter is organized as follows. Section 6.2 introduces a general space–time model for airborne radar systems, which is represented in a compressed sensing framework. The GSSMP algorithm and the compressed sensing–based multiple target detection algorithm are introduced in Sections 6.3 and 6.4, respectively. Numerical simulation and results are presented in Section 6.5, and conclusions are drawn in Section 6.6.

6.2   A general space–time model and its sparse representation

In this section, we consider an airborne radar system that transmits K coherent pulse trains and samples the returns on ULAs consisting of N elements. For each pulse, it collects Q temporal samples from each element receiver, where each time sample corresponds to a range cell. The collection of samples for the qth range cell is represented by an $N \times K$ data matrix $\mathbf{F}$ (snapshot) with elements $f(n, k)$ as follows:

$$\mathbf{F} = \begin{bmatrix} f(1,1) & f(1,2) & \cdots & f(1,K) \\ f(2,1) & f(2,2) & \cdots & f(2,K) \\ \cdots & \cdots & & \cdots \\ f(N,1) & f(N,2) & \cdots & f(N,K) \end{bmatrix}. \tag{6.1}$$

A test cell is assumed to be comprised of the target and clutter components. First, assuming that D targets are observed in the far field, the ith target is at a DOA angle of θ_i^t with Doppler frequency $f_{d_i}^t$. We can obtain an $NK \times 1$ complex vector $\mathbf{y}_t$ as follows:

$$\mathbf{y}_t = \sum_{i=1}^{D} \beta(\theta_i^t, f_{d_i}^t)[\mathbf{s}_S(\theta_i^t) \otimes \mathbf{s}_T(f_{d_i}^t)], \tag{6.2}$$

where $\beta(\theta_i^t, f_{d_i}^t)$ is the scattering coefficient of the ith target, and $\otimes$ represents the Kronecker product [27] of two vectors. The spatial steering vector $\mathbf{s}_S(\theta_i^t)$ and the Doppler filtering steering vector $\mathbf{s}_T(f_{d_i}^t)$ are represented by the following:

$$\mathbf{s}_S(\theta_i^t) = [1 \quad e^{j\frac{2\pi d}{\lambda} \sin \theta_i^t} \quad \cdots \quad e^{j(N-1)\frac{2\pi d}{\lambda} \sin \theta_i^t}]^T \tag{6.3}$$

and

$$\mathbf{s}_T(f_{d_i}^t) = [1 \quad e^{j\frac{2\pi f_{d_i}^t}{f_r}} \quad \cdots \quad e^{j(K-1)\frac{2\pi f_{d_i}^t}{f_r}}], \tag{6.4}$$

where d is the distance between the elements of the arrays, and λ and f_r denote the wavelength and pulse repetition frequency, respectively.

In addition to the target components, there also exists the clutter component $\mathbf{y}_c$, which can be considered a collection of independent scatters as follows:

$$\mathbf{y}_c = \sum_{i=1}^{N_c} \beta(\theta_i^c, f_{d_i}^c)[\mathbf{s}_S(\theta_i^c) \otimes \mathbf{s}_T(f_{d_i}^c)], \tag{6.5}$$

where N_C is the number of clutter scatters; θ_i^c and $f_{d_i}^c$ are the DOA angle and Doppler frequency for the ith clutter scatter, respectively; and $\beta(\theta_i^c, f_{d_i}^c)$ is the scattering coefficient for clutter. In Equation 6.5, $\mathbf{s}_S(\theta_i^c)$ and $\mathbf{s}_T(f_{d_i}^c)$ represent the spatial steering vector and the Doppler filtering steering vector, respectively.

Using the above modeling, the $NK \times 1$ complex vector of the test cell can be modeled as follows:

$$\mathbf{y}_{\text{test}} = \mathbf{y}_t + \mathbf{y}_c + \mathbf{e}, \tag{6.6}$$

where $\mathbf{e}$ is an $NK \times 1$ complex Gaussian noise vector.

In this chapter, compressed sensing is used to estimate the spectral distributions of the target and clutter scatters in the DOA–Doppler plane. To do so, the DOA–Doppler plane is divided into $V \times L$ grids, where V and L denote the number of rows (for Doppler frequency) and columns (for DOA angle), respectively. Each cell of the grid has the same size $\Delta\theta \times \Delta f_d$. Grid (i,j) represents a DOA angle of $\theta_i = \theta_0 + (i-1)\Delta\theta$ and a Doppler frequency of $f_{d_i} = f_{d_0} + (i-1)\Delta f_d$, where θ_0 and f_{d_0} represent the initial DOA angle and initial Doppler frequency. All the grids in the DOA–Doppler plane are mapped into a 2D vector $\mathbf{x}_{\text{test}}$ with the jth column put at the end of the $(j-1)$th column.

Because the airborne radar scene has a high CSR (>20 dB), the significant elements of spectral distribution focus along the clutter ridge in the DOA–Doppler plane. Therefore, it is reasonable to assume that the received data of the test cell is sparse in the DOA–Doppler plane [20]. A small number of grids are occupied by the targets and clutter scatters in the DOA–Doppler plane, and $\mathbf{x}_{\text{test}}$ is a sparse vector.

Based on the above derivation, a system for the test cell is built in a compressed sensing framework as follows:

$$\mathbf{y}_{\text{test}} = \mathbf{\Phi}\, \mathbf{x}_{\text{test}} + \mathbf{e}, \tag{6.7}$$

where $\mathbf{\Phi}$ is a sensing matrix with dimension $NK \times VL$ and is defined as $\mathbf{\Phi} = [\boldsymbol{\varphi}_1 \ \boldsymbol{\varphi}_2 \ \cdots \ \boldsymbol{\varphi}_{VL}]$, where $\boldsymbol{\varphi}_i$ is a column vector of dimension NK. The $((i-1)L+j)$th column of $\mathbf{\Phi}$ is defined as follows:

$$\boldsymbol{\varphi}_{(i-1)L+j} = \mathbf{s}_S(\theta_i) \otimes \mathbf{s}_T(f_{d_j}), \tag{6.8}$$

The sensing matrix $\mathbf{\Phi}$ has high coherence because V and D are set to large values to obtain a high resolution of the DOA–Doppler plane. Although in Equation 6.7 the radar vectors and matrices are complex valued in contrary to the original compressed sensing environment, it is easy to transfer it to real variables [27,28].

For simplicity, Equation 6.7 is rewritten in a classical format in compressed sensing with subscripts removed:

$$\mathbf{y} = \mathbf{\Phi}\mathbf{x} + \mathbf{e}. \tag{6.9}$$

6.3   The proposed GSSMP algorithm

In the recently proposed compressed sensing–based STAP algorithms [17–20], it was assumed that the sparse vector **x** can be reconstructed perfectly based on the sensing matrix $\mathbf{\Phi}$. However, the coherence of the sensing matrix is not low due to the high resolution of the DOA–Doppler plane, which does not guarantee a good reconstruction of the sparse vector with high probability [21–26]. Consequently, the direct estimation of the target amplitude may be unreliable using sparse representation when locating a moving target from the high clutter surrounding. Sun, Meng, and Wang [20] extracted only the prominent elements (clutter) from the sparse radar scene and used an additional adaptive filter to suppress the clutter and identify the target.

In this chapter, a novel method called *GSSMP* is proposed to reconstruct the sparse vector **x** representing the sparse radar scene. Our proposed algorithm can efficiently cope with the deterministic sensing matrix $\mathbf{\Phi}$ with high coherence. As a result, our proposed algorithm can estimate the weak elements (targets) as well as the strong elements (clutter) accurately and can distinguish the targets from clutter successfully in the DOA Doppler plane.

This section mainly introduces our proposed GSSMP algorithm. First, the intuition of our proposed algorithm is introduced in Section 6.3.1. A similarity analysis of the original sensing matrix is then performed in Section 6.3.2, and the construction process of the similar sensing matrix is introduced in Section 6.3.3. Section 6.3.4 introduces the detailed procedures for the complete GSSMP algorithm. Finally, the convergence analysis is provided in Section 6.3.5.

6.3.1   Intuition

In compressed sensing, the major challenge associated with K-sparse signal reconstruction is to identify in which subspace (generated by not more than K columns of the sensing matrix $\mathbf{\Phi}$) the measured signal **y** lies [28]. Once the correct subspace is determined, then the nonzero signal coefficients are calculated by applying the pseudo-inverse. The key procedure of greedy algorithms is the method used for finding the columns that span the correct subspace. An important prior for greedy algorithms is that the sensing matrix satisfy the restricted isometry property or have low coherence, which implies that the columns of the sensing matrix are locally near-orthogonal (incoherent with each other). However, in practice, it is challenging to design a deterministic sensing matrix having low coherence. Greedy algorithms always fail to reconstruct the K-sparse vector based on the deterministic sensing matrix with high coherence, since they cannot find the right columns from a group of coherent (similar) columns to form the correct subspace.

In this chapter, we consider the general condition where a deterministic sensing matrix has high coherence. A novel GSSMP algorithm is proposed to reconstruct the K-sparse signal, which builds a similar sensing matrix based on the original sensing matrix. Each condensed column of the similar sensing matrix represents the characteristics of a group of coherent columns or a single incoherent column in the original sensing matrix. It can be easily proved that the similar sensing matrix has a low coherence, provided that the minimum similar distance between any two condensed columns is large. An approximate estimate of the correct subspace is obtained based on the condensed columns of the newly built similar sensing matrix. Each condensed column of the estimated subspace indicates a group of coherent columns or a single incoherent column in the original sensing matrix, which contribute to the correct subspace and are termed *candidate columns*. A combinatorial search is then performed among all the candidate columns to find the columns forming the correct subspace. It is proved that under appropriate conditions, the GSSMP algorithm can identify the correct subspace quite well.

6.3.2 Similarity analysis of the original sensing matrix

The construction process of the similar sensing matrix is based on the similarity analysis of the original sensing matrix. In this chapter, similarity is defined as the absolute and normalized inner product between any two different columns in the original sensing matrix $\boldsymbol{\Phi}$:

$$\lambda(\boldsymbol{\varphi}_i, \boldsymbol{\varphi}_j) = \frac{|\boldsymbol{\varphi}_i^T \boldsymbol{\varphi}_j|}{||\boldsymbol{\varphi}_i||_2 \cdot ||\boldsymbol{\varphi}_j||_2}, \quad 1 \le i, j \le N \text{ and } i \ne j \tag{6.10}$$

The number of similarity values of $\boldsymbol{\Phi}$ is $N(N-1)/2$. It is evident that coherence is the largest similarity among the columns of a matrix. Any two columns with high similarity are coherent with each other and vice versa.

The relationship between the columns of the sensing matrix is efficiently represented using the similarity. A scenario is considered where each column is represented as a point in $\mathbb{R}^{M'}$, called the *similarity space*, where M' denotes the dimension of the similarity space, with $M' \le M$. The "distance" between any two points (columns) is equal to the "distance" related with similarity rather than the traditional Euclidean distance.

The detailed procedure is as follows. All the columns are mapped to a similarity space to obtain a number of column points, which are indicated as $\{P_{\boldsymbol{\varphi}_1}, \cdots, P_{\boldsymbol{\varphi}_i}, \cdots, P_{\boldsymbol{\varphi}_N}\}$, with $P_{\boldsymbol{\varphi}_i}$ denoting the column point corresponding to the column $\boldsymbol{\varphi}_i$. The similarity distance d_{similar} is defined as the distance between any two column points $P_{\boldsymbol{\varphi}_i}$ and $P_{\boldsymbol{\varphi}_j}$ as follows:

$$d_{\text{similar}}(P_{\boldsymbol{\varphi}_i}, P_{\boldsymbol{\varphi}_j}) = 1 - \lambda(\boldsymbol{\varphi}_i, \boldsymbol{\varphi}_j) \tag{6.11}$$

Therefore, the distribution range of the similarity distance is $[0, 1]$, and the closer the two column points are in the similarity space (with smaller distance), the more similar their corresponding columns are (with larger similarity).

A hierarchical clustering method is then used to cluster the column points. First, the similarity distance between every pair of column points is calculated. Second, the column points are grouped into a binary, hierarchical cluster tree, where pairs of column points that are in close proximity in similarity distance are linked to each other. As column points are paired into binary clusters, the newly formed clusters are grouped into larger clusters until a hierarchical tree is formed. Finally, the cutoff criterion is set to prune branches off the bottom of the hierarchical tree, and all the column points below each cutoff are assigned to a single cluster, resulting in a number of clusters. Therefore, the cluster number is determined once the cutoff criterion is set. The setting of the cutoff criterion is based on the minimum similarity distance between clusters (similar column groups), which is discussed in Section 6.3.5.

Each cluster contains one or more column points. For the cluster containing more than one column point, the similarity distance between any two column points is small. Accordingly, columns from the same cluster are similar (coherent) with each other, defined as a similar column group. It should be noted that a similar column group may only contain one column.

Figure 6.1 shows the classification results in a similarity space. Here we choose a two-dimensional representation of the similarity space for clarity and simplicity. A column point and a similar column group are indicated by a small square and a large circle, respectively. In Figure 6.1, d_{similar} indicates the similarity distance between any two column points. A similar column group may contain one or more columns. Any two columns in a similar

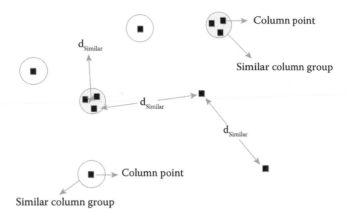

FIGURE 6.1
Classification results in a similarity space.

column group have a small similarity distance and are similar with each other. Next, we build the similar sensing matrix based on the similar column groups.

6.3.3   Construction of the similar sensing matrix

All the columns of the sensing matrix are mapped to a similarity space to obtain a number of column points $\{P_{\varphi_1}, \cdots, P_{\varphi_i}, \cdots, P_{\varphi_N}\}$. A hierarchical clustering method is then used to cluster the column points based on the similarity distance, and we can obtain a number of clusters containing one or more column points. The columns from a same cluster constitute a similar column group. All the columns of the sensing matrix are divided into D similar column groups, $\{\boldsymbol{\Gamma}_1, \boldsymbol{\Gamma}_2, \cdots, \boldsymbol{\Gamma}_i, \cdots, \boldsymbol{\Gamma}_D\}$. We assume that $M \ll D < N$. Each similar column group contains one or more columns, for example, $\boldsymbol{\Gamma}_i = \{\boldsymbol{\gamma}_1^i, \cdots, \boldsymbol{\gamma}_{N_{\Gamma_i}}^i\}$, where γ_j^i is the jth column of the similar column group $\boldsymbol{\gamma}_i$ and N_{Γ_i} is the number of columns in $\boldsymbol{\gamma}_i$.

The columns in a similar column group are similar to each other. We think of finding a single column from each similar column group, which represents the characteristics of all the columns in a group. The column is called a *condensed column*. There are different ways to calculate a condensed column. Here the average of the columns in a similar column group is chosen as the condensed column; for example, for the ith similar column group $\boldsymbol{\gamma}_i$, its condensed column $\boldsymbol{\gamma}_C^i$ is defined as follows:

$$\boldsymbol{\gamma}_C^i = \frac{1}{N_{\boldsymbol{\gamma}_i}} \sum_{j=1}^{N_{\boldsymbol{\gamma}_i}} \boldsymbol{\gamma}_j^i. \tag{6.12}$$

The similar sensing matrix is then built by combining the condensed columns from all the similar column groups as $\boldsymbol{\Psi} = [\boldsymbol{\gamma}_C^1 \quad \boldsymbol{\gamma}_C^2 \quad \cdots \quad \boldsymbol{\gamma}_C^i \quad \cdots \quad \boldsymbol{\gamma}_C^D]$, with size $M \times D$.

Remarks It should be noted that the original sensing matrix and the similar sensing matrix are connected via the condensed columns: each condensed column of the similar sensing matrix represents the characteristics of a similar column group in the original sensing matrix. The similar sensing matrix can be seen as a "compact" version of the original sensing matrix. Next, we show that the similar sensing matrix has low coherence provided that the similarity distance between any two of its condensed columns is large.

Property 6.3.1 *Let TH denote the minimum similarity distance between any two condensed columns of the similar sensing matrix $\boldsymbol{\Psi}$; for example, for any two condensed* columns, $\boldsymbol{\gamma}_C^i$ and $\boldsymbol{\gamma}_C^j$, the similarity distance between their corresponding column points is larger than TH.

$$d_{similar}(P_{\gamma_C^i}, P_{\gamma_C^j}) \geq TH \tag{6.13}$$

Therefore, the similarity between any two condensed columns, $\boldsymbol{\gamma}_C^i$ and $\boldsymbol{\gamma}_C^j$, is no larger than $1 - TH$.

$$\lambda(\boldsymbol{\gamma}_C^i, \boldsymbol{\gamma}_C^j) \leq 1 - TH \tag{6.14}$$

Consequently, we have the following proposition for the similar sensing matrix.

Proposition 6.3.2 *The coherence of the similar sensing matrix is less than or equal to* $1 - TH$.

Proof. The similar sensing matrix is built by combining the condensed columns from all the similar column groups, as $\boldsymbol{\Psi} = [\boldsymbol{\gamma}_C^1, \cdots, \boldsymbol{\gamma}_C^i, \cdots \boldsymbol{\gamma}_C^D]$. According to Property 6.3.1, the similarity between any two condensed columns is no larger than $1 - TH$. Therefore, the coherence of the similar sensing matrix, which is the largest similarity, is less than or equal to $1 - TH$:

$$\mu(\boldsymbol{\Psi}) \leq 1 - TH \tag{6.15}$$

6.3.4   The GSSMP algorithm

Our proposed GSSMP algorithm consists of six major steps, which are described next.

Algorithm 6.1: GSSMP algorithm.

Input: Sensing matrix $\boldsymbol{\Phi}$, measurement vector $\mathbf{y}$.
Output: The estimated signal $\hat{\mathbf{x}}$.

6.3.4.1   Constructing the similar sensing matrix

The process is same as that described in Section 3.3.

6.3.4.2   Obtaining an initial estimate of the correct subspace

The OMP algorithm is used to find an estimate of the support set based on the measurement vector $\mathbf{y}$ and the similar sensing matrix $\boldsymbol{\Psi}$. The estimated support set is represented as $\hat{\mathbf{a}} = \{\hat{a}^1(\boldsymbol{\gamma}_C^i), \hat{a}^2(\boldsymbol{\gamma}_C^j), \cdots, \hat{a}^{K'}(\boldsymbol{\gamma}_C^l)\}, K' \leq K$, where $\hat{a}^j(\boldsymbol{\gamma}_C^i)$ indicates that the jth element of $\hat{\mathbf{a}}$ corresponds to the ith condensed column $\boldsymbol{\gamma}_C^i$. We obtain an initial estimate of the correct subspace, $\hat{\mathbf{S}}_{\text{init}}$, spanned by the columns $\boldsymbol{\gamma}_C^i \cdots \boldsymbol{\gamma}_C^j, \cdots, \boldsymbol{\gamma}_C^l$, defined as $\hat{\mathbf{S}}_{\text{init}} = \text{SPAN}(\boldsymbol{\gamma}_C^i, \boldsymbol{\gamma}_C^j, \cdots, \boldsymbol{\gamma}_C^l)$.

6.3.4.3   Finding the candidate columns in the original sensing matrix

Each column of the estimated subspace $\hat{\mathbf{S}}_{\text{init}}$ corresponds to a similar column group of the original sensing matrix; for example, $\boldsymbol{\gamma}_C^i$ is the condensed column from the similar column group $\boldsymbol{\Gamma}_i$. We can then obtain a set $\hat{\boldsymbol{\Lambda}}$ containing the indices of K' similar column groups corresponding to the columns in the estimated subspace $\hat{\mathbf{S}}_{\text{init}}$, as $\hat{\boldsymbol{\Lambda}} = \{\boldsymbol{\Gamma}_i, \boldsymbol{\Gamma}_j, \cdots, \boldsymbol{\Gamma}_l\}$. All the columns in each similar column group from $\hat{\boldsymbol{\Lambda}}$ are listed and form a set of candidate columns $\hat{\mathbf{f}}$, where $\hat{\mathbf{f}} = \{\boldsymbol{\gamma}_1^i, \cdots, \boldsymbol{\gamma}_{N_{\Gamma_i}}^i, \cdots, \boldsymbol{\gamma}_1^j, \cdots, \boldsymbol{\gamma}_{N_{\Gamma_j}}^j, \cdots, \boldsymbol{\gamma}_1^l, \cdots \boldsymbol{\gamma}_{N_{\Gamma_l}}^l\}$. We assume that the total number of columns in $\hat{\mathbf{f}}$ is H_{cc}.

6.3.4.4   Calculating the candidate subspaces

List $\mathbf{C}_{H_{cc}}^{K}$ combinations based on the columns in $\hat{\mathbf{f}}$. Each combination of K columns spans a candidate subspace; for example, the pth candidate subspace is represented as $\Upsilon^p = SPAN(\boldsymbol{\gamma}_1^i, \boldsymbol{\gamma}_1^j, \cdots, \boldsymbol{\gamma}_1^l)$, $p = 1, 2, \cdots, N_{co}$, where N_{co} indicates the total number of candidate subspaces.

6.3.4.5   Calculating the candidate estimates

Based on each candidate subspace, our proposed algorithm solves a least-squares problem to approximate the nonzero entries of the original K-sparse vector $\mathbf{x}$ (Equation 6.16) and sets other entries to zero (Equation 6.17), resulting in a candidate estimate $\hat{\mathbf{x}}^p$, $p = 1, 2, \cdots, N_{co}$, as follows:

$$\hat{\mathbf{x}}_{\Upsilon^p}^p = (\Upsilon^p)^{\dagger} \mathbf{y} \tag{6.16}$$

$$\hat{\mathbf{x}}_{\{1,2,\cdots,N\}-\Upsilon^p}^p = \mathbf{0}, \tag{6.17}$$

where † indicates the pseudo-inverse operation, $\hat{x}_{\Upsilon^p}^p$ is composed of the entries of $\hat{x}^p$ indexed by $i \in \Upsilon^p$, and $\hat{\mathbf{x}}_{\{1,2,\cdots,N\}-\Upsilon^p}^p$ is composed of the entries of $\hat{x}^p$ indexed by $i \in \{1, 2, \cdots, N\} - \Upsilon^p$ [23,29].

6.3.4.6   Outputting the final estimates

For each candidate estimate $\hat{\mathbf{x}}^p (p = 1, \cdots, N_{co})$, calculate the residual $\mathbf{r}^p (p = 1, \cdots, N_{co})$ via Equation 6.18:

$$\mathbf{r}^p = \mathbf{y} - \boldsymbol{\Phi}\hat{\mathbf{x}}^p \tag{6.18}$$

The l_2 norm of $\mathbf{r}^p$ is indicated as $||\mathbf{r}^p||_2$. Among the residuals, $\mathbf{r}^p (p = 1, \cdots, N_{co})$, find the residual with the least l_2 norm, r_{min}, and its associate candidate estimate, which is denoted as $\hat{\mathbf{x}}_{min}$. Set $\hat{\mathbf{x}}_{min}$ and its associate subspace Υ_{min} as the final estimates of the K-sparse signal $\mathbf{x}$ and the correct subspace, respectively:

$$\hat{\mathbf{x}} = \hat{\mathbf{x}}_{min} \tag{6.19}$$

$$\hat{\mathbf{S}} = \Upsilon_{min} \tag{6.20}$$

6.3.5   Proof of the guaranteed reconstruction performance of the GSSMP algorithm

A sufficient condition for the perfect reconstruction of arbitrary K-sparse signal $\mathbf{x}$ is stated in the following theorem.

Theorem 6.3.3 *Let $\mathbf{x} \in \mathbb{R}^N$ be a K-sparse signal, and let its corresponding measurement be $\mathbf{y} = \boldsymbol{\Phi}\mathbf{x} + \mathbf{e} \in \mathbb{R}^M$, where $\mathbf{e}$ denotes the noise vector. Let $\boldsymbol{\Psi}$ be a similar sensing matrix built based on the original sensing matrix $\boldsymbol{\Phi}$. Suppose that the minimum similarity distance TH between any two condensed columns of the similar sensing matrix $\boldsymbol{\Psi}$ satis es the following:*

$$TH \geq 1 - 1/(4K' - 1), \tag{6.21}$$

where $K'(K' \leq K)$ is the sparsity level of the intermediate vector $\mathbf{x}'$ de ned in Proposition 6.3.4. Then the GSSMP algorithm can identify the correct subspace, and it reconstructs the K-sparse vector $\mathbf{x}$ perfectly with the reconstruction error satis ed as follows:

$$||\mathbf{x} - \hat{\mathbf{x}}||_2 \leq \frac{||\mathbf{e}||_2}{\sqrt{1 - \mu(\boldsymbol{\Phi})(K-1)}}. \tag{6.22}$$

We now commence the proof of Theorem 6.3.3. First, we prove that the measurement vector **y** can be represented using the condensed columns of the similar sensing matrix **Ψ**. At the same time, we can obtain an initial estimate of the correct subspace.

Proposition 6.3.4 *The measurement vector* **y** *can be represented using the condensed columns of the similar sensing matrix, as* $\mathbf{y} = \mathbf{\Psi}\mathbf{x}' + \mathbf{e}'$, *where* $\mathbf{x}'$ *is a* K'-*sparse vector* $(K' \leq K)$, *and* $\mathbf{e}'$ *is an equivalent measurement noise vector. At the same time, we can obtain the initial estimate of the correct subspace* $\hat{\mathbf{S}}_{init}$; *that is,* $\hat{\mathbf{S}}_{init} = SPAN(\boldsymbol{\gamma}_C^1, \cdots, \boldsymbol{\gamma}_C^i, \cdots, \boldsymbol{\gamma}_C^{K'})$, *where* $\boldsymbol{\gamma}_C^i$ *is de ned as the condensed column of the similar column group* $\mathbf{\Gamma}_i, i = 1, \cdots, K'$.

Proof. Let $\{\tau_1, \tau_2, \cdots, \tau_K\}$ be the K randomly distributed nonzero elements of the original K-sparse vector **x**. The vectors $\{\mathbf{b}_1, \mathbf{b}_2, \cdots, \mathbf{b}_K\}$ are defined as the columns of the sensing matrix **Φ**, which correspond to the nonzero elements $\{\tau_1, \tau_2, \cdots, \tau_K\}$. The correct subspace **S** is spanned by the vectors $\{\mathbf{b}_1, \mathbf{b}_2, \cdots, \mathbf{b}_K\}$ as $\mathbf{S} = SPAN(\mathbf{b}_1, \mathbf{b}_2, \cdots, \mathbf{b}_K)$.

We consider two cases: (1) the vectors $\{\mathbf{b}_1, \mathbf{b}_2, \cdots, \mathbf{b}_K\}$ are from K different similar column groups of the original sensing matrix; and (2) the vectors $\{\mathbf{b}_1, \mathbf{b}_2, \cdots, \mathbf{b}_K\}$ are from $K'(K' \leq K)$ similar column groups.

For Case 1, $\mathbf{b}_i$ is from $\mathbf{\Gamma}_i, i = 1, \cdots, K$. The vectors $\{\boldsymbol{\gamma}_C^1, \boldsymbol{\gamma}_C^2, \cdots, \boldsymbol{\gamma}_K\}$ are the condensed columns of the similar column groups $\{\mathbf{\Gamma}_1, \mathbf{\Gamma}_2, \cdots, \mathbf{\Gamma}_K\}$, respectively. The difference between $\mathbf{b}_i$ and $\boldsymbol{\gamma}_C^i$ is defined as $\Delta\mathbf{b}_i = \mathbf{b}_i - \boldsymbol{\gamma}_C^i$. According to the definitions of the vectors $\{\mathbf{b}_1, \mathbf{b}_2, \cdots, \mathbf{b}_K\}$, we have the following:

$$\mathbf{y} = \mathbf{\Phi}\mathbf{x} + \mathbf{e}$$
$$= \sum_{i=1}^{K} \tau_i \mathbf{b}_i + \mathbf{e}$$
$$= \sum_{i=1}^{K} \tau_i(\boldsymbol{\gamma}_C^i + \Delta\mathbf{b}_i) + \mathbf{e}$$
$$= \sum_{i=1}^{K} \tau_i \boldsymbol{\gamma}_C^i + \sum_{i=1}^{K} \tau_i(\Delta\mathbf{b}_i) + \mathbf{e}$$
$$= \sum_{i=1}^{K} \tau_i \boldsymbol{\gamma}_C^i + \mathbf{e}', \tag{6.23}$$

where $\mathbf{e}'$ is the equivalent measurement noise, $\mathbf{e}' = \sum_{i=1}^{K} \tau_i(\Delta\mathbf{b}_i) + \mathbf{e}$. Considering that $\{\boldsymbol{\gamma}_C^1, \boldsymbol{\gamma}_C^2, \cdots, \boldsymbol{\gamma}_C^K\}$ are from the similar sensing matrix **Ψ**, Equation 6.23 naturally results in

$$\mathbf{y} = \mathbf{\Psi}\mathbf{x}' + \mathbf{e}', \tag{6.24}$$

where $\mathbf{x}'$ is a $D \times 1$ sparse vector with $K'(K' = K)$ nonzero elements $\tau_1, \tau_2, \cdots, \tau_{K'}$. We obtain the initial estimate of the correct subspace $\hat{\mathbf{S}}_{init} = SPAN(\boldsymbol{\gamma}_C^1, \boldsymbol{\gamma}_C^2, \cdots, \boldsymbol{\gamma}_C^{K'})$.

For Case 2, without loss of generality, we assume that $K' = K - 1$ and that the two vectors $\mathbf{b}_1$ and $\mathbf{b}_2$ are both from a same group $\mathbf{\Gamma}_1$ with condensed column $\boldsymbol{\gamma}_C^1$. The difference between $\mathbf{b}_1$ and $\boldsymbol{\gamma}_C^1$ is defined as $\Delta\mathbf{b}_1 = \mathbf{b}_1 - \boldsymbol{\gamma}_C^1$, and the difference between $\mathbf{b}_2$ and $\boldsymbol{\gamma}_C^1$ is defined as $\Delta\mathbf{b}_2 = \mathbf{b}_2 - \boldsymbol{\gamma}_C^1$. Moreover, b_i is from $\mathbf{\Gamma}_{i-1}, i = 3, \cdots, K$. We have the following:

$$\mathbf{y} = \mathbf{\Phi}\mathbf{x} + \mathbf{e}$$
$$= \sum_{i=1}^{K} \tau_i \mathbf{b}_i + e$$

$$= \sum_{i=1}^{K} \tau_i (\gamma_C^i + \Delta \mathbf{b}_i) + \mathbf{e}$$

$$= \sum_{i=1}^{K} \tau_i \gamma_C^i + \sum_{i=1}^{K} \tau_i \Delta \mathbf{b}_i + \mathbf{e}$$

$$= \sum_{i=1}^{K} \tau_i \gamma_C^i + \mathbf{e}'$$

$$= (\tau_1 + \tau_2)\gamma_C^1 + \sum_{i=3}^{K} \tau_i \gamma_C^{i-1} + \mathbf{e}'$$

$$= \sum_{i=1}^{K-1} \tau_i' \gamma_C^i + e'$$

$$= \boldsymbol{\Psi}\mathbf{x}' + \mathbf{e}', \tag{6.25}$$

where $\mathbf{x}'$ is a $D \times 1$ sparse vector with $K'(K' = K - 1)$ nonzero elements $\tau_1', \tau_2', \cdots, \tau_{K'}'$. Further, we have $\tau_1' = \tau_1 + \tau_2$ and $\tau_i' = \tau_{i+1}, i = 2, \cdots, K'$. The vector $\mathbf{e}'$ is the equivalent measurement noise, which is defined as $\mathbf{e}' = \sum_{i=1}^{K} \tau_i(\Delta \mathbf{b}_i) + \mathbf{e}$. We obtain the initial estimate of the correct subspace $\hat{\mathbf{S}}_{\text{init}} = SPAN(\gamma_C^1, \gamma_C^2, \cdots, \gamma_C^{K'})$. This completes the proof.

Next, we show that the K'-sparse vector $\mathbf{x}'$ can be reconstructed perfectly based on the similar sensing matrix $\boldsymbol{\Psi}$ and the measurement vector $\mathbf{y}$, provided that the minimum similarity distance between any two condensed columns of the similar sensing matrix is large, and the amplitude of the equivalent measurement noise $\mathbf{e}'$ is small.

Proposition 6.3.5 *We have* $\mathbf{y} = \boldsymbol{\Psi}\mathbf{x}' + \mathbf{e}' \in \mathbb{R}^M$, *where* $\mathbf{x}' \in \mathbb{R}^D$ *is a* K'-*sparse signal,* $\mathbf{e}'$ *denotes the equivalent noise vector as* $\mathbf{e}' = \sum_{i=1}^{K} \tau_i(\Delta \mathbf{b}_i) + \mathbf{e}$, *according to Proposition 6.3.4. Suppose that the minimum similarity distance* T *of the similar sensing matrix* $\boldsymbol{\Psi}$ *satisfies the following:*

$$T \geq 1 - 1/(4K' - 1) \tag{6.26}$$

and the amplitude of the equivalent noise $\|e'\|$ *satisfies the following:*

$$\|\mathbf{e}'\|_2 \leq A(1 - \mu(\boldsymbol{\Psi})(2K' - 1))/2, \tag{6.27}$$

where A *is a positive lower bound on the magnitude of the nonzero entries of* $\mathbf{x}'$, *and* $\mu(\boldsymbol{\Psi})$ *denotes the coherence of the similar sensing matrix* $\boldsymbol{\Psi}$. *The* K'-*sparse vector* $\mathbf{x}'$ *can be reconstructed using the OMP algorithm, with reconstruction error:*

$$\|\mathbf{x}' - \hat{\mathbf{x}}'\|_2 \leq \frac{\|\mathbf{e}'\|_2}{\sqrt{1 - \mu(\boldsymbol{\Psi})(K' - 1)}}. \tag{6.28}$$

Proof. According to Proposition 6.3.2, the coherence of the similar sensing matrix is less than or equal to $1 - TH$.

$$\mu(\boldsymbol{\Psi}) \leq 1 - TH \tag{6.29}$$

Moreover, according to Theorem 3.1 in [28], in order to reconstruct the K'-sparse vector $\mathbf{x}'$ perfectly using the OMP algorithm, the coherence $\mu(\boldsymbol{\Psi})$ should satisfy

$$\mu(\boldsymbol{\Psi}) \leq 1/(4K' - 1) \tag{6.30}$$

The combination of Equations 6.29 and 6.30 results in Equation 6.31:

$$T \geq 1 - 1/(4K' - 1) \tag{6.31}$$

which is a sufficient condition for the perfect reconstruction of the K'-sparse vector $\mathbf{x}'$. Moreover, the reconstruction error satisfies the following (Theorem 3.1 in [28]):

$$||\mathbf{x}' - \hat{\mathbf{x}}'||_2 \leq \frac{||\mathbf{e}'||_2}{\sqrt{1 - \mu(\boldsymbol{\Psi})(K' - 1)}}. \tag{6.32}$$

This completes the proof.

Though we can reconstruct $\mathbf{x}'$ perfectly, we cannot obtain the exact value of the original signal $\mathbf{x}$. There is no direct transform from $\mathbf{x}'$ ($D \times 1$ vector) to $\mathbf{x}$ ($N \times 1$ vector). However, we can still identify the correct subspace based on the initial estimate of the correct subspace $\hat{\mathbf{S}}_{\text{init}}$, provided $\mathbf{x}'$ is perfectly reconstructed.

Proposition 6.3.6 *Based on the initial estimate of the correct subspace* $\hat{\mathbf{S}}_{init}$, *we can identify the correct subspace* $\mathbf{S}$.

Proof. According to Proposition 6.3.5, the K'-sparse signal x' can be reconstructed perfectly with its K' nonzero elements and their associate vectors $\{\boldsymbol{\gamma}_C^1, \boldsymbol{\gamma}_C^2, \cdots, \boldsymbol{\gamma}_C^{K'}\}$, which span the initial estimated subspace $\hat{S}_{ini} = SPAN(\boldsymbol{\gamma}_C^1, \boldsymbol{\gamma}_C^2, \cdots, \boldsymbol{\gamma}_C^{K'})$. The vectors $\{\boldsymbol{\gamma}_C^1, \boldsymbol{\gamma}_C^2, \cdots, \boldsymbol{\gamma}_C^{K'}\}$ correspond to a set of similar column groups in the original sensing matrix, $\{\boldsymbol{\Gamma}_1, \cdots, \boldsymbol{\Gamma}_i, \cdots, \boldsymbol{\Gamma}_{K'}\}$; that is, $\boldsymbol{\gamma}_C^i$ is the condensed column of $\boldsymbol{\gamma}_i, i = 1, \cdots K'$. According to Proposition 6.3.4, the vectors spanning the correct subspace, $\{\mathbf{b}_1, \mathbf{b}_2, \cdots, \mathbf{b}_K\}$, are contained in the similar column groups, $\{\boldsymbol{\Gamma}_1, \cdots, \boldsymbol{\Gamma}_i, \cdots, \boldsymbol{\Gamma}_{K'}\}$, which correspond to the K' condensed columns spanning the initial estimated subspace $\hat{\mathbf{S}}_{ini}$. As a result, a combinatorial search among all the columns contained in the similar column groups, $\{\boldsymbol{\Gamma}_1, \cdots, \boldsymbol{\Gamma}_i, \cdots, \boldsymbol{\Gamma}_{K'}\}$, can identify the vectors spanning the correct subspace, $\{\mathbf{b}_1, \mathbf{b}_2, \cdots, \mathbf{b}_K\}$. This completes the proof.

Proof of Theorem 6.3.3: We now complete the proof of Theorem 6.3.3. According to Proposition 6.3.6, the correct subspace $\mathbf{S}$ can be accurately identified based on the initial estimate of the correct subspace $\hat{\mathbf{S}}_{ini}$, provided x' is perfectly reconstructed. We can obtain an estimate of the original K-sparse vector $\mathbf{x}$, by solving a least-squares problem to approximate the nonzero entries of x (Equation 6.33) and set other entries as zero (Equation 6.34):

$$\hat{\mathbf{x}}_{\mathbf{S}} = (\mathbf{S})^{\dagger} \mathbf{y} \tag{6.33}$$

$$\hat{\mathbf{x}}_{\{1,2,\cdots,N\}-\mathbf{S}} = 0 \tag{6.34}$$

The reconstruction error satisfies the following (Theorem 3.1 in [28]):

$$||\mathbf{x} - \hat{\mathbf{x}}||_2 \leq \frac{||\mathbf{e}||_2}{\sqrt{1 - \mu(\boldsymbol{\Phi})(K - 1)}}. \tag{6.35}$$

Discussion: Accurate identification of the correct subspace relies on perfect reconstruction of the K'-sparse vector $\mathbf{x}'$. In order to reconstruct $\mathbf{x}'$ perfectly, the similar sensing matrix should have low coherence and the amplitude of the equivalent noise vector $\mathbf{e}'$ should be small enough.

According to Proposition 6.3.4, we have the following:

$$\mathbf{e}' = \sum_{i=1}^{K} \tau_i \Delta \mathbf{b}_i + \mathbf{e} \tag{6.36}$$

Let B be the maximum difference between any column $\mathbf{b}_i$ and its corresponding condensed column $\boldsymbol{\gamma}_C^i$ within a similar column group $\boldsymbol{\gamma}_i$. We then have the following:

$$\|\mathbf{e}'\|_2 = \|\sum_{i=1}^{K} \tau_i(\Delta\mathbf{b}_i) + \mathbf{e}\|_2$$

$$\leq \sum_{i=1}^{K} |\tau_i| \cdot \|(\Delta\mathbf{b}_i)\|_2 + \|\mathbf{e}\|_2$$

$$\leq \sum_{i=1}^{K} |\tau_i| \cdot B + \|\mathbf{e}\|_2$$

$$\leq B \cdot \sum_{i=1}^{K} |\tau_i| + \|\mathbf{e}\|_2,$$

$$\leq BKA + \|\mathbf{e}\|_2, \tag{6.37}$$

where A is a positive lower bound on the magnitude of the nonzero entries of $\mathbf{x}'$. Equation 6.37 implies that a small B results in a small amplitude of the equivalent noise $\|\mathbf{e}'\|_2$, provided that A and $\|\mathbf{e}\|_2$ are fixed. Therefore, a column $\mathbf{b}_i$ should be close to its corresponding condensed column $\boldsymbol{\gamma}_C^i$ to obtain a small value of B.

6.3.5.1   Bound on the coherence of the similar sensing matrix

In order to guarantee a perfect reconstruction of the sparse vector, the minimum similarity distance between any two condensed columns of the similar sensing matrix $\boldsymbol{\Psi}$ should satisfy $TII \geq 1 - 1/(4K' - 1)$ (refer to Equation 6.21 in Theorem 6.3.3). Moreover, according to Proposition 6.3.2, the coherence of the similar sensing matrix is less than or equal to $1 - TH$; that is, $\mu(\boldsymbol{\Psi}) \leq 1 - TH$. Therefore, we can obtain the bound on the coherence of the similar sensing matrix; that is, $\mu(\boldsymbol{\Psi}) \leq 1/(4K' - 1)$.

6.3.5.2   Bound on the high coherence of the original sensing matrix

Our proposed GSSMP algorithm can distinguish columns that are very similar to each other (with very high coherence) using combinatorial search procedures. Therefore, in theory, the bound on the high coherence of the original sensing matrix is less than 1; that is, $\mu(\boldsymbol{\Phi}) < 1$. In practice, the typical value of the bound on the original high coherence can be set as high as 0.99.

6.4   Compressed sensing–based multiple target detection algorithm

In recent work related with compressed sensing–based STAP [17–20], the coherence of the sensing matrix is not low due to the high resolution of the DOA–Doppler plane, which does not guarantee a good reconstruction of the sparse vector with large probability. Consequently, direct estimation of the target amplitude may be unreliable using sparse representation when locating a moving target from the surrounding strong clutter. The weak element (target) is always submerged in prominent elements (clutter). In [20], only the prominent elements were extracted from the sparse radar scene, and an additional adaptive filter was used to suppress the clutter to identify the target.

However, in this chapter, we can obtain a reconstructed radar scene with high accuracy based on the test cell using our proposed GSSMP algorithm. Both the prominent elements (clutter) and the weak elements (multiple targets) can be identified accurately with our proposed algorithm. Consequently, it is not difficult to distinguish the weak elements from the prominent elements in the reconstructed radar scene. In the following, a simple algorithm is proposed to detect multiple targets in the reconstructed radar scene.

Algorithm 6.2: Compressed sensing–based multiple target detection algorithm

---

i. Use our proposed GSSMP algorithm to obtain the estimate of the original sparse vector ($\mathbf{x}_{\text{test}}$), denoted as $\hat{\mathbf{x}}_{\text{test}}$, based on the original sensing matrix $\boldsymbol{\Phi}$, and the measurement vector $\mathbf{y}_{\text{test}}$ (the snapshot from the test cell). The entries corresponding to noise in the original sparse vector $\mathbf{x}_{\text{test}}$ are set to 0 in $\hat{\mathbf{x}}_{\text{test}}$ according to our proposed algorithm. The nonzero elements of the reconstructed sparse vector $\hat{\mathbf{x}}_{\text{test}}$ contain the prominent elements (clutter) and the weak elements (the targets).

ii. Distinguish the weak elements from the prominent elements to detect multiple targets.

   For each element of the estimated sparse vector, $\hat{\mathbf{x}}_{\text{test}}(i), i = 1, \cdots, VD,$

       if $|\hat{\mathbf{x}}_{\text{test}}(i)| > T_{\text{clutter}}$

           $\hat{\mathbf{x}}_{\text{test}}(i)$ corresponds to clutter.

       else if $|\hat{\mathbf{x}}_{\text{test}}(i)| > 0$

           $\hat{\mathbf{x}}_{\text{test}}(i)$ corresponds to a target.

       end

Here, T_{clutter} is a threshold set to distinguish the targets from the clutter. It is assumed that the CSR is sufficiently large (>20 dB). The clutter scatters have much higher amplitudes than the targets. The entries of $\hat{\mathbf{x}}_{\text{test}}$ can be arranged in descending order according to their amplitudes as follows:

$$|\hat{\mathbf{x}}_{test}(1)| > |\hat{\mathbf{x}}_{test}(2)| > \cdots > |\hat{\mathbf{x}}_{test}(k)| \gg |\hat{\mathbf{x}}_{test}(k+1)| > \cdots > |\hat{\mathbf{x}}_{test}(NK \times VL)|, \quad (6.38)$$

where $\hat{\mathbf{x}}_{\text{test}}(1)$ denotes the entry that with the largest amplitude, with similar definitions for $|\hat{\mathbf{x}}_{test}(2)|, \cdots, |\hat{\mathbf{x}}_{test}(NK \times VL)|$. The sudden change of the amplitude between $|\hat{\mathbf{x}}_{test}(k)|$ and $|\hat{\mathbf{x}}_{test}(k+1)|$ is caused by the large difference between the amplitudes of the clutter scatter and target element. Moreover, the value of $T_{clutter}$ is set proportional to the smallest amplitude of the clutter scatters, as $T_{clutter} = \kappa |\hat{\mathbf{x}}_{test}(k)|$. Our proposed algorithm can identify multiple targets directly from the reconstructed radar scene (the DOA–Doppler plane), which reduces the computing complexity significantly.

6.5 Simulation results and analysis

In this section, our proposed GSSMP algorithm is compared with the D3SR method [20] in reconstructing sparse radar scenarios. Furthermore, our proposed algorithm is also compared with several classic STAP algorithms, for example, the sample matrix inversion (SMI) method [14], the angle-Doppler compensation (ADC) method [29], and the D3SR method, using the improvement factor loss (IF_{loss}), which is a common metric in evaluating the performance of the STAP methods.

An airborne, side-looking radar system consisting of half-wavelength spaced ULAs is considered in this section. The radar system was comprised of 16 arrays and the data were organized in coherence processing intervals (CPI) of 16 pulses. The clutter was uniformly distributed between the directions of $-80° \sim 70°$ and was contained in both the training cells and the test cell. The DOA–Doppler plane was divided into 200×180 square grids, where the x-axis was for the DOA angle and the y-axis for the Doppler frequency. The initial DOA angle (θ_0), the DOA angle interval ($\Delta\theta$), the initial Doppler frequency (f_{d_0}), and the Doppler frequency interval (Δf_d) were $-90°$, $1°$, -400 Hz, and 4 Hz, respectively.

First, our proposed algorithm is compared with the D3SR method in reconstructing sparse radar scenarios in two examples. The environment setting of the first example is shown in Figure 6.2, where two pairs of targets (represented by four blue dots) were placed near the clutter ridge (represented by a red ridge). One pair of targets had the same Doppler frequency (-160 Hz), and their DOA angles were $-25°$ and $-20°$, respectively. The other pair of targets had the same DOA angle ($16°$), and their Doppler frequencies were 68 Hz and 56 Hz, respectively. The color represents the amplitude of the scattering coefficients (in dB). The clutter is represented by the color red, with a range of $[-5, 0]$ dB from the side color bar. The target point is blue, with a range of $[-30, -25]$ dB. Therefore, the simulation setting had a high CSR ranging from 20 to 30 dB.

The simulation results shown in Figures 6.2 through 6.4. Figure 6.2 gives the actual sparse radar scene. Figures 6.3 and 6.4 provide the sparse radar scenes reconstructed by the D3SR method and our proposed method, respectively. From Figure 6.3, it is evident that the D3SR method incorrectly recovered the elements near the clutter ridge in addition to the clutter scatters in the DOA–Doppler plane, where the targets were submerged in the incorrectly recovered nearby elements. This is due to the fact that the columns in the original sensing matrix corresponding to the elements near the clutter were highly coherent (similar) with the columns corresponding to the clutter scatters. The D3SR method cannot distinguish the highly coherent columns and assigned large values to their corresponding elements. However, our proposed method was able to estimate the weak elements (targets) as well as the prominent elements (clutter) accurately, and it distinguished the targets from clutter successfully in the DOA–Doppler plane (Figure 6.4). This verifies that our proposed similar sensing matrix pursuit algorithm can cope with the deterministic sensing matrix with high coherence efficiently.

The second example is used to compare the reconstruction performance of the two algorithms (the D3SR algorithm and our proposed algorithm) in estimating targets with

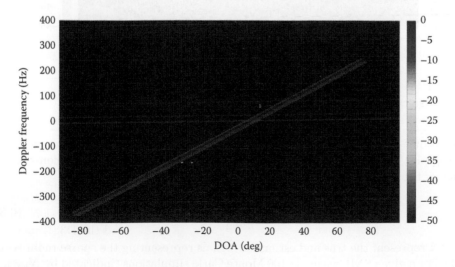

FIGURE 6.2
True sparse radar scene with four targets.

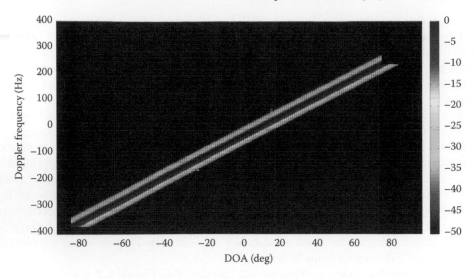

FIGURE 6.3
Estimated sparse radar scene with four targets using the direct data domain method using sparse representation (D3SR).

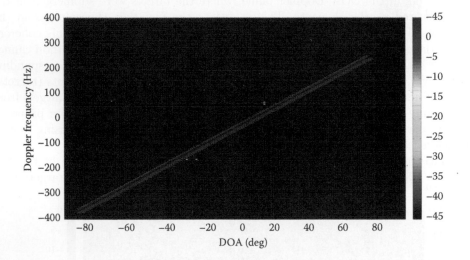

FIGURE 6.4
Estimated sparse radar scene with four targets using our proposed algorithm.

different positions under different noise levels. Ten targets were distributed randomly in the DOA–Doppler plane. The reconstruction error adopted to evaluate the reconstruction performance of the two algorithms was defined as follows:

$$\chi = \frac{||\hat{\mathbf{x}} - \mathbf{x}||_2^2}{||\mathbf{x}||_2^2}, \tag{6.39}$$

where $\mathbf{x}$ and $\hat{\mathbf{x}}$ represent the true and estimated signals representing the sparse radar scene, respectively. For a given SNR, we made 100 Monte Carlo simulations (indicated by N_{MC}). In each simulation, the locations of 10 targets were randomly distributed in the DOA–Doppler plane. Figures 6.5 through 6.7 show the simulation results from one simulation with an SNR

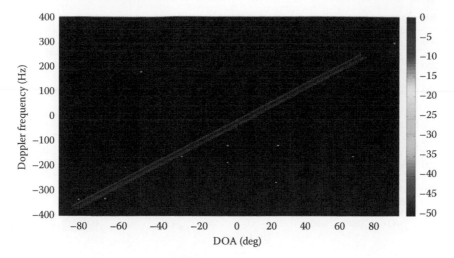

FIGURE 6.5
True sparse radar scene with 10 randomly distributed targets.

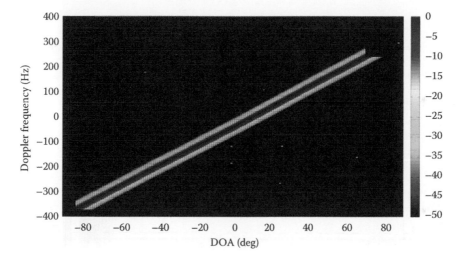

FIGURE 6.6
Estimated sparse radar scene with 10 randomly distributed targets using the D3SR algorithm.

of 20 dB. In Figure 6.6, the targets near the clutter ridge were submerged in the incorrectly reconstructed nearby elements by the D3SR algorithm, whereas our proposed algorithm was able to estimate the weak elements (targets) as well as the prominent elements (clutter) accurately (Figure 6.7). Figure 6.8 indicates the variation of average reconstruction error $(\bar{\chi} = \sum_{i=1}^{N_{MC}} \chi^i / N_{MC})$ with SNR varying from 0 dB to 30 dB, which shows that our proposed algorithm is resilient to the measurement noise. Our proposed algorithm was able to detect (< 0.1) 10 targets perfectly in the presence of measurement noise when the SNR was above 20 dB. On the contrary, a large reconstruction error (> 2) occurred for the D3SR algorithm, even when the SNR exceeded 20 dB.

Moreover, we compared our proposed algorithm with several classic STAP algorithms, for example, the SMI, ADC, and D3SR algorithms, using the improvement factor loss, which

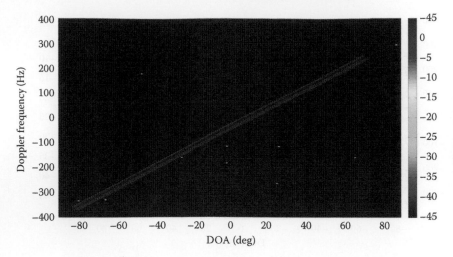

FIGURE 6.7
Estimated sparse radar scene with 10 randomly distributed targets using our proposed algorithm.

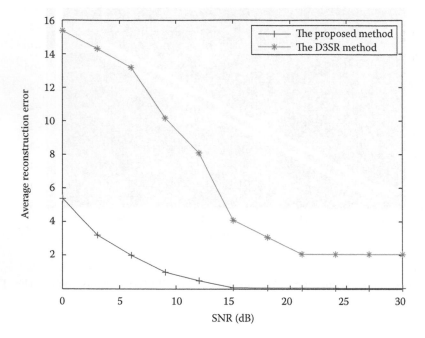

FIGURE 6.8
Reconstruction performance (average reconstruction error) of our proposed algorithm and the D3SR algorithm with varying signal-to-noise ratios.

is defined as follows [30]:

$$IF_{Loss} = \frac{SCR_{out}/SCR_{in}}{(SCR_{out}/SCR_{in})_{opt}}, \tag{6.40}$$

where SCR_{out} and SCR_{in} denote the output and signal-to-clutter ratio (SCR), respectively. A classic simulation setup in STAP simulations was adopted, where a moving target had

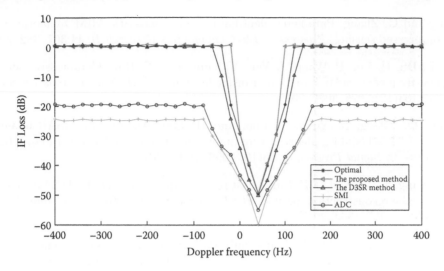

FIGURE 6.9
The IF$_{\text{Loss}}$ performance of different space–time adaptive processing methods at a direction-of-arrival angle of 20° as a function of the Doppler frequency.

a DOA angle of 20°. Different output SCRs were then considered with varying Doppler frequencies. Figure 6.9 gives the IF$_{\text{Loss}}$ performance of different STAP algorithms. Because the SCR improvement was mostly achieved in the subspace orthogonal to the clutter, all the STAP methods (the SMI, ADC, and D3SR algorithms) suffered considerable degradation near the clutter notch, regardless of the size of the total space (i.e., system degree of freedom [DOF]). However, our proposed similar sensing matrix pursuit algorithm detects the target directly based on the reconstructed radar scene, and it achieved comparable performance with the optimal filter.

6.6 Conclusion

In this chapter, a novel compressed sensing–based algorithm, the GSSMP, was proposed to detect unknown moving targets in high clutter directly in the test cell, which largely reduces the computing complexity. From simulation results, we observed that our proposed algorithm can estimate the weak elements (targets) as well as the strong elements (clutter) accurately, and it can distinguish the targets from clutter successfully in the sparse radar scene (the DOA–Doppler plane).

References

[1] R. Baraniuk and P. Steeghs, Compressive radar imaging, *Proceedings of the Radar Conference*, pp. 128–133, 2007.

[2] Z. Liu, X. Wei, and X. Li, Adaptive clutter suppression for airborne random pulse repetition interval radar based on compressed sensing, *Progress in Electromagnetics Research*, 128:291–311, 2012.

[3] M. Yang and G. Zhang, Parameter identifiability of monostatic MIMO chaotic radar using compressed sensing, *Progress in Electromagnetics Research B*, 44:367–382, 2012.

[4] D. Xu, L. Du, H. Liu, P. Wang, J. Yan, Y. Cong, and X. Han, Compressive sensing of stepped-frequency radar based on transfer learning, *IEEE Transactions on Signal Processing*, 63(12):3076–3087, 2015.

[5] Y. Chen, Q. Zhang, N. Yuan, Y. Luo, and H. Lou, An adaptive ISAR-imaging-considered task scheduling algorithm for multi-function phased array radars, *IEEE Transactions on Signal Processing*, 63(19):5096–5110, 2015.

[6] Z. Li, J. Wang, J. Wu, and Q. H. Liu, A fast radial scanned near-field 3-D SAR imaging system and the reconstruction method, *IEEE Transactions on Geoscience and Remote Sensing*, 53(3):1355–1363, 2015.

[7] X. Y. Pan, W. Wang, and G. Y. Wang, Sub-Nyquist sampling jamming against ISAR with CS-based HRRP reconstruction, *IEEE Sensors Journal*, 16(6):1597–1602, 2016.

[8] C. Gangodagamage, E. Foufoula-Georgiou, S. P. Brumby, R. Chartrand, A. Koltunov, D. Liu, M. Cai, and S. L. Ustin, Wavelet-compressed representation of landscapes for hydrologic and geomorphologic applications, *IEEE Geoscience and Remote Sensing Letters*, 13(4):480–484, 2016.

[9] S. Tomei, A. Bacci, E. Giusti, M. Martorella, and F. Berizzi, Compressive sensing-based inverse synthetic radar imaging from incomplete data, *IET Radar, Sonar & Navigation*, 10(2):386–397, 2016.

[10] J. Ward, *Space–Time Adaptive Processing for Airborne Radar*, Massachusetts Institute of Technology Lincoln Laboratory, Lexington, KY, 1994.

[11] M. L. Tounsi, R. Touhami, A. Khodja, and M. C. E. Yagoub, Analysis of the mixed coupling in bilateral microwave circuits including anisotropy for MICS and MMICS applications, *Progress in Electromagnetics Research*, 62:281–315, 2006.

[12] S. Asadi and M. C. E. Yagoub, Efficient time-domain noise modeling approach for millimeter-wave FETS, *Progress in Electromagnetics Research*, 107:129–146, 2010.

[13] M. A. Habib, A. Bostani, A. Djaiz, M. Nedil, M. C. E. Yagoub, and T. A. Denidni, Ultra wideband CPW-FED aperture antenna with WLAN band rejection, *Progress in Electromagnetics Research*, 106:17–31, 2010.

[14] J. Guerci, *Space–Time Adaptive Processing for Radar*, Boston: Artech House, 2003.

[15] R. Klemm, *Applications of Space–Time Adaptive Processing*, London: Inspec/IEE, 2004.

[16] W. Melvin, A STAP overview, *IEEE Aerospace and Electronic Systems Magazine*, 19(1):19–35, 2004.

[17] H. Zhang, G. Li, and H. Meng, A class of novel STAP algorithms using sparse recovery technique, arXiv: 0904.1313 [cs.IT], 2009.

[18] I. W. Selesnick, S. U. Pillai, K. Y. Li, and B. Himed, Angle-Doppler processing using sparse regularization, *Proceedings of the IEEE International Conference on Acoustics, Speech and Signal Processing*, pp. 2750–2753, 2010.

[19] J. Parker and L. Potter, A Bayesian perspective on sparse regularization for STAP post-processing, *IEEE Radar Conference*, pp. 1471–1475, 2010.

[20] K. Sun, H. Meng, and Y. Wang, Direct data domain STAP using sparse representation of clutter spectrum, *Signal Processing*, 91(9):2222–2236, 2011.

[21] D. L. Donoho and M. Elad, Optimally sparse representation in general (nonorthogonal) dictionaries via l^1 minimization, *Proceedings of the National Academy of Sciences of the United States of America*, 100(5):2197–2202, 2003.

[22] R. Gribonval and M. Nielsen, Sparse representations in unions of bases, *IEEE Transactions on Information Theory*, 49(12):3320–3325, 2003.

[23] J. A. Tropp, Greed is good: Algorithmic results for sparse approximation, *IEEE Transactions on Information Theory*, 50(10):2231–2242, 2004.

[24] D. Donoho, Compressed sensing, *IEEE Transactions on Information Theory*, 52(4):1289–1306, 2006.

[25] Z. Ben-Haim, Y. C. Eldar, and M. Elad, Coherence-based performance guarantees for estimating a sparse vector under random noise, *IEEE Transactions on Signal Processing*, 58(10):5030–5043, 2010.

[26] M. F. Duarte and Y. C. Eldar, Structured compressed sensing: From theory to applications, *IEEE Transactions on Signal Processing*, 59(9):4053–4085, 2011.

[27] R. A. Horn and C. R. Johnson, *Topics in Matrix Analysis*, Cambridge University Press, Cambridge, UK, 1991.

[28] W. Dai and O. Milenkovic, Subspace pursuit for compressive sensing signal reconstruction, *IEEE Transactions on Information Theory*, 55(5):2230–2249, 2009.

[29] B. Himed, Y. Zhang, and A. Hajjari, STAP with angle-Doppler compensation for bistatic airborne radars, *Proceedings of the IEEE National Radar Conference*, pp. 22–25, Long Beach, CA, 2002.

[30] R. Klemm, *Principles of Space–Time Adaptive Processing*, London: IEE Press, 2002.

7

Bayesian Sparse Estimation of Radar Targets in the Compressed Sensing Framework

Stéphanie Bidon, Marie Lasserre, and François Le Chevalier

CONTENTS

7.1   Introduction

Pulse-Doppler radars classically transmit a train of pulses with a narrow bandwidth at a constant pulse repetition frequency (PRF) to detect moving targets. By design, they are subject to the well-known problem of range and/or velocity ambiguities. Indeed, the ambiguous velocity v_a and range ambiguity R_a are linked through

$$R_a v_a = \frac{c\lambda_c}{4} \quad \text{with} \quad v_a = \frac{\lambda_c F_r}{2} \quad \text{and} \quad R_a = \frac{cT_r}{2}, \qquad (7.1)$$

where λ_c is the wavelength, c is the speed of light, and $F_r = 1/T_r$ is the PRF. Equation 7.1 shows that (1) the unambiguous range–velocity coverage of a pulsed radar is independent of the PRF and (2) the ambiguous range and velocity depend on the PRF in an opposite way. Consequently, decreasing aliasing in one dimension by changing the PRF will inevitably increase it in the alternate. For instance, in a low PRF mode, there are no ambiguities in the range domain but many in the velocity domain. It prevents not only from estimating unambiguously the velocity of exo-clutter targets but also from detecting targets located in the so-called blind velocities which are the clutter ambiguous velocities. Hence, in practice for pulsed radars, the Nyquist sampling criterion cannot be satisfied in both dimensions.

To obtain a nonambiguous mode in real radar systems, a series of bursts having different PRFs are usually transmitted successively [56]. Combining adequately the received signal, radar ambiguities can be removed and targets can be detected in the blind velocities. Most of the techniques described in the literature rely on the Chinese remainder theorem [67]. Nonetheless, using multiple PRFs to remove ambiguities usually consumes large amounts of radar timeline and entails several practical difficulties (e.g., ghosts might appear with not enough PRFs [48]).

A promising alternative is the use of a wideband waveform with a low PRF. Wideband waveforms have several merits like discretion due to their spectral spreading and an increased discrimination power due to their high range resolution property [44,47]. Indeed, increasing the instantaneous bandwidth B improves the radar range resolution since the size of a range cell is inversely proportional to it

$$\delta_R = \frac{c}{2B}.$$

For instance, a range resolution of 15 cm is obtained with a 1 GHz bandwidth. With such a waveform, fast moving targets migrate from range gate to range gate during the coherent processing interval (CPI). Conventional detection algorithms are not designed to handle range walk so that augmented techniques are required [32,35,53]. Among them, the most famous is certainly the Keystone transform that allows the received signal to be summed coherently while compensating for range walk [53]. However, these techniques mostly consider the range migration as a nuisance. On the contrary, unlike Doppler phase measurement, range migration can give unambiguous velocity measurement as illustrated in Figure 7.1 [42]. A nonambiguous radar mode can thus be obtained where detection in the blind velocities becomes possible. Accordingly, several estimation algorithms of the target scene have been proposed but endure some limitations [17]. Particularly, the main challenge lies in the strong sidelobes of the wideband ambiguity function that are located at each multiple of the ambiguous velocity apart from the main peak. To enable a nonambiguous mode, they need to be carefully dealt with.

In this chapter, while assuming reasonably that the signal of interest (i.e., the target echoes) is sparse or at least compressible in a given dictionary, the compressed sensing (CS) paradigm is invoked to allow the target scene to be unambiguously reconstructed.

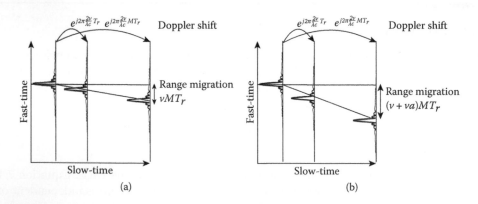

FIGURE 7.1
Range migration during the CPI. Doppler phase measurement is identical for scatterer with velocity v (a) and $v + v_a$ (b), whereas range walk is more significant for the fastest scatterer.

CS is a recent part of signal processing stating that sparse signals can be recovered with measurements sampled at a rate lower than that of Nyquist [13,19]. It is a two part problem in which a sampling scheme needs first to be designed and then a sparse recovery algorithm is implemented. CS has been lately of interest in many domains of application including radar imaging and detection [3,6,21,24,27,55]. Numerous SSR techniques have been developed accordingly such as ℓ_1-penalized least squares (e.g., Basis Pursuit [68], Lasso [64]), greedy algorithms (e.g., CLEAN method [29], OMP [66]), and hard- and soft-thresholding techniques [12]. We refer the reader to the first chapter of this book for a more comprehensive overview of CS and SSR techniques.

Herein, considering a wideband waveform with low PRF, the sampling scheme is not questioned. Nonetheless, as explained earlier, a low PRF leads to sub-Nyquist observations in the velocity domain. Using range migration, SSR methods offer then the possibility to recover the target scene unambiguously, namely without velocity sidelobes. In what follows, a Bayesian[*] approach is favored to solve the sparse estimation problem. Specifically, several hierarchical Bayesian models are presented where sparsity is induced in the target amplitude vector via its prior probability density function (pdf). Starting from a basic model describing on-grid targets in a white noise background, we demonstrate how this model can be augmented to include more realistic phenomena such as colored noise (to deal with the diffuse clutter component), off-grid targets, and target scene with high dynamic range. To obtain optimal estimators, algorithms based on Monte-Carlo Markov chains (MCMC) are implemented and discussed for each proposed hierarchical Bayes model. Though computationally intensive, the MCMC approach has several merits, a significant one being to obtain empirical posterior pdfs which entail much more information than a single-point estimate. Performance of the algorithms is shown successively on synthetic, semi-, and full-experimental radar data and compared with a more conventional SSR technique, namely an ℓ_1-penalized least-squares approach. A short discussion finally highlights some remaining issues to be tackled before implementing the proposed CS approach in practical/real-time radar scenarios.

7.2   Signal model

7.2.1   Expression of the received signal

Let us first describe the signature of a single point target. For conventional narrowband radars, a target remains confined in its range resolution cell during the whole CPI. Its signature is simply expressed in the slow-time domain by a cisoid [56], that is, for $m = 0, \dots, M - 1$[†]

$$[\boldsymbol{a}]_m = \exp\left\{ j2\pi \frac{2vF_c}{c} T_r m \right\} \qquad \text{(1D-cisoid)}, \qquad (7.2)$$

where M is the number of pulses, m is the pulse index, and $F_c = c/\lambda_c$ is the carrier frequency. In Equation 7.2, one recognizes the well-known Doppler frequency

$$f_d = \frac{2v}{\lambda_c} T_r \qquad \text{(normalized frequency)}.$$

---

[*]Note that deterministic approaches are currently developed specifically for the same purpose but will not be discussed here (e.g., [54]).

[†]In radar, slow-time refers to the pulse index, whereas fast-time refers to the range dimension.

For wideband radars, a moving target migrates in range and can travel several cells during the CPI as illustrated in Figure 7.1. The signature in Equation 7.2 fails at describing this phenomenon. Instead, one needs to consider a low-range resolution segment of K range cells that contain the migrating target [35,43]. To express the target signature, it is easier then to transpose the signal in the fast-frequency domain by applying a K-point fast Fourier transform (FFT) at each pulse return (cf. Figure 7.2). The initial K range cells are thus transformed into K subbands that cover the whole bandwidth B. The target signature in the fast-frequency/slow-time domain can then be expressed as [17]

$$[\boldsymbol{A}]_{k,m} = \exp\left\{j2\pi\left(-\tau_0\frac{B}{K}k + \frac{2v}{\lambda_c}T_r m\right)\right\} \qquad \text{(2D-cisoid)}$$

$$\times \exp\left\{j2\pi\frac{2v}{c}\frac{B}{K}T_r k m\right\} \qquad \text{(cross-coupling terms)}, \qquad (7.3)$$

where $k = 0,\ldots,K-1$ is the subband index, and τ_0 is the initial round-trip delay of the target. To obtain (Equation 7.3), it was assumed that range migration is significant during the whole CPI but remains imperceptible in the pulse duration denoted by T_p, viz

$$vMT_r \gg \delta_R \quad \text{and} \quad vT_p \ll \delta_R.$$

In the signature of Equation 7.3, two main terms can be distinguished. The first term is a conventional bidimensional cisoid that models the target signature for narrowband radars. It is the counterpart of Equation 7.2 expressed in the fast-frequency/slow-time domain where the additional range frequency appears

$$f_r = \frac{\tau_0 B}{K} \qquad \text{(normalized frequency)}.$$

The second term is a complex exponential that entails cross-coupling terms between the range and velocity dimensions. It models the range migration *per se*.

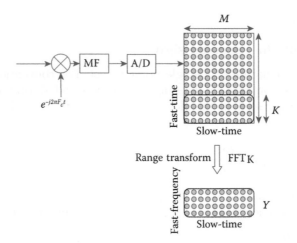

FIGURE 7.2
Preprocessing for wideband radars: conventional preprocessing (demodulation, range-matched filtering, sampling) followed by K-point FFT applied to a selection of K range cells.

In practice, the received signal may entail several scatterer echos plus a noise component. The whole signal can then be expressed in the fast-frequency/slow-time domain by a $K \times M$ matrix

$$Y = \sum_{\ell=1}^{L} \alpha_\ell A_\ell + N,$$

where L is the number of scatterers within the K selected range cells; α_ℓ and A_ℓ are the complex amplitude and signature of the ℓth scatterer, respectively; and N is the noise matrix modeling at least the internal noise of the receiver. Note that given the high-range resolution of the radar, a target can actually be extended in range. Therefore, according to the radar scenario, some of the L scatterers can be part of the same target. In the remainder of the chapter, we use interchangeably the row-vectorized notation

$$y = \sum_{\ell=1}^{L} \alpha_\ell a_\ell + n, \tag{7.4}$$

where each KM-length vector involved can be clearly identified with its matrix counterpart.

7.2.2 Wideband ambiguity function

The radar ambiguity function is an important analysis tool that allows to characterize a waveform jointly with the filter matched to the received signal [56]. Several wideband ambiguity functions have been studied in the literature (e.g., [1,45,60]). Here, given the signal model, it is obtained by the following coherent sum (or equivalently by the matched filter in white noise)

$$a^H y, \tag{7.5}$$

when there is only one scatterer with signature a_ℓ in the signal y, that is, $y = a_\ell$. The resulting wideband ambiguity function has been described in [43] and is represented in Figure 7.3a. The scatterer response is characterized by a main peak at its unambiguous location and velocity sidelobes that are arranged in a typical "butterfly" shape. The pth

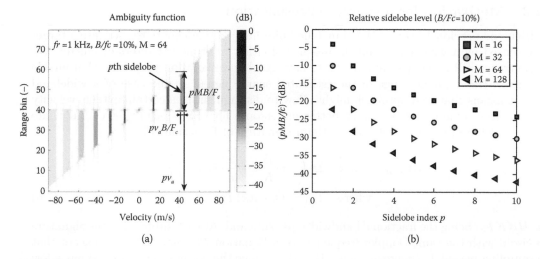

FIGURE 7.3
Wideband ambiguity function. (a) View in range–velocity domain. (b) Velocity sidelobe level as a function of pulse number M.

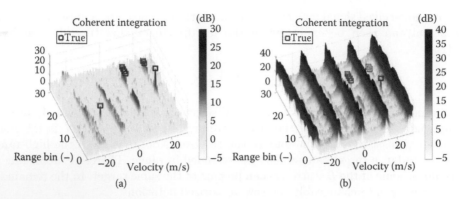

(a)   (b)

FIGURE 7.4
Coherent integration of Equation 7.5 for wideband radar signal. Square markers indicate true target locations. (a) Multiple scatterers in white noise. (b) Multiple scatterers in white noise plus clutter.

velocity sidelobe ($p \in \mathbb{Z}^*$) can be approximately described by a cube whose features are described in Figure 7.3. Interestingly, its velocity and range spreads as well as its relative sidelobe level do not depend on the target's range and velocity. Typical order of values of the sidelobe level are given in Figure 7.3b.

In practice, these sidelobes are high enough to limit the performance of conventional linear processing techniques based on matched filtering. Two outputs are depicted in Figure 7.4 to illustrate that point. First, a scenario with multiple scatterers in white noise is considered in Figure 7.4a. Clearly, it is hard to distinguish sidelobes of strong scatterers from main peaks of low scatterers. Considering then a scenario with an additionnal clutter component at each range gate in Figure 7.4b, one can notice that clutter sidelobes tend to add up at the location of the usual blind velocities so that dectection remains there infeasible (even if range walk is compensated for in the matched filtering operation).

7.2.3   Motivation for a sparse representation

As discussed previously, conventional matched filtering is not adequate to estimate a target scene sensed by a wideband radar in presence of either multiple scatterers or clutter. On the contrary, SSR techniques are known to have good deconvolution properties that enable to increase the dynamic range of the estimated sparse signal. In the case of a wideband radar, it means that SSR approach might allow to distinguish, to a certain extent, velocity sidelobes from true target peaks. To better understand how SSR techniques can alleviate velocity sidelobes, let us rewrite the scatterer signature of Equation 7.3 as

$$[\mathbf{A}]_{k,m} = \exp\left\{ j\, 2\pi \left(-\tau_0 \frac{B}{K}k + \frac{2v}{\lambda_c}T_{r,k}m \right) \right\}$$
$$\text{where} \quad T_{r,k} = (1 + \mu k)\, T_r, \tag{7.6}$$

$\mu = B/(KF_c)$ being the fractional bandwidth per subband. At each subband k, the signature is a cisoid with the same Doppler frequency as in Equation 7.3, that is, $2v/\lambda_c$, except that the sampling period $T_{r,k}$ increases linearly with k. Since the waveform considered has a low PRF, aliasing already occurs in the slow-time domain at the first subband $k = 0$ and then increases with k. This phenomenon is illustrated in Figure 7.5. The greater the subband index, the slower the variation of the sine function actually observed with the samples.

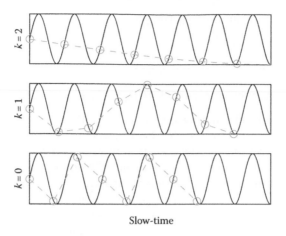

Slow-time

FIGURE 7.5

Representation of aliasing in the slow-time dimension for a wideband radar signal. Plain line is the analog version of a cisoid with frequency $2v/\lambda_c$. Circle markers represent the observed data with sampling frequency $T_{r,k}$.

Therefore, the frequency diverse waveform allows the same signal to be observed several times with a different aliasing factor. If the frequency diversity is sufficient (or equivalently if the range walk phenomenon is significant), an SSR technique might then produce a target scene estimate that lacks the sidelobe structure of conventional matched filtering output.

To formulate the target scene estimation problem as a sparse recovery problem, the signal model in Equation 7.4 is recast as

$$y = Hx + n, \tag{7.7}$$

where H is the sparsifying dictionary that stems from a discretization of the range and velocity dimensions and x is the target amplitude vector associated with it. In practice, one selects an analysis grid that covers the following range-bin/velocity domain

$$[0, K) \times \left[-n_{va}\frac{v_a}{2}, n_{va}\frac{v_a}{2}\right),$$

where n_{va} is the unfolding factor in velocity. This bidimensional domain is discretized respectively in $\bar{K} = n_{zp}^r K$ and $\bar{M} = n_{zp}^v n_{va} M$ points of reconstruction, where $n_{zp}^r, n_{zp}^v \in \mathbb{N}^* \times \mathbb{N}^*$ are the zero-padding factors in range and velocity, respectively (see Figure 7.11). The dictionary H is thus of size $KM \times \bar{K}\bar{M}$. By design, the $\bar{i}$th column of H, denoted by $h_{\bar{i}}$, corresponds to the (normalized) steering vector that points in the range–velocity direction of the $\bar{i}$th point of the analysis grid, that is,

$$[h_{\bar{i}}]_i = \frac{1}{\sqrt{KM}} \exp\{-j2\pi\frac{\bar{k}}{\bar{K}}k\} \exp\{j2\pi n_{va}\frac{\bar{\bar{m}}}{\bar{M}}m(1+\mu k)\}, \tag{7.8}$$

where $i = m + kM$, $\bar{i} = \bar{m} + \bar{k}\bar{M}$, and $\bar{\bar{m}}$ is the modulus of $\bar{m}$ centered around zero (e.g., if $\bar{M}$ is even, $\bar{\bar{m}} = \bar{m}$ if $\bar{m} < M/2$ and $\bar{\bar{m}} = \bar{m} - \bar{M}/2$ otherwise). Indeed, unlike in a conventional Doppler processing, one needs to consider a signed velocity index in the target signature of Equation 7.8 to ensure a correct description of the range walk by the cross-coupling terms.

The sparsifying dictionary being described, the remaining SSR problem consists in estimating the target amplitude vector x. However, given the velocity unfolding (i.e., $n_{va} > 1$), the matrix H has in practice more columns than rows, meaning that the problem in Equation 7.7 is ill-posed. To regularize it, a conventional approach consists in explicitly enforcing sparsity in the estimation problem. To do so, a full Bayesian approach is favored in this chapter where sparsity is induced by the prior pdf assigned to the target amplitude vector $\boldsymbol{x}$.

7.3 Bayesian sparse estimation of migrating targets

This section describes a hierarchical Bayesian model detailed in [10] that provides the basis for augmented models reported later in this chapter and summarized in Figure 7.6. First, a simple scenario is considered where scatterers are embedded in white noise and are on the grid, that is, located precisely at a point of the analysis grid. To enrich the data model, prior information is added statistically to the observations $\boldsymbol{y}$ and concerns the sparse nature of the signal as well as the target and noise power levels. Following the common practice, conjugate priors are chosen to facilitate subsequently the computation of Bayesian estimators [57]. Indeed, posteriors are then in the same family as that of the priors. Nonetheless, their parameters are also carefully selected to ensure some physical sense. The hierarchical Bayesian model is represented in Figure 7.6a and detailed hereafter.

7.3.1 Hierarchical Bayesian model

Usually, the first stage of a hierarchical Bayes model describes the likelihood function of the observations. Here, the noise component $\boldsymbol{n}$ is supposed to be a white Gaussian vector with power σ^2, which is denoted by $\boldsymbol{n} \sim \mathcal{CN}\left(\mathbf{0}, \sigma^2 \boldsymbol{I}_{KM}\right)$, where $\boldsymbol{I}_\xi$ is the $\xi \times \xi$ identity matrix.

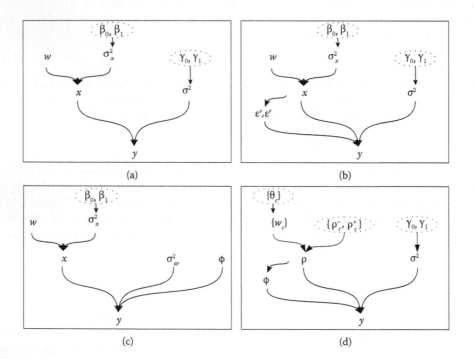

FIGURE 7.6
Graphs of the proposed hierarchical Bayesian models for the estimation of migrating targets. Parameters in dotted circles are chosen by the radar operator. Arrows represent statistical dependence. (a) On-grid targets in white noise. (b) Off-grid targets in white noise. (c) On-grid targets in AR noise. (d) On-grid targets in white noise with high dynamic range.

The likelihood function can thus be written

$$f(\boldsymbol{y}|\boldsymbol{x}, \sigma^2) \propto \frac{1}{\sigma^{2KM}} \exp\left\{-\frac{\|\boldsymbol{y} - \boldsymbol{Hx}\|_2^2}{\sigma^2}\right\}, \tag{7.9}$$

where $\propto$ means proportional to. In practice, a white noise model embodies two types of scenarios: (1) clutter* is simply not present in the raw data (e.g., in look-up mode) or has been removed beforehand in a prewhitening operation and (2) clutter only consists of a discrete component that can be entirely represented by a finite number of scatterers at zero velocity in $\boldsymbol{x}$. This point will be further discussed by comparing performance of a white noise model with that of a colored noise on experimental data.

In the likelihood function of Equation 7.9, two parameters $\boldsymbol{x}$ and σ^2 are unknown. As done conventionnally in a Bayesian framework, they are considered as random variables so that a second stage is added to the hierarchical model. Let us first discuss the prior pdf of the noise power σ^2. An inverse-gamma distribution is chosen since it is conjugate to the likelihood function of Equation 7.9. The pdf is denoted by $\sigma^2|\gamma_0, \gamma_1 \sim \mathcal{IG}(\gamma_0, \gamma_1)$ and expressed by

$$\pi(\sigma^2|\gamma_0, \gamma_1) \propto \frac{e^{-\gamma_1/\sigma^2}}{(\sigma^2)^{\gamma_0+1}} \mathbb{I}_{\mathbb{R}^+}(\sigma^2), \tag{7.10}$$

where $\mathbb{I}_A()$ is the indicator function of the set A (i.e., $\mathbb{I}_A(x) = 1$ if $x \in A$ and $\mathbb{I}_A(x) = 0$ otherwise) and (γ_0, γ_1) are the so-called shape and scale parameters, respectively. Both hyperparameters (γ_0, γ_1) can be chosen by the radar operator so that the shape of the prior in Equation 7.10 conveys, as far as possible, the information known about σ^2 [25]. It may be easier to tune them as a function of the mean m_{σ^2} and the variance v_{σ^2} of Equation 7.10, if they exist,

$$\gamma_0 = \frac{m_{\sigma^2}^2}{v_{\sigma^2}} + 2 \quad \text{and} \quad \gamma_1 = m_{\sigma^2}\left(\frac{m_{\sigma^2}^2}{v_{\sigma^2}} + 1\right)$$

since they have a more comprehensible sense in practice. In a radar framework, thermal noise is mostly due to the receiver's internal noise so that its value is generally well known [59,61]. Using an informative prior seems reasonable; nonetheless, a noninformative prior, obtained with $(\gamma_0, \gamma_1) \rightarrow (0,0)$ [25], has also been tested successfully [9]. Actually, it has been observed that the tuning of the σ^2-hyperparameters is not too critical.

Choosing the prior of $\boldsymbol{x}$ is a delicate step in the design of the hierarchical Bayesian model. Indeed, it has to induce sparsity in the vector $\boldsymbol{x}$, thereby regularizing the estimation problem of Equation 7.7. Additionally, it must ensure sufficient mathematical tractability to allow Bayesian estimators to be derived. Recalling that the data are complex valued and given the likelihood function of Equation 7.9, a complex Bernoulli–Gaussian pdf offers a good compromise [14,16,37]. Hence, the elements of $\boldsymbol{x}$, denoted by $x_{\bar{i}} = [\boldsymbol{x}]_{\bar{i}}$ for $\bar{i} = 0, \dots, \bar{K}\bar{M}-1$, are assumed *a priori* independent and identically distributed (iid), viz

$$\pi(x_{\bar{i}}|w, \sigma_x^2) = (1-w)\,\delta(|x_{\bar{i}}|) + w\,\frac{1}{\pi\sigma_x^2}\exp\left\{-\frac{|x_{\bar{i}}|^2}{\sigma_x^2}\right\}, \tag{7.11}$$

where the hyperparameters w, σ_x^2 are subject to $w \in [0,1]$ and $\sigma_x^2 > 0$. The pdf is denoted by $x_{\bar{i}}|w, \sigma_x^2 \sim \mathcal{BerCN}(w, 0, \sigma_x^2)$. Hence, with a probability $(1-w)$ the $\bar{i}$th range–velocity bin is empty, and with a probability w it is occupied by a scatterer whose amplitude is complex Gaussian distributed with zero mean and variance σ_x^2. Using the iid hypothesis, the prior pdf of $\boldsymbol{x}$ is

$$\pi(\boldsymbol{x}|w, \sigma_x^2) = (1-w)^{n_0}\left(\frac{w}{\pi\sigma_x^2}\right)^{n_1}\exp\left\{-\frac{\|\boldsymbol{x}\|_2^2}{\sigma_x^2}\right\}\prod_{\bar{i}/x_{\bar{i}}=0}\delta(|x_{\bar{i}}|),$$

---

*We only consider ground clutter in this chapter.

where $\delta()$ is the Delta Dirac function and n_1 is the number of nonzero elements in $\boldsymbol{x}$

$$n_1 = \|\boldsymbol{x}\|_0 \quad \text{and} \quad n_0 = \bar{K}\bar{M} - \|\boldsymbol{x}\|_0.$$

It is worth noticing that the Bernoulli–Gaussian pdf in Equation 7.11 allows the sparsity level and the target power to be monitored separately by the hyperparameters w and σ_x^2, respectively. Without the atom at zero, the lone hyperparameter σ_x^2 would have to monitor both criteria though antagonistic for high-power targets. As a matter of fact, mixed-type distributions having an atom at zero have been used a number of times to enforce sparsity in signal recovery problems [18,50].

Finally, in a hierarchical Bayesian model, the unknown hyperparameters can be in turn considered as random variables. Hyperpriors are then assigned to the former, thereby defining an additional level to the model. At this stage, two hyperparameters are unknown, namely w and σ_x^2. Firstly, a uniform distribution is chosen to describe statiscally w, that is,

$$\pi(w) = \mathbb{I}_{[0,1]}(w). \tag{7.12}$$

The pdf of Equation 7.12 is denoted by $w \sim \mathcal{U}_{[0,1]}$ and means that *a priori* the sparsity level is unknown. Note that a more informative prior could be chosen when a certain number of targets is expected. Secondly, an inverse-gamma pdf is chosen for σ_x^2 since it is conjugate to the prior Equation 7.11. It is denoted by $\sigma_x^2|\beta_0,\beta_1 \sim \mathcal{IG}(\beta_0,\beta_1)$ and expressed by

$$\pi(\sigma_x^2|\beta_0,\beta_1) \propto \frac{e^{-\beta_1/\sigma_x^2}}{\sigma_x^{2\,(\beta_0+1)}}\mathbb{I}_{\mathbb{R}^+}(\sigma_x^2), \tag{7.13}$$

where (β_0,β_1) are the shape and scale parameters, respectively. Both are located at the last stage of the hierachical model. Similarly to the hyperparameters of σ^2, their values need to be chosen by the radar operator so that the shape of the prior in Equation 7.13 conveys the information known *a priori* about σ_x^2. Considering both the physical and SSR data models (Equations 7.4 through 7.7), one obtains the approximation $\sigma_x^2 \approx KM/L \left(\sum_\ell |\alpha_\ell|^2\right)$. It means that σ_x^2 represents the average target power after processing; that is, it entails the processing gain KM. Examples of numerical values for (β_0,β_1) are given later in Section 7.3.3 for practical scenarios. In any event, note that unlike the σ^2-hyperparameters, the tuning of the σ_x^2-hyperparameters turns out to be quite determining in the reconstruction performance.

Remark 7.3.1 It is worth mentioning that the "true" prior of $\boldsymbol{x}$ is a hierarchical pdf given by

$$\pi(\boldsymbol{x}|\beta_0,\beta_1) = \int\int \pi(\boldsymbol{x}|w,\sigma_x^2)\pi(w)\pi(\sigma_x^2|\beta_0,\beta_1)dw d\sigma_x^2$$

$$= \prod_{\bar{i}} \frac{1}{2}\left\{\delta(x_{\bar{i}}) + \frac{\beta_0/\beta_1}{\pi}\left[1 + \frac{|x_{\bar{i}}|^2}{\beta_1}\right]^{-(\beta_0+1)}\right\}. \tag{7.14}$$

One recognizes a mixed-type prior with an equal probability between the atom at zero and a complex Student-t distribution. Without the atom at zero, the prior would result in that used in the problem known as *relevance vector machine* or *sparse Bayesian learning* [34,65,70]. Alternatively, choosing a gamma prior for σ_x^2 would lead to a Laplace prior [4,5,23,51,64].

7.3.2 Bayesian estimator

According to the previously described Bayesian model, classical Bayesian estimates of the target amplitude vector $\boldsymbol{x}$ can be derived. Because of its superior performance, the focus is placed on the minimum mean square error (MMSE) estimator [36,57]. (Note that implementation and performance of the maximum a posteriori [MAP] estimator have also been reported in [10].) The MMSE estimator is defined as the mean of the posterior distribution, that is,

$$\hat{\boldsymbol{x}}_{\mathrm{mmse}} = \int \boldsymbol{x} f(\boldsymbol{x}|\boldsymbol{y}) d\boldsymbol{x}, \tag{7.15}$$

where it can be shown that [10]

$$f(\boldsymbol{x}|\boldsymbol{y}) \propto \frac{B(1+n_1, 1+n_0)\, \Gamma(\beta_0+n_1) \prod_{\bar{i}/x_{\bar{i}}=0} \delta(|x_{\bar{i}}|)}{(\beta_1 + \|\boldsymbol{x}\|_2^2)^{\beta_0+n_1} \left[\|\boldsymbol{y}-\boldsymbol{H}\boldsymbol{x}\|_2^2 + \gamma_1 \right]^{KM+\gamma_0}}, \tag{7.16}$$

where $B(.)$ and $\Gamma()$ are the beta and gamma functions, respectively.

Given the expression of the posterior in Equation 7.16, deriving in closed form the estimator of Equation 7.15 does not seem feasible. Furthermore, the pdf does not belong to any class of known distributions that could help to sample it straightforwardly. Instead, one can turn to a numerical approach called an MCMC method. In particular here, a Gibbs sampler can be simply implemented [57]. The latter generates iteratively samples, denoted by $(\sigma^{2\,(t)}, w^{(t)}, \sigma_x^{2\,(t)}, \boldsymbol{x}^{(t)})$, distributed according to their respective *conditional posterior distribution*. For instance, for the parameter σ^2, the pdf sampled is $f(\sigma^2|\boldsymbol{y}, w^{(t-1)}, \sigma_x^{2\,(t-1)}, \boldsymbol{x}^{(t-1)})$. Interestingly, after a burn in period of N_{bi} samples, each subchain $\theta^{(t)}$ generates samples that are distributed according to their respective *posterior distribution* $f(\theta|\boldsymbol{y})$ (θ designates successively σ^2, w, σ_x^2, and $\boldsymbol{x}$). With a sufficient number of samples N_r, the MMSE estimator of $\boldsymbol{x}$ can thus be obtained empirically as

$$\hat{\boldsymbol{x}}_{\mathrm{mmse}} \triangleq \frac{1}{N_r} \sum_{t=1}^{N_r} \boldsymbol{x}^{(t+N_{bi})}.$$

The conditional posterior distributions used in the Gibbs sampler are explicited hereafter in Equation 7.18. They are obtained using the joint posterior distribution of $\sigma^2, \boldsymbol{x}, w, \sigma_x^2|\boldsymbol{y}$. Applying several times Bayes theorem, one obtains

$$f(\sigma^2, w, \boldsymbol{x}, \sigma_x^2, |\boldsymbol{y}) \propto f(\boldsymbol{y}|\boldsymbol{x}, \sigma^2)\pi(\boldsymbol{x}|w, \sigma_x^2)\pi(w)\pi(\sigma_x^2)\pi(\sigma^2)$$

$$\propto \frac{e^{-[\|\boldsymbol{y}-\boldsymbol{H}\boldsymbol{x}\|_2^2+\gamma_1]/\sigma^2}}{\sigma^{2(KM+\gamma_0+1)}}\mathbb{I}_{\mathbb{R}+}(\sigma^2) \times \frac{e^{-[\beta_1+\|\boldsymbol{x}\|_2^2]/\sigma_x^2}}{(\sigma_x^2)^{\beta_0+n_1+1}}\mathbb{I}_{\mathbb{R}+}(\sigma_x^2)$$

$$\times (1-w)^{n_0}\, w^{n_1}\mathbb{I}_{\mathbb{R}+}(w) \times \prod_{\bar{i}/x_{\bar{i}}=0} \delta(|x_{\bar{i}}|). \tag{7.17}$$

Since the prior pdfs (Equations 7.10 through 7.13) have been chosen among a family of conjugate distributions, implementation of the Gibbs sampler moves proves to be simple. More specifically, one can notice that each conditional posterior belongs to the same family as that of the corresponding prior distribution except they have different parameters. Indeed, the latter merge the information brought by the observation and the prior information.

The Gibbs sampler consists of the following four iterative moves:

$$\sigma^2|\boldsymbol{y}, w, \boldsymbol{x}, \sigma_x^2 \sim \mathcal{IG}\left(KM + \gamma_0, \|\boldsymbol{y}-\boldsymbol{H}\boldsymbol{x}\|_2^2 + \gamma_1\right), \tag{7.18}$$

$$w|\boldsymbol{y}, \boldsymbol{x}, \sigma_x^2, \sigma^2 \sim \mathcal{Be}\left(1+n_1, 1+n_0\right), \tag{7.19}$$

$$\sigma_x^2|\boldsymbol{y}, \boldsymbol{x}, \sigma^2, w \sim \mathcal{IG}\left(\beta_0 + n_1, \beta_1 + \|\boldsymbol{x}\|_2^2\right), \tag{7.20}$$

$$\forall\bar{i}, \; x_{\bar{i}}|\boldsymbol{y}, \sigma^2, w, \boldsymbol{x}_{-\bar{i}}, \sigma_x^2 \sim \mathcal{BerCN}\left(w_{\bar{i}}, \mu_{\bar{i}}, \eta_{\bar{i}}^2\right). \tag{7.21}$$

The three first pdfs (Equations 7.18 through 7.20) are obtained straightforwardly from Equation 7.17. The last move (Equation 7.21) aims at generating samples distributed according to $f(\boldsymbol{x}|\boldsymbol{y}, \sigma^2, w, \sigma_x^2)$. However, since the distribution does not belong to any class of known distributions, the vector $\boldsymbol{x}$ is instead sampled element-by-element as suggested in [14,18]. Particularly, the $\bar{i}$th element is distributed according to a Bernoulli–Gaussian distribution with parameters $(w_{\bar{i}}, \mu_{\bar{i}}, \eta_{\bar{i}}^2)$, that is,

$$f(x_{\bar{i}}|\boldsymbol{y}, \sigma^2, w, \boldsymbol{x}_{-\bar{i}}, \sigma_x^2) = (1 - w_{\bar{i}})\delta(|x_{\bar{i}}|) + w_{\bar{i}}\, \mathrm{g}\left(x_{\bar{i}}|\mu_{\bar{i}}, \eta_{\bar{i}}^2\right),$$

where $\boldsymbol{x}_{-\bar{i}}$ is the vector $\boldsymbol{x}$ withtout the element $x_{\bar{i}}$ and the function $\mathrm{g}(.|\mu_{\bar{i}}, \eta_{\bar{i}}^2)$ is the univariate complex Gaussian pdf with mean $\mu_{\bar{i}}$ and variance $\eta_{\bar{i}}^2$. Parameters are given by

$$\eta_{\bar{i}}^2 = \left\{ \frac{1}{\sigma_x^2} + \frac{\|\boldsymbol{h}_{\bar{i}}\|_2^2}{\sigma^2} \right\}^{-1}, \quad \mu_{\bar{i}} = \frac{\eta_{\bar{i}}^2}{\sigma^2}\boldsymbol{h}_{\bar{i}}^H \boldsymbol{e}_{\bar{i}}, \quad \text{and} \quad w_{\bar{i}} = \frac{w\frac{\eta_{\bar{i}}^2}{\sigma_x^2}\exp\left\{\frac{|\mu_{\bar{i}}|^2}{\eta_{\bar{i}}^2}\right\}}{(1-w) + w\frac{\eta_{\bar{i}}^2}{\sigma_x^2}\exp\left\{\frac{|\mu_{\bar{i}}|^2}{\eta_{\bar{i}}^2}\right\}},$$

where $\boldsymbol{e}_{\bar{i}}$ is the observation vector $\boldsymbol{y}$ minus the contribution of the bins $\bar{j} \neq \bar{i}$

$$\boldsymbol{e}_{\bar{i}} = \boldsymbol{y} - \sum_{\bar{j} \neq \bar{i}} x_{\bar{j}}\boldsymbol{h}_{\bar{j}}.$$

For an efficient implementation of Equation 7.21, a similar approach as that of [18] is recommended.

7.3.3   Performance

To confirm the ability of the proposed sparse Bayesian technique to alleviate velocity ambiguity, two low scatterers are placed in the first velocity sidelobes of a strong scatterer. The range–velocity map obtained is depicted in Figure 7.7. For comparison purposes, the output of an ℓ_1-penalized least squares is also shown, namely the so-called ML1 estimate discussed in [46]. Clearly, SSR techniques manage to disambiguate the target scene with different level of qualities. The MMSE estimation offers the most appropriate reconstruction: scatterer amplitudes are properly restored while only very low false estimation occurs outside the support of $\boldsymbol{x}$ (nonvisible here due to colorbar setting). Moreover, the proposed Bayesian algorithm allows to obtain the empirical posterior distributions of σ^2, w, σ_x^2, and $\boldsymbol{x}$. In Figure 7.8, one can appreciate the tightening of each posterior with respect to its prior around the MMSE estimate. Trends observed with this single-run is reinforced in Figure 7.9 where the mean squared error (MSE) of the sparse vector $\boldsymbol{x}$ is represented as a function of the bandwidth B. These first numerical examples prefigure the high potential of wideband waveform in conjunction with sparse Bayesian recovery to enable a nonambiguous radar mode.

The estimators are further tested on semiexperimental radar data. To fit as much as possible to the hierarchical model, on-grid synthetic targets are injected in experimental data* in a region most probably devoid of true targets. Data are then prewhitened with an *ad hoc* filter to remove clutter. A rough estimate of the thermal noise power (obtained beforehand) serves as a normalization factor so that $\sigma^2 \approx 1$. The target scene mimics vehicles on a freeway. Range–velocity maps obtained are depicted in Figure 7.10 and show the ability of SSR techniques to alleviate velocity ambiguity of exo-clutter targets embedded in a *real* thermal noise component. In particular, as in the pure synthetic case, the MMSE estimator exhibits high potential for sparse recovery. However, at this stage, the proposed sparse Bayesian algorithm needs to be augmented to deal with several phenomena possibly encountered in full-experimental data. In the next section, three extended versions of the sampler described in Equation 7.18 are presented to that end.

---

*Experimental data have been collected by the PARSAX radar of the Delft University of Technology [38], while it was pointing on a freeway.

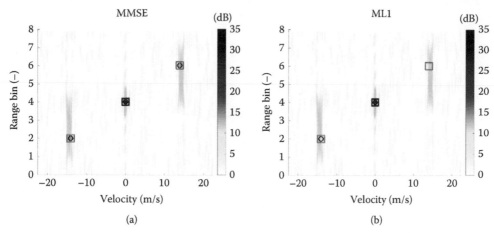

FIGURE 7.7

Sparse recovery of target scene for wideband radar. Synthetic data. $F_c = 10$ GHz, $B = 1$ GHz, $F_r = 1$ kHz ($\delta_R = 15$ cm, $v_a = 15$ m/s), $M = 32$, $K - 8$, (γ_0, γ_1) such that $m_{\sigma^2} = 0$ dB and $\sqrt{v_{\sigma^2}} = 5$ dB, (β_0, β_1) such that $m_{\sigma_x^2} = 20$ dB and $\sqrt{v_{\sigma_x^2}} = 20$ dB, $n_{va} = 3$, $n_{zp}^r = 1$, $n_{zp}^v = 1$. Square markers indicate true scatterer location. Diamond markers indicate estimated scatterers. Coherent integration of Equation 7.5 depicted as transparent background. (a) proposed Bayesian algorithm of Equation 7.18. (b) ML1 of [46] with noise confidence interval of 99%.

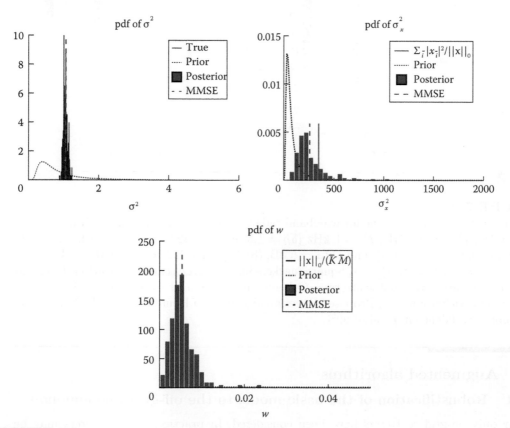

FIGURE 7.8

Prior pdf and empirical posterior pdf obtained with the samples of Equation 7.18.

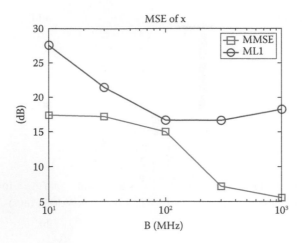

FIGURE 7.9
MSE of x as a function of the bandwidth B. Scenario is the same as that of Figure 7.7 except for the varying bandwidth.

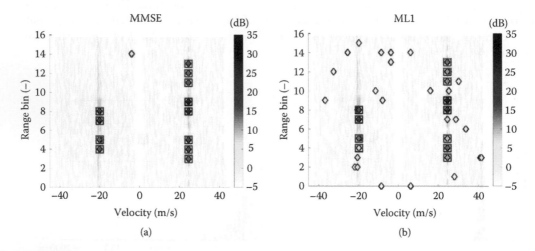

FIGURE 7.10
Sparse recovery of target scene for wideband radar. Semiexperimental PARSAX data. $F_c = 3.315$ GHz, $B = 100$ MHz, $F_r = 1$ kHz ($\delta_R = 1.5$ m, $v_a = 42.25$ m/s), $M = 64$, $K = 16$, (γ_0, γ_1) such that $m_{\sigma^2} = 0$ dB and $\sqrt{v_{\sigma^2}} = 5$ dB, (β_0, β_1) such that $m_{\sigma_x^2} = 20$ dB and $\sqrt{v_{\sigma_x^2}} = 20$ dB, $n_{va} = 2$, $n_{zp}^r = 1$, $n_{zp}^v = 1$. Square markers indicate true scatterer location. Diamond markers indicate estimated scatterers. Coherent integration of Equation 7.5 depicted as transparent background. (a) Proposed Bayesian algorithm of Equation 7.18. (b) ML1 of [46] with noise confidence interval of 99%.

7.4   Augmented algorithms

7.4.1   Robustification of the basic model to the off-grid phenomenon

So far only on-grid scatterers have been considered. In practice, real scatterers may be located between points of the analysis grid (cf. Figure 7.11) so that the signal of interest is not sparse anymore in the dictionary described by Equation 7.8. Performance degradation

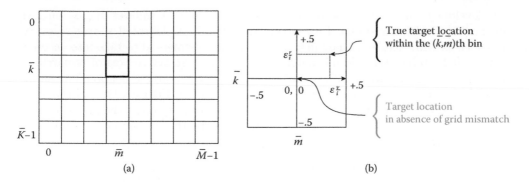

FIGURE 7.11

Representation of grid mismatch phenomenon. (a) Range–velocity grid of analysis. (b) Zoom on $(\bar{k}, \bar{m})$th bin.

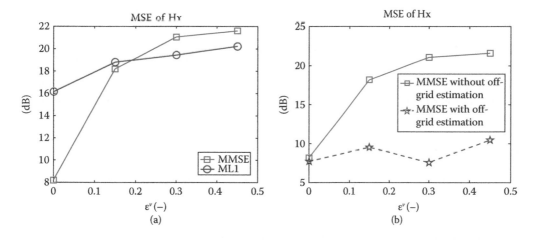

FIGURE 7.12

Robustness of SSR techniques toward grid mismatch. Performance metrics is the MSE of the reconstructed target scene in the dictionary, that is, $\mathcal{E}\left\{\|H(\varepsilon^v, \varepsilon^r)x - H(\hat{\varepsilon}^v, \hat{\varepsilon}^r)\hat{x}\|_2^2\right\}$. Each scatterer of Figure 7.7 is displaced off-the grid with a quantity ε^v. (a) Non-robust estimation with $(\hat{\varepsilon}^v, \hat{\varepsilon}^r) = (0, 0)$. (b) Robust estimation.

of the SSR algorithms is illustrated in presence of grid mismatch in Figure 7.12a. Obviously, robustification toward this phenomenon is necessary.

The grid mismatch problem is well known in the SSR literature. It has been described in [46] and its effect has been measured several times, for example, in [15,20,28,58]. One way to counter its deleterious effect is to iteratively refine the grid of analysis [20,22,46]. Nonetheless, coherence of the dictionary might increase which is in turn detrimental to the stability of the recovery [20]. Another strategy consists in estimating the mismatch jointly with the sparse vector. A parametric model is used in [30] where the sparsifying dictionary depends nonlinearly on the grid errors that are estimated via an expectation-maximization (EM) method. To obtain less complex estimator, several authors have proposed a simplified linear model where grid mismatch is described by an additive perturbation matrix to the dictionary [63,73]. Usually, the latter stems from a first-order Taylor-series expansion with respect to the grid errors [26,31,33,62,71,72]. Accordingly, several estimation techniques have been developed: an optimization method based on total least squares in [73], an EM algorithm in [71], and a variational Bayes technique in [31]. Celebrated sparse recovery techniques such as OMP [66] and Lasso [64] have been also augmented to take into account grid mismatch [62]. Today, the grid mismatch problem is still an active area of research.

In the framework of hierarchical Bayes inference, the basic model of Section 7.3 can be simply modified to incorporate this phenomenon as depicted in Figure 7.6b and reported in [30,39,41]. Two perturbation vectors $\boldsymbol{\varepsilon}^v$ and $\boldsymbol{\varepsilon}^r$ are added to represent the grid errors in the velocity and range dimensions, respectively. Accordingly, the sparsifying dictionary is redefined as $\boldsymbol{H} \triangleq \boldsymbol{H}(\boldsymbol{\varepsilon}^v, \boldsymbol{\varepsilon}^r)$ with

$$[\boldsymbol{h}_{\bar{\imath}}(\varepsilon_{\bar{\imath}}^v, \varepsilon_{\bar{\imath}}^r)]_i = \frac{1}{\sqrt{KM}} \exp\left\{-j2\pi \frac{\bar{k} + \varepsilon_{\bar{\imath}}^r}{\bar{K}} k\right\} \exp\left\{j2\pi n_{va} \frac{\bar{\bar{m}} + \varepsilon_{\bar{\imath}}^v}{\bar{M}} m(1 + \mu k)\right\},$$

where $\varepsilon_{\bar{\imath}}^v, \varepsilon_{\bar{\imath}}^r$ are confined to the interval $[-0.5, 0.5)$ to avoid designing a unit-coherence dictionary. To complete the model, a prior pdf is assigned to the additional parameters $\boldsymbol{\varepsilon}^v$ and $\boldsymbol{\varepsilon}^r$. A simple yet natural choice is to consider iid grid errors with noninformative prior when a scatterer is present in the bin; otherwise mismatch need not be estimated, that is,

$$\pi(\boldsymbol{\varepsilon}^v, \boldsymbol{\varepsilon}^r | \boldsymbol{x}) = \prod_{\bar{\imath}} \pi(\varepsilon_{\bar{\imath}}^v, \varepsilon_{\bar{\imath}}^r | x_{\bar{\imath}}), \quad \text{where} \quad \pi(\varepsilon_{\bar{\imath}}^v, \varepsilon_{\bar{\imath}}^r | x_{\bar{\imath}}) = \begin{cases} \delta(\varepsilon_{\bar{\imath}}^v)\delta(\varepsilon_{\bar{\imath}}^r) & \text{if } x_{\bar{\imath}} = 0 \\ \mathbb{I}_{[-.5,.5]}(\varepsilon_{\bar{\imath}}^v)\mathbb{I}_{[-.5,.5]}(\varepsilon_{\bar{\imath}}^r) & \text{otherwise.} \end{cases}$$

Finally, using the same approach as that described in Section 7.3, a fifth move is added to the Gibbs sampler Equation 7.18 where samples $\varepsilon_{\bar{\imath}}^{v(t)}, \varepsilon_{\bar{\imath}}^{r(t)}$ are generated jointly according to the conditional posterior distribution $f(\varepsilon_{\bar{\imath}}^v, \varepsilon_{\bar{\imath}}^r | \boldsymbol{y}, \boldsymbol{x}, \boldsymbol{\varepsilon}^v_{-\bar{\imath}}, \boldsymbol{\varepsilon}^r_{-\bar{\imath}}, \sigma^2, w, \sigma_x^2)$. It can be shown that this pdf does not belong to any known class of distributions. A Metropolis-Hastings move can be implemented instead; details of the method are reported in [41]. Benefits of this robustification are clearly visible in Figure 7.12b. This is at the cost of an increased computational load, though the latter is in a certain way contained since mismatch is estimated only for bins where scatterers are present.

7.4.2   Extension of the basic model to colored noise

Another phenomenon arising with full-experimental data is the possible presence of diffuse clutter. The basic algorithm of Section 7.3, which assumes only white noise, is not capable to deal properly with such component. Indeed, it has been reported in [8] that residual sidelobes appear then in the blind velocities (see also Figure 7.13a). In light of this observation, mainly two strategies can be thought of: diffuse clutter can be removed beforehand via conventional filtering which requires then an adequate training interval, or the basic hierarchical model can be augmented to incorporate this additional component. In what follows, a first attempt at the second strategy is presented. For the sake of simplicity, the noise is assumed Gaussian distributed, iid from subband to subband, and correlated in the slow-time dimension according to a finite-order autoregressive (AR) model. Note that AR models have been successfully used in several radar applications (e.g., [2]). The approach can be seen as a regularization technique similarly to the famous diagonal loading [7,69].

According to the previous hypotheses, the noise distribution can be expressed as

$$\boldsymbol{n} | \boldsymbol{R} \sim \mathcal{CN}_{KM}(\boldsymbol{0}, \boldsymbol{R}) \quad \text{with} \quad \boldsymbol{R} = \boldsymbol{I}_K \otimes \boldsymbol{\Gamma},$$

where the slow-time covariance matrix $\boldsymbol{\Gamma}$ is of size $M \times M$ and has a P-banded inverse matrix [49]; P is the order of the AR model and is supposed to be known. Particularly, $\boldsymbol{\Gamma}^{-1}$ has a specific Cholesky factorization

$$\boldsymbol{\Gamma}^{-1} = \sigma_{\mathrm{ar}}^{-2}(\boldsymbol{I}_M - \boldsymbol{\Phi})^H(\boldsymbol{I}_M - \boldsymbol{\Phi}),$$

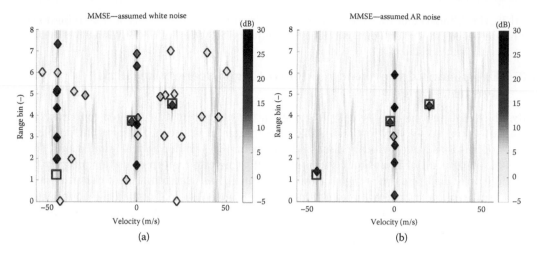

FIGURE 7.13

Sparse recovery of target scene for wideband radar. Full experimental PARSAX data with additional synthetic targets (square markers) including one target at $-v_a$ to demonstrate possible estimation in first-blind velocity. Parameters mostly the same as that of Figure 7.10 ($\mathrm{m}_{\sigma_x^2}$ increased and $\mathrm{v}_{\sigma_x^2}$ decreased to favor sparse recovery instead of target split). Diamond markers indicate estimated scatterers. Coherent integration of Equation 7.5 depicted as transparent background. (a) Estimation of off-grid migrating targets in assumed white noise. (b) Estimation of off-grid migrating targets in assumed AR noise (robustication toward grid mismatch is also incorported).

where σ_{ar}^2 is the variance of the white input to the AR model ($\sigma_{\mathrm{ar}}^2 \triangleq \sigma^2$ only when $P = 0$) and $\boldsymbol{\Phi}$ is a lower triangular Toeplitz matrix containing the P AR parameters denoted by $\phi_1, \ldots, \phi_P$. It is defined as

$$\boldsymbol{\Phi} = \text{Toeplitz} \left\{ \left[0, \boldsymbol{\phi}^T, 0, \ldots, 0\right] \right\} \quad \text{with} \quad \boldsymbol{\phi} = \left[\phi_1, \ldots, \phi_P\right]^T.$$

The likelihood function becomes

$$f(\boldsymbol{y}|\boldsymbol{x}, \sigma_{\mathrm{ar}}^2, \boldsymbol{\phi}) = \frac{1}{\pi^{KM} \sigma_{\mathrm{ar}}^{2KM}} \exp \left\{ -\sigma_{\mathrm{ar}}^{-2} \| \left[\boldsymbol{I}_K \otimes \left(\boldsymbol{I}_M - \boldsymbol{\Phi}\right)\right] (\boldsymbol{y} - \boldsymbol{Hx}) \|_2^2 \right\},$$

where compared to that of Section 7.3, the noise is now described by two parameters σ_{ar}^2 and $\boldsymbol{\phi}$. To complete the new hierarchical model, prior distributions need to be assigned to them. Without further knowledge about noise, flat priors can be favored such as [49]

$$\pi(\sigma_{\mathrm{ar}}^2) \propto \frac{1}{\sigma_{\mathrm{ar}}^2} \mathbb{I}_{\mathbb{R}^+}(\sigma_{\mathrm{ar}}^2) \quad \text{and} \quad \pi(\boldsymbol{\phi}) \propto 1,$$

The resulting hierarchical Bayesian model is represented in Figure 7.6c. Again, following the same approach as that described in Section 7.3, a new Gibbs sampler can be defined where steps 1 and 4 of Equation 7.18 are modified to sample σ_{ar}^2 and $\boldsymbol{x}$, respectively, in presence of AR noise and where an additional step allows to sample $\boldsymbol{\phi}$. Details of the method are reported in [8]. Benefits of the robustification toward diffuse noise is shown in Figure 7.13 with a single-run output on full-experimental data. Interestingly enough, clutter is estimated as the sum of a discrete and diffuse component around zero velocity which is in accordance with previous observations [11,43]. A significant result here is the estimation of targets located in the blind-velocities. This supports the feasibility of designing a nonambiguous radar mode based on a wideband waveform in conjunction with sparse Bayesian recovery.

7.4.3 Extension of the basic model to target scene with high dynamic range

Finally, the third and last real-world phenomenon considered in this chapter concerns the simultaneous presence of very low and high power targets in the scene, a recurrent problem in radar applications. As previously discussed and examplified in numerical simulations, the sparsity-inducing prior of Equation 7.14 requires adjusting two hyperparameters, namely β_0 and β_1. In a very high dynamic range scenario, no tuning has been found that can provide a satisfying sparse recovery of low targets [40] (see also Figure 7.14b). To push the boundaries of performance, a novel prior for x can be thought of to augment the basic algorithm of Section 7.3. To that end, the sparse vector is parametrized in terms of modulus ρ and angle φ as $x \triangleq \rho \odot e^{i\varphi}$, where $\odot$ is the Hadamard product, that is, $x_{\bar{\imath}} = \rho_{\bar{\imath}} e^{i\varphi_{\bar{\imath}}}$. To complete the hierarchical Bayes scheme, priors need to be assigned to both vectors. In practice, the phase of each scatterer present in the scene is unknown so that a noninformative prior seems to be the most appropriate

$$\pi(\varphi_{\bar{\imath}}|\rho_{\bar{\imath}}) = \begin{cases} \delta(\varphi_{\bar{\imath}}) & \text{if } \rho_{\bar{\imath}} = 0 \\ \frac{1}{2\pi}\mathbb{I}_{[0,2\pi]}(\varphi_{\bar{\imath}}) & \text{otherwise.} \end{cases}$$

Concerning the modulus of the target amplitude, the prior is designed while taking into account the finite dynamic range of the radar sensor. In particular, to cover more efficiently

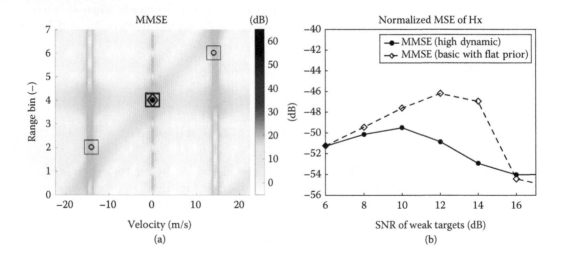

FIGURE 7.14
Comparison between the basic sparse Bayesian recovery (diamond markers) and its new version extended to high dynamic range (circle markers). Synthetic data under consideration. Square markers indicate true target locations. (a) Range–velocity map. Scenario is the same as that of Figure 7.7 except for the hyperparameters (γ_0, γ_1) and (β_0, β_1) adjusted to improve estimation in high dynamic range scenarios. (b) Normalized MSE of the reconstructed target scene in the dictionary. Target scene similar to that of (a): one strong target with two low targets with same SNR. The latter is varied here.

its span, the new sparse-promoting prior considered consists of an atom at zero and a finite mixture of uniform distributions

$$\pi(\rho_{\tilde{i}}|\boldsymbol{w}) = w_0 \, \delta(\rho_{\tilde{i}}) + \sum_{c=1}^{C} w_c \, \frac{1}{\rho_c^+ - \rho_c^-} \, \mathbb{I}_{[\rho_c^-,\rho_c^+)}(\rho_{\tilde{i}}), \tag{7.22}$$

where C is the number of target power classes, $\boldsymbol{w} = [w_c]_{c=0,\ldots,C}$ is the vector of class probabilities with $w_0 = 1 - \sum_{c=1}^{C} w_c$, and ρ_c^-, ρ_c^+ are the class boundaries. Using Equation 7.22 means that $\rho_{\tilde{i}}$ is null with probability w_0, or belongs to class c with probability w_c, and is uniformly distributed within this class. Ultimately, a last stage is added to the model, to describe statistically $\boldsymbol{w}$. A multivariate Dirichlet distribution is selected (which is a generalization of Bernoulli distribution) with concentration parameters $\theta_0, \ldots, \theta_C$

$$\pi(\boldsymbol{w}|\theta_0, \ldots, \theta_C) \propto w_0^{\theta_0-1} \mathbb{I}_{[0,1]}(w_0) \times \prod_{c=1}^{C} w_c^{\theta_c-1} \mathbb{I}_{[0,1]}(w_c).$$

The full Bayesian model is depicted in Figure 7.6d. Accordingly, following the same approach as that presented in Section 7.3, a Gibbs sampler can be simply designed. Details of implementation and typical values for $C, \rho_c^-, \rho_c^+, \theta_c$ can be found in [40]. Numerical examples are shown in Figure 7.14 on synthetic data and give a glimpse of the potential of designing nonambiguous radar mode even in a challenging high dynamic range scenario.

7.5 Concluding remarks

Pulse-Doppler radars with constant PRF naturally induce a subsampling of the received signal in range and/or velocity. This is detrimental to the target scene estimation, particularly in the blind velocities. In this chapter, we investigated the use of a wideband waveform jointly with sparse recovery techniques to unambiguously reconstruct the target signal. Using a frequency diverse waveform brings indeed additional information. In particular, with the use of a low PRF, range migration of moving targets provides nonambiguous measurement of their velocity. Disambiguation, *per se*, can then be performed via sparse reconstruction.

For this purpose, several hierarchical Bayes models were presented where sparsity of the target scene is induced via a sparse-promoting prior. We showed that a full Bayes approach allows one to design models where parameters can preserve a strong physical sense, which offers an ideal framework to incorporate prior information about the radar scene. Furthermore, hierarchical models can be easily modified to take into account additional real-world phenomena. Resulting Bayes estimates showed promising performance that supports the feasibility of designing a nonambiguous radar mode with a wideband waveform in conjunction with sparse recovery techniques.

However, several limitations remain concerning the proposed approach. We underline here two of them. First, the computational complexity remains an hindrance to real-time implementation. Indeed, Bayes estimators pertaining to hierarchical models rarely have analytical solutions. As a consequence, computationally intensive numerical approaches are used instead. Second, for the purpose of radar detection, sparse recovery techniques only estimate the target scene and do not give any decision about the presence or not of a target. To overcome these two main problems, ongoing work can be found in the literature. The Bayesian community is actively working on decreasing the complexity of hierarchical inference (e.g., variational method and stochastic optimization [52]) while first attempts at designing CS radar detector are suggested [3].

Acknowledgments

The authors would like to thank O. Krasnov at TU Delft for kindly providing the PARSAX experimental data.

References

[1] T. J. Abatzoglou and G. O. Gheen. Range, radial velocity, and acceleration MLE using radar LFM pulse train. *IEEE Trans. Aerosp. Electron. Syst.*, 34(4):1070–1084, 1998.

[2] Y. I. Abramovich, N. K. Spencer, and M. D. E. Turley. Time-varying autoregressive (TVAR) models for multiple radar observations. *IEEE Trans. Signal Process.*, 55(4):1298–1311, 2007.

[3] L. Anitori, A. Maleki, M. Otten, R. G. Baraniuk, and P. Hoogeboom. Design and analysis of compressed sensing radar detectors. *IEEE Trans. Signal Process.*, 61(4):813 –827, 2013.

[4] S. D. Babacan, R. Molina, and A. K. Katsaggelos. Bayesian compressive sensing using Laplace priors. *IEEE Trans. Image Process.*, 19(1):53–63, 2010.

[5] S. D. Babacan, S. Nakajima, and M. N. Do. Bayesian group-sparse modeling and variational inference. *IEEE Trans. Signal Process.*, 62(11):2906–2921, 2014.

[6] C. Berger, S. Zhou, and P. Willett. Signal extraction using compressed sensing for passive radar with OFDM signals. In *Proceedings of International Conference on Information Fusion*, pages 1–6, June 30–July 3, 2008.

[7] P. J. Bickel and E. Levina. Regularized estimation of large covariance matrices. *Ann. Stat.*, 36(1):199–227, 2008.

[8] S. Bidon, O. Besson, J.-Y. Tourneret, and F. Le Chevalier. Bayesian sparse estimation of migrating targets in autoregressive noise for wideband radar. In *Proceedings of IEEE International Radar Conference*, Cincinnati, OH, May 19–23, 2014.

[9] S. Bidon, J.-Y. Tourneret, and L. Savy. Sparse representation of migrating targets in low PRF wideband radar. In *Proceedings IEEE International Radar Conference*, Atlanta, GA, May 7–11, 2012.

[10] S. Bidon, J.-Y. Tourneret, L. Savy, and F. Le Chevalier. Bayesian sparse estimation of migrating targets for wideband radar. *IEEE Trans. Aerosp. Electron. Syst.*, 50(2): 871–886, 2014.

[11] J. B. Billingsley and J. F. Larrabee. *Multifrequency Measurements of Radar Ground Clutter at 42 Sites*. Technical report, MIT Lincoln Laboratory, Lexington, MA, 1991.

[12] T. Blumensath and M. E. Davies. Iterative hard thresholding for compressed sensing. *Appl. Comput. Harmon. Anal.*, 27(3):265–274, 2009.

[13] E. J. Candès and M. B. Wakin. An introduction to compressive sampling. *IEEE Signal Process. Mag.*, 25(2):21–30, 2008.

[14] Q. Cheng, R. Chen, and T. H. Li. Simultaneous wavelet estimation and deconvolution of reflection seismic signals. *IEEE Trans. Geosci. Remote Sens.*, 34(2):377–384, 1996.

[15] Y. Chi, L. L. Scharf, A. Pezeshki, and A. R. Calderbank. Sensitivity to basis mismatch in compressed sensing. *IEEE Trans. Signal Process.*, 59(5):2182–2195, 2011.

[16] G. Z. Dai and J. M. Mendel. Maximum a posteriori estimation of multichannel Bernoulli-Gaussian sequences. *IEEE Trans. Inf. Theory*, 35(1):181–183, 1989.

[17] F. Deudon, S. Bidon, O. Besson, and J.-Y. Tourneret. Velocity dealiased spectral estimators of range migrating targets using a single low-PRF wideband waveform. *IEEE Trans. Aerosp. Electron. Syst.*, 49(1):244–265, 2013.

[18] N. Dobigeon, A. O. Hero, and J.-Y. Tourneret. Hierarchical Bayesian sparse image reconstruction with application to MRFM. *IEEE Trans. Image Process.*, 18(9): 2059–2070, 2009.

[19] D. L. Donoho. Compressed sensing. *IEEE Trans. Inf. Theory*, 52:1289–1306, 2006.

[20] M. F. Duarte and R. G. Baraniuk. Spectral compressive sensing. *Appl. Comput. Harmon. Anal.*, 35:111–129, 2013.

[21] J. H. G. Ender. On compressive sensing applied to radar. *Signal Process.*, 90(5): 1402–1414, 2010.

[22] A. C. Fannjiang and W. Liao. Coherence-pattern guided compressive sensing with unresolved grids. *SIAM J. Imaging Sci.*, 5:179–202, 2012.

[23] M. A. T. Figueiredo. Adaptive sparseness for supervised learning. *IEEE Trans. Pattern Anal. Mach. Intell.*, 25(9):1150–1159, 2003.

[24] J.-J. Fuchs and F. Le Chevalier. Détection d'une cible mobile en présence de fouillis à l'aide d'un radar large bande. In *Proceedings of GRETSI*, pages 531–534, Vannes, 1999.

[25] S. J. Godsill and P. J. W. Rayner. Statistical reconstruction and analysis of autoregressive signals in impulsive noise using the Gibbs sampler. *IEEE Trans. Speech Audio Process.*, 6(4):352–372, 1998.

[26] A. Gretsistas and M. D. Plumbley. An alternating descent algorithm for the off-grid DoA estimation problem with sparsity constraints. In *Proceedings of the 20th European Signal Processing Conference (EUSIPCO 2012)*, pages 874–878, Bucharest, Romania, 2012.

[27] M. A. Herman and T. Strohmer. High-resolution radar via compressed sensing. *IEEE Trans. Signal Process.*, 57(6):2275–2284, 2009.

[28] M. A. Herman and T. Strohmer. General deviants: An analysis of perturbations in compressed sensing. *IEEE J. Sel. Top. Signal Process.*, 4(2):342–349, 2010.

[29] J. A. Högbom. Aperture synthesis with a non-regular distribution of interferometer baselines. *Astron. Astrophy. Suppl.*, 15:417–426, 1974.

[30] L. Hu, Z. Shi, J. Zhou, and Q. Fu. Compressed sensing of complex sinusoids: An approach based on dictionary refinement. *IEEE Trans. Signal Process.*, 60(7): 3809–3822, 2012.

[31] L. Hu, J. Zhou, Z. Shi, and Q. Fu. A fast and accurate reconstruction algorithm for compressed sensing of complex sinusoids. *IEEE Trans. Signal Process.*, 61(22):5744–5754, 2013.

[32] D. E. Iverson. Coherent processing of ultra-wideband radar signals. *Proc. IEE Radar, Sonar Navig.*, 141(3):171–179, 1994.

[33] R. Jagannath and K. V. S. Hari. Block sparse estimator for grid matching in single snapshot DoA estimation. *IEEE Signal Process. Lett.*, 20(11): 1038–1041, 2013.

[34] S. Ji, Y. Xue, and L. Carin. Bayesian compressive sensing. *IEEE Trans. Signal Process.*, 56(6):2346–2356, 2008.

[35] N. Jiang, R. Wu, and J. Li. Super resolution feature extraction of moving targets. *IEEE Trans. Aerosp. Electron. Syst.*, 37(3):781–793, 2001.

[36] S. M. Kay. *Fundamentals of Statistical Signal Processing: Estimation Theory.* Prentice Hall, Englewood Cliffs, NJ, 1993.

[37] J. J. Kormylo and J. M. Mendel. Maximum likelihood detection and estimation of Bernoulli-Gaussian processes. *IEEE Trans. Inf. Theory*, 28(3):482–488, 1982.

[38] O. A. Krasnov, G. P. Babur, Z. Wang, L. P. Ligthart, and F. van der Zwan. Basics and first experiments demonstrating isolation improvements in the agile polarimetric FM-CW radar—PARSAX. *Int. J. Microw. Wirel. Technol.*, 2(Special Issue 3–4):419–428, 2010.

[39] M. Lasserre, S. Bidon, O. Besson, and F. Le Chevalier. Bayesian sparse Fourier representation of off-grid targets with application to experimental radar data. *Signal Process.*, 111:261–273, 2015.

[40] M. Lasserre, S. Bidon, and F. Le Chevalier. New sparse-promoting prior for the estimation of a radar scene with weak and strong targets. *IEEE Trans. Signal Process.*, PP(99):1–1, 2016.

[41] M. Lasserre, S. Bidon, and F. Le Chevalier. Velocity ambiguity mitigation of off-grid range migrating targets via Bayesian sparse recovery. In *Proceedings of Statistical Signal Processing Workshop (SSP)*, Palma de Mallorca, June 26–29, 2016.

[42] F. Le Chevalier. Radar non ambigü à large bande, 1996. French Patent 9 608 509.

[43] F. Le Chevalier. *Principles of Radar and Sonar Signal Processing.* Artech House, Norwood, MA, 2002.

[44] F. Le Chevalier. Space-time coding for active antenna systems. In W. L. Melvin and J. A. Scheer, editors, *Principle of Modern Radar. Advanced Techniques*, pages 499–258. SciTech, Edison, NJ, 2013.

[45] D. C. Lush and D. A. Hudson. Ambiguity function analysis of wideband radars. In *Proceedings of 1991 IEEE National Radar Conference*, pages 16–20, Los Angeles, CA, March 12–13, 1991.

[46] D. M. Malioutov. A sparse signal reconstruction perspective for source localization with sensor arrays. Master's thesis, MIT, Cambridge, MA, 2003.

[47] A. R. Moore. Practical phased array radar system design (capability versus compromise). In *Tutorial European Radar Conference*, Manchester, UK, 2011.

[48] G. V. Morris and L. Harkness, editors. *Airborne Pulse Doppler Radar*, 2nd ed. Artech House, Boston, MA, 1996.

[49] S. Ni and D. Sun. Bayesian estimates for vector autoregressive models. *J. Bus. Econ. Stat.*, 23(1): 105–117, 2005.

[50] P. Schnitter, L. Potter, and J. Ziniel. Fast Bayesian matching pursuit. In *Proceedings of the Workshop on Information Theory and Applications*, La Jolla, CA, January 2008.

[51] T. Park and G. Casella. The Bayesian Lasso. *J. Am. Stat. Assoc.*, 103(482):681–686, 2008.

[52] M. Pereyra, P. Schniter, E. Chouzenoux, J. C. Pesquet, J. Y. Tourneret, A. O. Hero, and S. McLaughlin. A survey of stochastic simulation and optimization methods in signal processing. *IEEE J. Sel. Top. Signal Process.*, 10(2):224–241, 2016.

[53] R. P. Perry, R. C. DiPietro, and R. L. Fante. SAR imaging of moving targets. *IEEE Trans. Aerosp. Electron. Syst.*, 35(1):188–200, 1999.

[54] N. Petrov and F. Le Chevalier. Iterative adaptive approach for unambiguous wideband radar target detection. In *Proceedings of European Radar Conference (EURAD)*, Paris, September 6–11, 2015.

[55] L. C. Potter, E. Ertin, J. T. Parker, and M. Çetin. Sparsity and compressed sensing in radar imaging. *Proc. IEEE*, 98(6):1006–1020, 2010.

[56] M. A. Richards, J. A. Scheer, and W. A. Holm, editors. *Principle of Modern Radar. Basic Principles*. SciTech, Raleigh, NC, 2010.

[57] C. P. Robert and G. Casella. *Monte Carlo Statistical Methods*. Springer Science, New York, 2004.

[58] M. Rosenbaum and A. B. Tsybakov. Sparse recovery under matrix uncertainty. *Ann. Stat.*, 38(5):2620–2651, 2010.

[59] M. I. Skolnik. *Radar Handbook*. The McGraw-Hill, New York, 1970.

[60] J. M. Speiser. Wide-band ambiguity functions. *IEEE Trans. Inf. Theory*, 13(1): 122–123, 1967.

[61] M. J. Steiner and K. Gerlach. Fast converging adaptive canceller for a structured covariance matrix. *IEEE Trans. Aerosp. Electron. Syst.*, 36(4):1115–1126, 2000.

[62] Z. Tan and A. Nehorai. Sparse direction of arrival estimation using co-prime arrays with off-grid targets. *IEEE Signal Process. Lett.*, 21(1):26–29, 2014.

[63] Z. Tan, P. Yang, and A. Nehorai. Joint sparse recovery method for compressed sensing with structured dictionary mismatches. *IEEE Trans. Signal Process.*, 62(19): 4997–5008, 2014.

[64] R. Tibshirani. Regression shrinkage and selection via the Lasso. *J. Roy. Stat. Soc.*, 58(1):267–288, 1996.

[65] M. E. Tipping. Sparse Bayesian learning and the relevance vector machine. *J. Mach. Learn. Res.*, 1:211–244, 2001.

[66] J. A Tropp and A. C. Gilbert. Signal recovery from random measurements via orthogonal matching pursuit. *IEEE Trans. Inf. Theory*, 53(12):4655–4666, 2007.

[67] G. Trunk and S. Brockett. Range and velocity ambiguity resolution. In *Proceedings IEEE National Radar Conference*, pages 146–149, Lynnfield, MA, April 20–22, 1993.

[68] E. Van den Berg and M. P Friedlander. Probing the Pareto frontier for basis pursuit solutions. *SIAM J. Sci. Comput.*, 31(2):890–912, 2008.

[69] A. Wiesel, O. Bibi, and A. Globerson. Time varying autoregressive moving average models for covariance estimation. *IEEE Trans. Signal Process.*, 61(11):2791–2801, 2013.

[70] D. P. Wipf and B. D. Rao. Sparse Bayesian learning for basis selection. *IEEE Trans. Signal Process.*, 52(8):2153–2164, 2004.

[71] Z. Yang, L. Xie, and C. Zhang. Off-grid direction of arrival estimation using sparse Bayesian inference. *IEEE Trans. Signal Process.*, 61(1):38–43, 2013.

[72] Y. Zhang, Z. Ye, X. Xu, and N. Hu. Off-grid DoA estimation using array covariance matrix and block-sparse Bayesian learning. *Signal Process.*, 98:197–201, 2014.

[73] H. Zhu, G. Leus, and G. B. Giannakis. Sparsity-cognizant total least-squares for perturbed compressive sampling. *IEEE Trans. Signal Process.*, 59(5):2002–2016, 2011.

8

Virtual Experiments and Compressive Sensing for Subsurface Microwave Tomography

Martina Bevacqua, Lorenzo Crocco, Loreto Di Donato,
Tommaso Isernia, and Roberta Palmeri

CONTENTS

8.1   Introduction

The development of imaging methods able to provide information on surveyed areas as reliable as possible, not subject to user's interpretation, is an important and still open challenge for ground-penetrating radar (GPR) technology [1–6]. Nowadays, microwave tomographic approaches represent a powerful way to achieve such a goal, but they require coping with the solution of a nonlinear and ill-posed inverse scattering problem [7]. Such a nontrivial task entails the need of appropriate regularization methods [8] to restore the well-posedness of the problem and avoid physically meaningless solutions. On the other hand, defeating nonlinearity requires suitable strategies to counteract the occurrence of "false solutions" [4,5], that is, estimated solutions that match the data (within the expected accuracy) but are indeed different from the ground truth. These difficulties are further worsened when, as in GPR surveys, it is not possible to probe the targets from all possible directions, which results in an obvious reduction of the available data and a consequent deterioration of the imaging results. Last, but not least, the existing solution procedures based on global or local optimization strategies are usually computationally demanding and represent a huge drawback when real-time analyses are required [4–6].

To partially overcome the difficulties arising from the nonlinearity of the problem, tomographic inversion techniques based on Born (BA) or Kirchhoff (KA) approximations have been proposed and exploited to process GPR data [2,3,5]. These methods rely on first-order scattering approximations which allow to deal with linear, but still ill-posed, problems. However, linearized methods can be rigorously applied in a very limited range of cases, and, in practical instances, can provide only information on location and extent of buried targets [3]. In addition, even when the model's approximation is fulfilled, quantitative results cannot be achieved anyway for aspect limited data due to specific properties of the mathematical operator which rules the relationship between data and unknowns [9]. Finally, neglecting higher order interactions may give rise to artifacts, or "ghosts," in the tomographic images. These artifacts may in turn be erroneously interpreted as "false" targets. Conversely, weakly scattering objects may look like artifacts and therefore be neglected. As such, linearized approximations can only limitedly contribute to the quest for processing tools capable to provide images of the underground that are as objective as possible.

In recent years, the authors of this chapter have introduced a new framework to tackle the microwave imaging problem in a convenient and effective way: the virtual experiments (VE) framework [10–16]. Such a framework stands on an original, target-oriented preprocessing of the data problem, able to "transform," through simple recombinations, and without additional measurements, the actual experiments into a set of equivalent *virtual* ones wherein it is possible to exploit the advantages of linearized approaches in a range of cases much wider than standard weak scattering approximations [10]. As a matter of fact, the design of the VE implicitly takes into account the geometrical and electromagnetic features of the unknown scatterers, so that the resulting linearized inversion approach has a wider range of validity than the usual diffraction tomography based on the BA.

The other main drawback in microwave imaging is the ill-posedness of the problem, which entails that only an approximated, typically smoothed, version of the unknown profile can be reliably recovered [8]. This drawback can be circumvented, at least in principle, by adopting imaging recovery approaches based on compressive sensing (CS) [17,18], which offers a powerful framework to achieve nearly optimal reconstruction of "sparse images" as well as super-resolution. However, to date, it has been completely well assessed only in the case of linear recovery problems [17,18]. As a result, in subsurface microwave imaging, CS has been so far exploited only in conjunction with weak scattering approximations [19].

In this chapter, we describe how to take advantage together from the CS theory and the VE-based linear inversion approach. In particular, we describe an imaging strategy which consists of three steps. The first step provides the information needed to design the VE and to set the "scenario-aware" linearized scattering model. The second step exploits the properties of the VE to achieve the approximation of the field inside the imaging domain needed to cast the linearized model. Finally, the last step returns the unknown permittivity and conductivity profiles of the surveyed region, by exploiting CS as regularization strategy for the VE based inversion. Different from most of the adopted CS approaches dealt with the pixel representation, herein we exploit CS to promote sparsity of solution in a representation basis different from the usual one. In fact, exploiting the fact that many scattering structures have piecewise constant electromagnetic profiles, they can be effectively represented in terms of few step functions. Notably, although piecewise constant profiles do not represent the most general case, in many circumstances involved with GPR surveys, such as mines, stones, or pipes, the scatterers can be assumed to be sparse in such a kind of basis.

While the VE-CS imaging strategy has a general applicability, it has a peculiar advantage if applied to GPR configurations, since it offers the unique possibility of counteracting "aspect limitation" of data, which significantly affects the imaging performances of tomographic imaging. Indeed, exploiting the VE framework allows the prediction of the field which would be measured in locations wherein actual measurements do not or can not occur. As a result, it is possible to exploit a number of additional equations corresponding to specific "fictitious" measurements [14], which are further enhanced in the framework of CS.

The adopted method requires data gathered under multiview–multistatic configurations. Although multichannel acquisitions are not the most common configuration in conventional GPR systems, it is worth noting that several facilities have been introduced and developed in recent years with ongoing interest in this field [20–22]. Furthermore, in some cases, these arrangements can be simply implemented also through a pair of antennas which gather the GPR echoes for multiple positions of the transmitter and the receiver. Some examples may include a couple of antennas mounted on unmanned vehicles, or moved along boreholes. On the other hand, since the method works fine with single-frequency data, the extra time paid for multichannel acquisitions is traded with no need of properly modeling the frequency dispersivity of the soil.

The chapter is organized as follows. In Section 8.2, the mathematical formulation of the inverse scattering problem underlying the microwave imaging is given for the two dimensional (2D) scalar problem. In Section 8.3, the VE framework is addressed with emphasis on the adopted internal field approximation, as well as on the possibility to counteract difficulties arising from aspect limitation of the measurement setup. Then, basics of CS for microwave imaging formulation are given in Section 8.4, while Section 8.5 presents in detail the method's implementation. Finally, Section 8.6 contains reconstruction results toward simulated data in two different measurements setups, the surface GPR and the cross-borehole configurations, corresponding to reflected and transmitted field measurements, respectively. Finally, an example concerned with a stratified soil is also addressed.

8.2   Formulation of the tomographic imaging problem

In the following, the time harmonic factor $exp\{j\omega t\}$ is assumed and dropped and a 2D geometry with the y-axis as invariance direction for both the inhomogeneities and for the electric field is considered. This allows to deal with an electromagnetic scalar problem. The investigated domain Ω, embedded in a soil with complex permittivity $\varepsilon_b(\underline{r}, \omega) = \varepsilon_b'(\underline{r}) - j\sigma_b(\underline{r})/(\omega\varepsilon_0)$, and placed at a given depth below the air–soil interface, is assumed to host the cross section $\Sigma = \Sigma_1 \cup \Sigma_2 \cup \Sigma_3$ of one or more penetrable scatterers (or anomalies) with complex permittivity $\varepsilon_s(\underline{r}, \omega) = \varepsilon_s'(\underline{r}) - j\sigma_s(\underline{r})/(\omega\varepsilon_0)$, being $\underline{r} = (x, z)$ the coordinate spanning the generic point in Ω and ω the angular frequency of the electromagnetic wave. Sketches of the two scenarios considered in the remainder of the chapter are shown in Figure 8.1.

The region Ω is probed by means of filamentary currents, located on a curve Γ_{TX}, and the field is gathered by all the receivers (located on Γ_{RX}) when, in turn, only one antenna is transmitting, that is, we are assuming a multiview–multistatic acquisition.

Under the above assumptions, the scattering phenomenon can be simply cast through a couple of integral equations, which read:

$$E_s(\underline{r}_r, \underline{r}_t) = k_b^2 \int_\Omega g_m(\underline{r}', \underline{r}_r)E_t(\underline{r}', \underline{r}_t)\chi(\underline{r}')d\underline{r}' \qquad \underline{r}_t \in \Gamma_{TX}, \ \underline{r}_r \in \Gamma_{RX}, \qquad (8.1)$$

$$E_t(\underline{r}, \underline{r}_t) - E_i(\underline{r}, \underline{r}_t) = k_b^2 \int_\Omega g_d(\underline{r}', \underline{r})E_t(\underline{r}', \underline{r}_t)\chi(\underline{r}')d\underline{r}' \qquad \underline{r} \in \Omega, \ \underline{r}_t \in \Gamma_{TX}, \qquad (8.2)$$

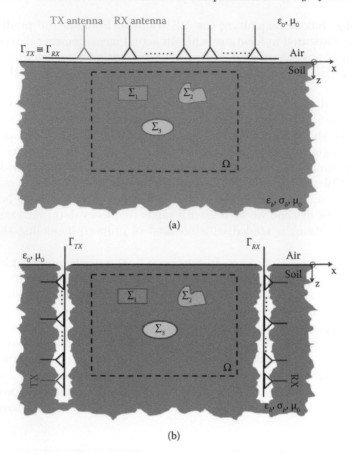

FIGURE 8.1

The geometry of the 2D problem and the adopted measurement configuration for (a) surface GPR and (b) cross-borehole measurement setup.

wherein:

- $\underline{r}_r$ and $\underline{r}_t$ denote the receiver and transmitter position, respectively;

- E_s is the scattered field at the receiver position;

- E_i and E_t denote the incident and the total field in Ω, respectively;

- $\chi(\underline{r}) = \varepsilon_s(\underline{r})/\varepsilon_b(\underline{r}) - 1$ is the unknown contrast function;

- $k_b = \omega\sqrt{\mu_0\varepsilon_0\varepsilon_b}$ is the complex wave number in the soil, μ_0 and ε_0 being the magnetic permeability and electric permittivity in the vacuum, respectively;

- $g_m(\underline{r}',\underline{r}_r)$ is the Green's function pertaining the adopted measurement configuration;

- $g_d(\underline{r}',\underline{r})$ is the Green's function pertaining to the surveyed domain.

In particular, for the two configurations depicted in Figure 8.1:

$$g_m(\underline{r}',\underline{r}_r) = \begin{cases} g_{12}(\underline{r}',\underline{r}_r) & \text{for surface GPR acquisition} \\ g_{22}(\underline{r}',\underline{r}_r) & \text{for cross-borehole acquisition} \end{cases} \qquad (8.3)$$

while for both surface GPR and cross-borehole acquisition:

$$g_d(\underline{r}', \underline{r}) = g_{22}(\underline{r}', \underline{r}), \tag{8.4}$$

wherein:

- $g_{12}(\underline{r}', \underline{r}_r)$ is the external Green's function of the half space, that is, the field radiated in the medium 1 (the air) by an elementary source placed in the medium 2 (the soil);

- $g_{22}(\underline{r}', \underline{r})$ is the internal Green's function of the half space, that is, the field radiated in the soils by an elementary source placed in the same medium.

Equation 8.1 is also referred to as "data equation," as it expresses the functional relationship between the scattered field data and the current induced (proportional to the contrast source χE_t) in the scatterer support Σ. On the other hand, Equation 8.2 is often referred to as "state equation" (or "object equation") as it relates the scattered field in Ω to the induced current.

The inverse scattering problem amounts to retrieve the unknown contrast function, which encodes the geometrical and electromagnetic properties of the targets, solving the couple of Equations 8.1 and 8.2 for given incident fields and corresponding (measured) scattered fields.

The problem introduced above shows two huge difficulties. The first one is the ill-posedness, which descends from the compactness of the integral operator in the data equation [8], and entails that only a finite number of unknown parameters can be recovered from processing the scattered fields [23]. Similarly, by virtue of reciprocity, only a finite number of independent experiments can be performed [23], so that the overall amount of independent information that can be retrieved in a multi-view configuration is essentially finite and the overall problem is ill-posed. As such, it needs to be properly regularized to avoid unstable (i.e., meaningless) solutions.

The second drawback is related to nonlinearity of the problem. Indeed from Equation 8.2, it can be seen that the total field depends on the unknown contrast function as well, so that, combining Equations 8.1 and 8.2, one easily discovers that the operator which relates the scattered field data to the contrast unknown is a nonlinear one. A very rough way to avoid nonlinearity is making use of the BA, which amounts to replace the total field with the incident one in Ω, hence coming to a linear relationship between data and unknowns. Such approximation is valid only for the weak scattering regime, that is, when the scatterers do not significantly perturb the incident field [24].

Finally, it is worth to underline that when data are gathered by means of arrays of limited aperture, as in any GPR system, another severe limitation is concerned with the unavoidable resolution loss, even when the scattering model assumptions are fulfilled. Mathematically, this is due to the spectral properties of the scattering operator when data are gathered only on a side rather than all around the imaging domain. For this reason, any linearized approach is prevented from achieving a full quantitative characterization of the surveyed area [9,14].

On the other hand, minor difficulties arise from the need of computing the half space Green's functions g_{12} and g_{22}. Notably, such a difficulty can be circumvented solving the relevant Green–Sommerfeld integrals by means of a one-dimensional discrete Fourier transform. Obviously, a sufficiently fine discretization of the spectral domain is needed and care has to be taken to ensure accuracy [25]. Also note that in some cases, such as possible deep boreholes, the scattering phenomenon can be modeled by using the Green's functions of a homogeneous background, neglecting the effect of the air–soil interface.

8.3   Basics of the VE framework

By the sake of clarity, this section is organized in three different parts. In the first part, the idea underling the VE and a simple way to design them are explained. Then, in the second part, a new linear approximation for the total field is introduced. Finally, the third part is focused on the strategy to overcome the resolution loss arising from the acquisition and processing of aspect limited data in standard GPR surveys.

8.3.1   The VE framework

Solving Equations 8.1 and 8.2 is a nontrivial task as discussed in Section 8.2. Leaving aside for the time being issues related to ill-posedness, note that nonlinearity can be faced in many different ways. In this respect, different approaches have been addressed in the last years concerned with both full nonlinear as well as approximated (first or higher orders) models [2–6].

Full nonlinear approaches do not introduce approximations in the physical model and are generally addressed as optimization problems exploiting local or global solution strategies [4,6]. However, both these strategies entail the possible occurrence of "false solutions" (or lack of convergence to the ground truth) [4,5] and require a large computational effort. For these reasons, approximated models have been also suggested in the past years as a trade-off between full nonlinear models and the basic BA. Some examples of nonlinear approximated methods are the extended Born approximation (EBA) [26], the quadratic approach [27], and the distorted Born iterative method [28,29]. The latter faces the problem via successive linearizations of the scattering problem to achieve more accurate reconstructions than BA.

Although many approaches have been addressed to manage the nonlinearity of the problem, their applicability is still limited in subsurface imaging, so that the BA remains the most popular way to tackle tomographic imaging in GPR surveys. However, approaches based on BA allow to perform only qualitative imaging, that is detection and imaging of the location and shape of buried targets, without possibility to achieve quantitative reconstructions in terms of dielectric parameters.

As a contribution toward fast and possibly accurate subsurface imaging, we consider in the following the use of the VE.

To understand the basics of such a method, let us observe that for a given linear combination (with known coefficients ξ) of the incident fields, a scattered field would be measured which is nothing else than a linear combination (with the same weight coefficients ξ) of the original scattered fields. In formulas, if we consider a superposition of the incident fields $E_i(\underline{r}, \underline{r}_t)$ radiated by some antennas placed at N_t transmitting positions, the incident field

$$\Phi_i(\underline{r}) = \sum_{t=1}^{N_t} \xi(\underline{r}_t) E_i(\underline{r}, \underline{r}_t) \tag{8.5}$$

gives rise to a total field

$$\Phi_t(\underline{r}) = \sum_{t=1}^{N_t} \xi(\underline{r}_t) E_t(\underline{r}, \underline{r}_t), \tag{8.6}$$

and, finally, this latter will produce the scattered field (measured at N_r receiving positions)

$$\Phi_s(\underline{r}_r) = \sum_{t=1}^{N_t} \xi(\underline{r}_t) E_s(\underline{r}_r, \underline{r}_t), \quad r = 1, ..., N_r, \tag{8.7}$$

where we denote with the capital greek letter Φ the "rearranged" electric fields. Note that the knowledge of the fields (Equations 8.5 and 8.7) do not require any further experiment, as all quantities can be computed via software processing. As a consequence, the logical chain from Equation 8.5 to Equation 8.7 can be interpreted as a "virtual" experiment.

According to the above, the data equation for a virtual experiment should read:

$$\Phi_s(\underline{r}_r) = k_b^2 \int_\Omega g_m(\underline{r}',\underline{r}_r)\Phi_t(\underline{r}')\chi(\underline{r}')d\underline{r}', \quad \underline{r}_r \in \Gamma_{RX}, \tag{8.8}$$

while the corresponding state equation

$$\Phi_t(\underline{r}) - \Phi_i(\underline{r}) = k_b^2 \int_\Omega g_d(\underline{r}',\underline{r})\Phi_t(\underline{r}')\chi(\underline{r}')d\underline{r}', \quad \underline{r} \in \Omega. \tag{8.9}$$

At a first glance, no advantage has been achieved comparing Equations 8.1 and 8.2 with Equations 8.8 and 8.9. However, by properly choosing the ξ coefficients, one can require given expected properties on the internal field Φ_t to simplify the solution of the problem. To pursue such a goal, the *leit-motive* of the VE is related to the following question: "What kind of linear combinations of the original incident fields is required to obtain a given behavior of the total field in a virtual experiment?"

A possible answer to this question can be given considering that one can enforce a recombined scattered field which seems to resemble the field of a filamentary current "emerging" from a given point belonging to the scatterers' support. Such a design equation is cast as

$$\sum_{t=1}^{N_t} E_s(\underline{r}_r,\underline{r}_t)\xi(\underline{r}_t,\underline{r}_s) = g_m(\underline{r}_r,\underline{r}_s), \tag{8.10}$$

where E_s is the measured scattered field, g_m is the Green's function pertaining to the measurement configuration, and ξ is the excitations coefficient sought for each *sampling point* $\underline{r}_s$ of an arbitrary grid which discretizes the imaging domain. Equation 8.10 has two interesting properties.

First, by means of a simple analytical prolongation of the field from the measurement line to the investigated area, it provides an alternative approximation of the scattered field inside the target, and hence for the total field in the whole imaging domain Ω (see the following subsection).

Second, the equation incidentally coincides with the discretized version of the *far eld equation* [30], a well-known equation in the inverse scattering community, being the cornerstone of the so-called linear sampling method (LSM). This equation has been originally conceived by Colton and co-workers [30] as an effective and simple tool to retrieve the shape of unknown scatterers.

Since Equation 8.10 is ill-posed [30], it has to be regularized. Then, according to theoretical results, the norm of the regularized solution attains low values when belongs to the support of the unknown scatterers and large values elsewhere [30]. Hence, the behavior of this *energy indicator* provides an estimate of the unknown targets' support. Moreover, since bounded energy solutions to Equation 8.10 can be achieved only for points belonging to the scatterers' support [31–33], the far field equation can be exploited as design equation for the VE only when considered for $\underline{r}_s \in \Sigma$.

It is interesting to note that, while in a standard scattering experiment, the incident field is always the same (but for the different incidence directions) regardless of the probed target, in the VE the opposite situation occurs. In fact, the incident field depends on the target, being achieved by considering the solutions $\xi(\underline{r}_t,\underline{r}_s)$. Conversely the scattered field has always the same structure, that is, $\Phi_s(\underline{r}_r,\underline{r}_s) = g_m(\underline{r}_r,\underline{r}_s)$, regardless the features of the scattering system.

8.3.2   Effective internal field approximation for nonweak scatterers

The peculiar exchange of roles between the incident and the scattered field occurring in the introduced VE can be exploited for solving the inverse scattering problem in an original fashion. Once the sampling points belonging to the scatterer's support are chosen and VE are built accordingly, the total field can be still expressed as the sum of the incident fields through $\xi(\underline{r}_t, \underline{r}_s)$, and a corresponding scattered field given by the field radiated by a filamentary current positioned in the considered sampling point. In this way, the scattered field arises from a kind of prolongation of the elementary field pattern from Γ_{RX} to Ω. We refer to all the points where the above decomposition (and the relevant approximation) can be considered as "pivot points" and denote them by $\underline{r}_p$. Obviously, for the reasons above, they should belong to the support of the scatterer (but for special cases [32,33]).

Accordingly, for each pivot point $\underline{r}_p$, the total field in the investigated region Ω can be conveniently approximated by means of the following expression:

$$\Phi_t(\underline{r}) \approx \Phi_i(\underline{r}) + LP\left[g_d(\underline{r}, \underline{r}_p)\right], \tag{8.11}$$

wherein, at the right hand side, the first addendum is the incident field arising in the VE and the second addendum is a low-pass filtered version of the Green's function pertaining to the surveyed domain. Note this operation avoids singularity for the scattered field in the pivot point [10]. It is worth to underline that this approximation takes implicitly into account the nature of the scatterer through the "weight" function ξ. Since, in this kind of VE, the target is forced by the primary sources Φ_i to behave like a point scatterer, we refer to it as the *point source   eld approximation* (PSFA).

Obviously, the above linear framework has a limited range of validity, as it is based on an approximation of the internal field, as well as on the validity range of the LSM exploited as design equation for the VE [30]. This notwithstanding, the use of the far field equation (Equation 8.10) as preprocessing step ensures the possibility to cope with a class of dielectric profiles (in terms of dimension and contrast values) widely exceeding the BA, as discussed in [34]. Such a property is related to the nature of the field approximation (Equation 8.11), that is, "scatterer-aware" through the recombination of the data provided by ξ. In particular, due to the nature of the qualitative preprocessing step, the method is applicable to compactly supported (including possibly not simply connected) targets, having electric size comparable or slighting exceeding the probing wavelength [34].

8.3.3   Auxiliary equations for aspect limited data

As discussed in the previous sections, the VE framework and the PSFA allow to provide a field approximation able to take into account the nature of the scatterer (shape and constitutive parameters). Then, by considering the field approximation (Equation 8.11) for a proper set of pivot points, we can turn the original multiview experiments into virtual multipivot ones without any significant loss of information [10].

However, the VE also turn out to be useful in dealing with the further difficulty which is specific of subsurface sensing, that is, the constrain that data has to be gathered only under aspect-limited configuration. In fact, the peculiar feature of the VE makes possible to devise an effective strategy also to counteract the relevant image degradations. Such a possibility arises from the fact that the PSFA (Equation 8.11) is valid in the whole surrounding background, since it expresses the field radiated by an elementary source embedded in the soil. As a result, one can take advantage of this expression to foresee the value of the (virtual) scattered field at locations where physical measurements are not carried out. By exploiting these   *ctitious* measurements one can enlarge the set of available equations to improve the reconstruction capability of the tomographic imaging [14]. In particular,

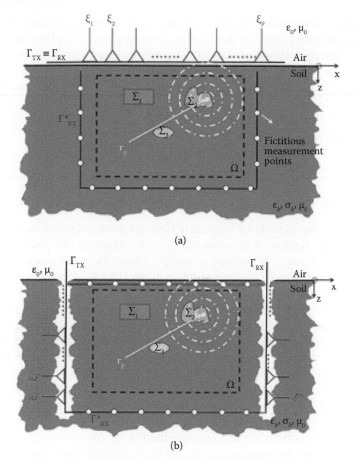

FIGURE 8.2
Virtual experiments and fictitious measurements strategy in (a) GPR surface and (b) cross-borehole survey. The incident field radiated by means of simultaneous excitation of the transmitting antennas (in blue) induces a scattered field (green dashed line) that seems to emerge from a filamentary current placed in the pivot point. The circle markers on the red lines represents the fictitious measurements which "capture" the cylindrical wave in locations where real antennas can not be positioned.

this can be attained by adding a number of fictitious measurements located on a curve Γ^*_{RX} (complementary to Γ_{RX}) in such a way to "restore" the full aspect configuration; see Figure 8.2a and b for a sketch of such fictitious arrangement. For these cases, the value of the virtual scattered field at the fictitious locations would be obviously given by $g_{22}(\underline{r}^*_r, \underline{r}_p)$.

According to the above, the data equations for the considered pivot point can be recast as

$$\int_\Omega \tilde{g}(\underline{r}', \underline{\rho}_r)\Phi_t(\underline{r}', \underline{r}_p)\chi(\underline{r}')d\underline{r}' = \begin{cases} \Phi_s(\underline{\rho}_r, \underline{r}_p) & \text{if } \underline{\rho}_r = \underline{r}_r \in \Gamma_{RX} \\ g_{22}(\underline{\rho}_r, \underline{r}_p) & \text{if } \underline{\rho}_r = \underline{r}^*_r \in \Gamma^*_{RX}, \end{cases} \tag{8.12}$$

where $\underline{r}_r$ and $\underline{r}^*_r$ denote the location of the actual and the fictitious receivers, respectively.

Accordingly, taking into account the different kind of measurements, that is, the actual and the fictitious ones, the Green's function in 8.12 reads

$$\tilde{g}(\underline{r}, \underline{\rho}_r) = \begin{cases} g_m(\underline{r}, \underline{\rho}_r) & \text{if } \underline{\rho}_r = \underline{r}_r \in \Gamma_{RX} \quad \text{as defined in Equation 8.3.} \\ g_{22}(\underline{r}, \underline{\rho}_r) & \text{if } \underline{\rho}_r = \underline{r}^*_r \in \Gamma^*_{RX} \end{cases} \tag{8.13}$$

In the exploitation of the above field conditioning, it could seem that one can place fictitious measurements arbitrarily, for instance as close as possible to the targets, inside Ω or even inside the scatterers. However, analytical prolongation (which allows for fictitious measurements) works fine outside the scatterers, while it becomes more critical when going inside the scatterers due to the effect of poorly radiating source [23,33], which cannot be taken into account by the VE design equation. Hence, a convenient and more reliable choice amounts to place the fictitious probes approximately at the same distance existing between the investigated domain and the actual probes.

8.4   Basics of the compressive sensing for microwave imaging

The CS theory provides several tools for reconstructing sparse signals from highly incomplete sets of measurements by means of constrained ℓ_1 minimizations. In particular, it is, in principle, capable to provide nearly optimal reconstructions of the unknown as long as such an unknown is sparse with respect to some suitable basis [17,18].

In order to recall the basics of the CS theory, let us consider a generic linear problem $\boldsymbol{y} = \boldsymbol{Ax}$, where $\boldsymbol{y}$ is the $N_y \times 1$ data vector, $\boldsymbol{x}$ is the $N_x \times 1$ vector that represents the unknown function, while $\boldsymbol{A}$ is the $N_y \times N_x$ matrix which relates the unknown vector to the data vector. By assuming the usual CS terminology, $\boldsymbol{A}$ represents the sensing matrix.

Let us now suppose to adopt a convenient representation matrix $\boldsymbol{\Psi}$ ($\boldsymbol{x} = \boldsymbol{\Psi s}$), so that the representation coefficients $\boldsymbol{s}$ are sparse (i.e., only few coefficients are different from zero). According to the CS theory, by taking advantage from the sparsity of the unknown coefficients $\boldsymbol{s}$, it is possible to solve the inverse problem even if N_y is (much) less than N_x, but it is anyway sufficiently larger than the number S of coefficients different from zero (with $S < N_y < N_x$). In particular, the correct number of measurements N_y, necessary to obtain a faithful solution, is lower-bounded and has to satisfy the inequality $N_y \geq N_y'$, where N_y' is proportional to S and to $logN_x$ [17]. Note that the exact reconstruction of a sequence of S-sparse signal can be achieved only when the number N_y of measurements is sufficiently larger than S. Numerical analyses of canonical cases have suggested that a successful reconstruction is achieved, in more than 50% of the cases, when $N_y \geq 4S$, and, in more than 90% of the cases, when $N_y \geq 8S$ [17,18].

A very relevant nonintuitive circumstance is that it is not just the number of measurements which plays a role, since also the kind of performed measurements plays a role as well. Intuitively, the matrix $\boldsymbol{\Theta} = \boldsymbol{A\Psi}$ has to be deeply different from a subset of the identity matrix and should not cancel out any information about the original signal. In particular, the image of the columns of $\boldsymbol{\Psi}$ should be spread out in the domain defined by the rows of $\boldsymbol{A}$. As a matter of fact, the larger this incoherence, the better the possibility to retrieve sparse signals by compressed measurements. Moreover, the matrix $\boldsymbol{\Theta}$ should approximately preserve the Euclidean length of S-sparse signals, so that each S-sparse vector cannot be mapped in the null space of $\boldsymbol{\Theta}$, and each S-sparse vector has to show a nonnegligible image in the space of data. The exact requirements can be formalized by means of the so-called restricted isometry property (RIP) [17,18], which guarantees conditions for an exact recovery.

Provided the above conditions are fulfilled, it is possible to solve the inverse problem $\boldsymbol{y} = \boldsymbol{Ax}$ by means of the following optimization constrained problem [17,18]:

$$\min_{\boldsymbol{s}}\{\|\boldsymbol{s}\|_{\ell_1}\}$$
$$\text{subject to } \|\boldsymbol{\Theta s} - \boldsymbol{y}\|_{\ell_2} < \delta, \tag{8.14}$$

where δ is a positive parameter which depends on the level of required accuracy as well as on the modeling and measurement errors.

The problem (Equation 8.14) is commonly known as the basis pursuit denoising (BPDN) or least absolute shrinkage and selection operator (LASSO) problem [35]. In Equation 8.14, the minimization of the ℓ_1 norm promotes the search of sparse solutions, while the constraint enforces the data consistency. In other words, among all solutions which are consistent with the measured data within a given error, the sparsest one is sought. Note that, while the optimization problem should consider indeed the so-called ℓ_0 norm [18], the relaxation into ℓ_1 norm adopted in Equation 8.14 reduces the problem to a convex programming one, and the two formulations are equivalent for a wide range of cases [17,36].

8.5 Implementation procedure of the VE-CS imaging approach

The implementation procedure of the inversion approach can be basically divided in three main steps. With reference to Figure 8.1, let us assume that the transmitting antenna is moved in some N_t positions, while all the others are receiving at N_r positions. Once collected the data, the first step of the procedure consists in the solution of the design equation (Equation 8.10).

To apply the LSM, we have to sample with an arbitrary grid of points the region under test Ω. The spatial sampling rate is arbitrary, since it will affect only the final visualization of the indicator function without increasing the computational burden. Accordingly, for the sake of simplicity, we select an even grid of N_s sampling points, spaced at least $\lambda_b/4$, λ_b being the wavelength in the host medium, i.e., the soil. For each sampling point, we solve the discretized version of Equation 8.10, which in a compact matrix-vector form reads

$$\mathbf{E_s}\xi = g_m, \tag{8.15}$$

where the exact meaning of the quantities in Equation 8.15 for the two different measurement setups is given in the following.

Surface GPR acquisition *(reflection measurement configuration).* When we deal with the surface GPR configuration, the N_t and N_r antennas are usually located at the interface between air and soil, possibly on the same line ($\Gamma_{TX} \equiv \Gamma_{RX}$) (see Figure 8.1a). In this case, $\mathbf{E_s}$ is an $N_r \times N_t$ data matrix having the following structure:

$$\mathbf{E_s} = \begin{pmatrix} Reflection \\ Data \end{pmatrix} = \begin{pmatrix} e_{11} & \cdots & e_{1N_t} \\ \vdots & \ddots & \vdots \\ e_{N_r 1} & \cdots & e_{N_r N_t} \end{pmatrix}. \tag{8.16}$$

Moreover, ξ is the $N_t \times 1$ vector containing the samples of the unknown function ξ and g_m is an $N_r \times 1$ vector containing the samples of the half space Green's function g_{12} for the considered sampling point $\underline{r}_s$ (see Equation 8.3).

Cross-borehole acquisition *(transmission measurement configuration).* When we deal with the cross-borehole configuration, the N_t antennas are located along the first borehole, say it Γ_{TX}, while the N_r ones are positioned along the second borehole Γ_{RX}, as shown in Figure 8.1b. In the latter case, for the sake of simplicity, we also assume $N_r = N_t$. In this case, $\mathbf{E_s}$ is a $2N_r \times 2N_t$ block matrix which has the following structure:

$$\mathbf{E_s} = \begin{pmatrix} 0 & \begin{array}{c} Reciprocal \\ data \end{array} \\ \begin{array}{c} Transmission \\ data \end{array} & 0 \end{pmatrix} = \begin{pmatrix} 0 & \cdots & 0 & e_{11} & \cdots & e_{N_r 1} \\ \vdots & \ddots & \vdots & \vdots & \ddots & \vdots \\ 0 & \cdots & 0 & e_{1N_t} & \cdots & e_{N_r N_t} \\ e_{11} & \cdots & e_{1N_t} & 0 & \cdots & 0 \\ \vdots & \ddots & \vdots & \vdots & \ddots & \vdots \\ e_{N_r 1} & \cdots & e_{N_r N_t} & 0 & \cdots & 0 \end{pmatrix}. \tag{8.17}$$

To build such a matrix, we exploit the available $N_r \times N_t$ data points collected by moving the transmitter along Γ_{TX} and the receiver along Γ_{RX}. Accordingly, the generic entry e_{ij} of the bottom-left block is the complex scattered field value at the ith receiver for the jth transmitter's position. The upper-right block of the matrix (Equation 8.17) is then properly filled by reciprocity ($e_{ji} = e_{ij}$), while the other two blocks are filled with zero entries corresponding to those pairs (i, j) for which no measurement is actually performed. As a matter of fact, reciprocity and the use of zero entries to build the matrix represent an effective way to handle aspect-limited data for the LSM [37]. Accordingly, $\boldsymbol{\xi}$ is the $2N_t \times 1$ vector containing the values of the unknown function ξ and $\boldsymbol{g_m}$ is a $2N_r \times 1$ vector containing the samples of the Green's function g_{22}, as in Equation 8.3, for the considered sampling point.

In order to deal with a common formulation for both the above measurement setups, we denote with R and T the total number of receivers and transmitters, which correspond to the dimension of the rows and columns of the matrix $\mathbf{E_s}$, respectively. Accordingly, $\boldsymbol{\xi}$ is an $T \times 1$ vector, and $\boldsymbol{g_m}$ is an $R \times 1$ vector.

Solution of Equation 8.15 can be suitable achieved by means of the singular value decomposition (SVD) of the data matrix $\mathbf{E_s}$. For the generic sampling point, it reads

$$\boldsymbol{\xi} = \mathbf{V}\boldsymbol{\Sigma}\mathbf{U}^+ \boldsymbol{g_m}, \tag{8.18}$$

wherein $\mathbf{U}$ is an $R \times R$ matrix whose columns are the left singular vectors of $\mathbf{E_s}$, $\mathbf{V}$ is a $T \times T$ matrix whose rows are the right singular vectors of $\mathbf{E_s}$, and $^+$ denotes the conjugate transpose.

$$\boldsymbol{\Sigma} = diag\left(\frac{\sigma_1}{\sigma_1^2 + \gamma^2}, \frac{\sigma_2}{\sigma_2^2 + \gamma^2}, \cdots, \frac{\sigma_N}{\sigma_N^2 + \gamma^2}\right), \tag{8.19}$$

with $N = min\{T, R\}$, σ_n denoting the nth singular value of $\mathbf{E_s}$ and γ the Tikhonov regularization parameter fixed according to [38], being the same for all the sampling points. Then, an effective visualization of the target's support can be obtained by means of the normalized indicator function defined as [39]

$$\Upsilon_N = \frac{log_{10}\mathcal{I} - (log_{10}\mathcal{I})_{max}}{min\left[log_{10}\mathcal{I} - (log_{10}\mathcal{I})_{max}\right]}, \tag{8.20}$$

wherein $\mathcal{I} = ||\boldsymbol{\xi}||/||\boldsymbol{g_m}||$, which continuously varies between 0 and 1 and assumes largest values in points belonging to the scatterers' support. Finally, it is worth noting that solving Equation 8.15 is an efficient computational process, as the required SVD of $\mathbf{E_s}$ is the same for all the sampling points.

The second step of the procedure deals with the implementation of the quantitative inversion. For that purpose, one has first to pick, within the N_S sampling points, a subset of P pivot points. As discussed in [10], the guidelines to select the pivot points are basically two:

1. P can be chosen in the order of the actual degrees of freedom of the scattered fields [23], which can be estimated from the electric size of the target's support as obtained via LSM indicator.

2. The pivot points have to be evenly spaced where Υ_N approaches 1 within the estimated target's support. However, since the only aim is to provide a sufficient "diversity" among the VEs, no meaningful differences are expected for slight changes of the pivot points' position.

Then, the corresponding LSM solutions are exploited to build the approximated total fields $\Phi_t(\underline{r}_1), ..., \Phi_t(\underline{r}_P)$, using (Equation 8.11). In this respect, assuming that the domain Ω_I to be imaged (which may correspond to the whole region Ω or a portion of it) is discretized into N_c square cells, the discretized data equation 8.12 accounting for the overall actual and fictitious measurements reads as

$$\begin{bmatrix} \boldsymbol{L}^a \\ \boldsymbol{L}^f \end{bmatrix} \boldsymbol{\chi} = \begin{bmatrix} \boldsymbol{\Phi}_s \\ \boldsymbol{g}_{22} \end{bmatrix}, \tag{8.21}$$

wherein:

- $\boldsymbol{L}^a$ is the $(R \times P) \times N_c$ block of the scattering operator accounting for the actual measurements, whose generic element is given by $l^a_{ij,k} = \Phi_t(\underline{r}_p, \underline{r}) g_m(\underline{r}_r, \underline{r})$;

- $\boldsymbol{L}^f$ is the $(R^* \times P) \times N_c$ block of the scattering operator accounting for the fictitious measurements, whose generic element is given by $l^f_{ij,k} = \Phi_t(\underline{r}_p, \underline{r}) g_{22}(\underline{r}_r^*, \underline{r})$, R^* being the number of fictitious measurements;

- $\boldsymbol{\Phi}_s$ is the $R \times P$ vector containing the values of the recombined scattered field according to Equation 8.7;

- $\boldsymbol{g}_{22}$ is the $R^* \times P$ value of the Green's function mimicking the virtual scattered field at the fictitious measurements $\underline{r}_r^*$;

- $\boldsymbol{\chi}$ is the $N_c \times 1$ unknown vector containing the value of the contrast in the kth cell of the imaging domain.

Accordingly, in a more compact matrix-vector notation, the imaging problem is cast as solution of the global matrix equation:

$$\boldsymbol{L}\boldsymbol{\chi} = \boldsymbol{f}, \tag{8.22}$$

wherein $\boldsymbol{L}$ has final dimension of $[(R + R^*) \times P] \times N_c$ and the data vector $\boldsymbol{f}$ has dimension $[(R + R^*) \times P] \times 1$.

Once the problem is linearized, the third and final steps of the imaging procedure consists in the solution of Equation 8.22.

In particular, we adopt the LASSO scheme described in Section 8.4 as customized to the case of profiles having a sparse gradient.

Hence, the optimization problems reads

$$\min_{\boldsymbol{\chi}} \{ \|\boldsymbol{\Delta}_x \boldsymbol{\chi}\|_{\ell_1} + \|\boldsymbol{\Delta}_z \boldsymbol{\chi}\|_{\ell_1} \}$$
$$\text{subject to} \quad \|\boldsymbol{L}\boldsymbol{\chi} - \boldsymbol{f}\|_{\ell_2} < \delta, \tag{8.23}$$

where $\boldsymbol{\Delta}_x$ and $\boldsymbol{\Delta}_z$ are the discretized version of the gradient along the two coordinate directions x and z. In other words, $\boldsymbol{\Delta}_x \boldsymbol{\chi}$ and $\boldsymbol{\Delta}_z \boldsymbol{\chi}$ are the $N_c \times 1$ vectors containing the forward finite differences of the unknown function $\boldsymbol{\chi}$.

Finally, in Equation 8.23, as assessed in [11], the tolerance parameter δ has to be set as a trade-off between the reconstruction accuracy and the feasibility of the optimization task. To avoid the trivial solution, that is, the null vector, the parameter is conveniently chosen lower than $\|\mathbf{f}\|_{\ell_2}$, i.e. $\delta = \hat{\delta}\|\mathbf{f}\|_{\ell_2}$, with $\hat{\delta} < 1$.

8.6   Numerical examples

To give a proof of the VE-CS methods's performances we present some numerical examples dealing with scenarios mimicking those faced in subsurface surveys. Our aim is to show how the effectiveness of field conditioning pursued through the VE, joint with the use of fictitious measurements and CS (exploiting sparseness of the unknown's gradient), allows to appraise the morphologic and constitutive properties of buried objects and not only their location and shape. The imaging strategy has been tested against data gathered under the two different measurement configurations, i.e. the GPR surface and the cross-borehole configurations.

The scattered field data have been simulated by means of a 2D full wave finite element solver (COMSOL Multiphysics®) and corrupted by means of white gaussian noise at a given noise-to-signal ratio (SNR).

To appraise the accuracy of the results, we have used different reconstruction error metrics. The first one is the usual mean square error:

$$MSE = \frac{\|\chi - \tilde{\chi}\|^2}{\|\chi\|^2}, \tag{8.24}$$

where χ is the actual contrast profile and $\tilde{\chi}$ the estimated one. As this metrics is a synthetic parameter, it is not always suitable to evaluate the accuracy of the reconstruction. For this reason, other metrics have been introduced to compensate this drawback, i.e.

$$Err_{max} = \frac{\|\chi\|_\infty - \|\tilde{\chi}\|_\infty}{\|\chi\|_\infty}, \tag{8.25}$$

where the $\| \cdot \|_\infty$ is the uniform norm, and

$$Err_{loc} = \frac{\sqrt{(\tilde{x} - x)^2 + (\tilde{z} - z)^2}}{\lambda_b}, \tag{8.26}$$

where $(\tilde{x}, \tilde{z})$ and (x, z) are the coordinates of the center of the retrieved and the actual objects, respectively.

Finally, to show the performance achievable by the method, we have compared the results with those obtained with the standard BA in conjunction with the same CS recovery approach of Equation 8.23. The numerical implementation of Equation 8.23 exploits the CVX MATLAB® toolbox [40].

8.6.1   Surface GPR configuration

In this section, we present a numerical example concerned with a GPR surface configuration, where the transmitting and receiving antennas are placed in a different medium (usually air) with respect to the investigated domain (see Figure 8.1a). Three rectangular targets, each one with dimension of about $\lambda_b/2 \times \lambda_b/4$, and different dielectric properties, are buried in a dry soil that exhibits a random variation of $\pm 5\%$ around the average values of $\varepsilon_b' = 4$

and $\sigma_b = 1$ mS/m. The first target is a void ($\varepsilon' = 1$), the second target mimics a plastic mine ($\varepsilon' = 2.5, \sigma = 0$), and the third has dielectric properties of a stone ($\varepsilon' = 6$, $\sigma = 1$ mS/m), at the working frequency of 400 MHz. The imaging domain, placed just below the air–soil interface, is 1.5×1 m^2 large (about $4\lambda_b \times 2.5\lambda_b$) and is discretized into 96×64 cells. The probing array above the air–soil interface is 2.5 m long and consists of $N_t = N_r = 14$ evenly spaced antennas (see Figure 8.3a and b). Finally, the simulated scattered field data have been corrupted with a SNR = 30dB.

The LSM indicator Υ_N obtained after solving Equation 8.15 is shown in Figure 8.3c. As it can be seen, this indicator allows to detect the three objects and to retrieve qualitatively their position and shape. Obviously, at this stage nothing can be inferred about their electric properties, but we can use this information to perform the following reconstruction step. In this respect, we are able to choose $P = 28$ pivot points needed to build the VE.

As a countermeasure to aspect-limited data, we add three dummy arrays with R^* fictitious receivers on the curve Γ^*_{RX}. The first one, with 14 antennas, is 2.5 m long and is placed at depth -1 m from the interface. The second and the third virtual arrays, 1 m long, are placed perpendicular to the actual array and each one consists of five antennas. The overall number of fictitious measurements is $R^* = 24$, and the resulting configuration is shown in Figure 8.3d. For this example, the number of nonnull coefficients in the step functions representation basis is $S = 53$.

The reconstruction results are shown in Figure 8.3e and f, wherein it is possible to observe that the VE-CS method allows to obtain nearly accurate results, while the BA does not allow to achieve a satisfactory imaging Figures 8.3g and h. In Table 8.1, error metrics and $\hat{\delta}$ regularization parameters are reported for the two imaging approaches.

Note that results in Figures 8.3e and f have been obtained by exploiting the explicit information arising from Υ_N, that is, by considering as imaging domain Ω_I only a portion of the whole surveyed area, see Figure 8.3d. This can be actually achieved, solving only the subset of equations in the system Equation 8.22 pertaining to those pixels where the targets are expected to belong. In this way, more accurate results and faster inversion procedures can be performed.

As a last comment, it is worth noting that the number of processed data in the VE-CS inversion can be reduced without significant loss in the reconstruction accuracy. Such an issue has been investigated in [15] where a random selection procedure, with different sampling rates, has been suggested to further reduce the computational burden of the CS-based recovery approach.

8.6.2 Cross-borehole configuration

The numerical example reported in this section concerns with a cross-borehole configuration, as depicted in Figure 8.1b. The surveyed area is a limestone region hosting an oil-shale and an empty cavity (see Figure 8.4a and b). At the working frequency of 18 MHz, the oil-shale's parameters are $\varepsilon'_s = 4$ and $\sigma_s = 2.5$ mS/m [41–43], while for limestone we assume a random variation of $\pm 5\%$ around the average values $\varepsilon'_b = 6$ and $\sigma_b = 0.1$ mS/m. The investigated domain Ω is a 30 m $\times$ 30 m square region discretized into 78×78 cells. Moreover, its top side is 12 m below the air–soil interface. The two boreholes (30 m long) are positioned at a distance of 31 m each from the other, hosting 11 evenly spaced probes. In particular, the N_t transmitters are moved along the leftmost borehole, while the N_r receivers are moved on the rightmost borehole. Note that, despite data have been simulated by taking into account the presence of the air–soil interface, in the inversion procedure we have neglected the presence of the interface adopting the Green's function of the homogeneous background [25] in solving Equation 8.22. This entails, in addition to the random variations of the background medium, another model error which allows to validate the robustness of the method against

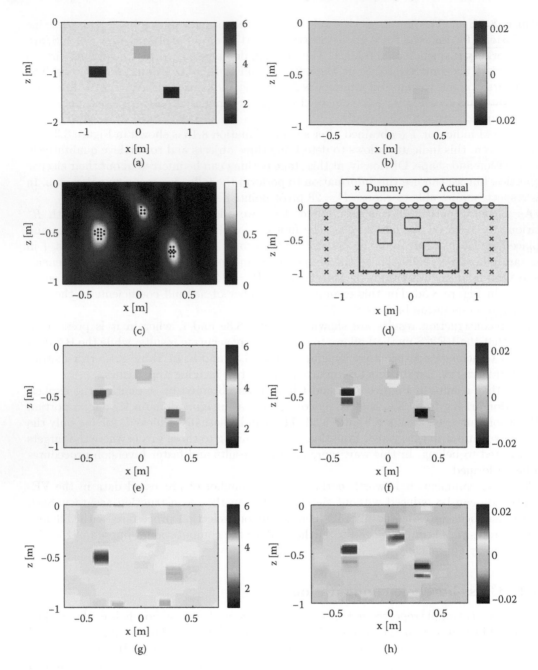

FIGURE 8.3

Imaging results with data gathered under a surface GPR configuration. (a) Permittivity and (b) conductivity distribution of the reference profile. (c) LSM support indicator with superimposed pivot points. (d) Sketch of the fictitious measurements and the contour plot of the imaging domain Ω_I (dashed line) considered in the CS reconstruction approach. Permittivity and conductivity of the retrieved profile by means of (e)–(f) the VE-CS and (g)–(h) the BA-CS approach.

TABLE 8.1

Error Metrics for the Surface GPR Scenario

	MSE	$\mathrm{Err_{max}}$	$\mathrm{Err_{loc}(stone)}$	$\mathrm{Err_{loc}(void)}$	$\mathrm{Err_{loc}(mine)}$	$\hat{\delta}$
VE-CS	0.63	0.24	0.041	0.083	0.042	0.35
BA-CS	0.63	0.51	0.060	0.080	0.060	0.14

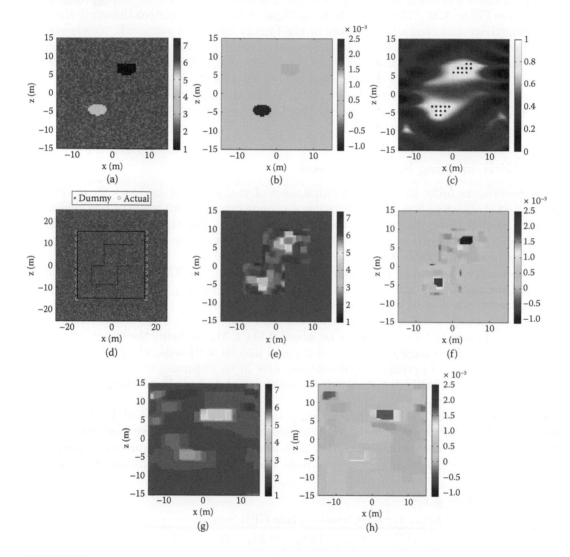

FIGURE 8.4

Imaging results with data gathered under a cross-borehole configuration. (a) Permittivity and (b) conductivity distribution of the reference profile. (c) LSM support indicator with superimposed pivot points. (d) Sketch of the fictitious measurements and the contour plot of the imaging domain Ω_I (dashed line) considered in the CS reconstruction approach. Permittivity and conductivity of the retrieved profile by means of (e)–(f) the VE-CS and (g)–(h) the BA-CS approach.

possible uncertainties. In this case, the number of nonnull coefficients is $S = 42$ while the scattered field data have been corrupted with a SNR=30dB.

In Figure 8.4c, the map of the normalized indicator Υ_N is reported. It allows to clearly appraise the presence, positions and dimensions of two disjoint anomalies. Then, we consider $P = 22$ pivot points evenly spaced within the estimated targets' support to built the VE and two fictitious arrays of 30 m placed perpendicularly to the actual ones at a distance of ± 18 m from the center of the investigated domain. The overall number of fictitious probes is $R^* = 18$, (see Figure 8.4d). The results shown in Figure 8.4e and f are quite satisfactory since both permittivity and conductivity values of the targets have been retrieved. Conversely, BA-CS inversion (see Figure 8.4g and h), only yields a rough estimate of the objects' shape and it does not provide any information about their electromagnetic properties. In Table 8.2, error metrics and $\hat{\delta}$ regularization parameters are reported for the two approaches.

As in the previous example, also in this case, the results shown in Figure 8.4e and f have been obtained by exploiting the information about the targets' location (see Figure 8.4d for the considered imaging domain).

8.6.3   GPR imaging in stratified soil

In this Section, in order to test the performances of the proposed approach in a more realistic and challenging scenario, we consider a nonhomogeneous soil made of non planar stratified medium. For this reason, the Green's functions and the approximations in Equations 8.22–8.24 have been numerically evaluated via software simulation and exploiting reciprocity arguments (Table 8.3).

The scenario considered in this example is shown in Figure 8.5a. The inhomogeneous soil is made of two materials: sand ($\varepsilon'_{b1} = 4$, $\sigma_{b1} = 1mS/m$) and clay ($\varepsilon'_{b2} = 5$, $\sigma_{b2} = 10mS/m$). The oval target is a void ($\varepsilon'_{s1} = 1$) and is located in the clay, while the rectangular target having properties $\varepsilon'_{s2} = 2.5$, $\sigma = 0$ is buried in the sand. The working frequency is $400MHz$. The imaging domain is $1.5 \times 1\,m^2$ large, i.e. about $4.5\lambda_b \times 3\lambda_b$, λ_b being the wavelength in the clay medium. The imaging domain is discretized into 96×64 cells. The probing array is made of $N_t = N_r = 14$ evenly spaced antennas over in an alignment of $2.5\,m$ located at the air-soil interface. Finally, the simulated scattered field data have been corrupted with a SNR = 20dB.

The LSM indicator Y_N with the superimposed pivot points ($P = 6$) is shown in Figure 8.5b. In order to apply the method of fictitious measurements we consider three "dummy" arrays, as shown in Figure 8.5c. The first one, with 14 measurement points, is long $2.5\,m$ and is placed at depth $-1\,m$ from the interface; the second and the third

TABLE 8.2
Error Metrics for the Cross-Borehole GPR Scenario

	MSE	Err$_{max}$	Err$_{loc}$(void)	Err$_{loc}$(oil)	$\hat{\delta}$
VE-CS	0.53	0.30	0.056	0.08	0.38
BA-CS	0.63	0.56	0	0.11	0.2

TABLE 8.3
Error Metrics for the Example in Stratified Soils

	MSE	Err$_{max}$	Err$_{loc}$(void)	Err$_{loc}$(oil)	$\hat{\delta}$
VE-CS	0.45	0.06	0.07	0.02	0.38
BA-CS	0.49	0.37	0.013	0.09	0.2

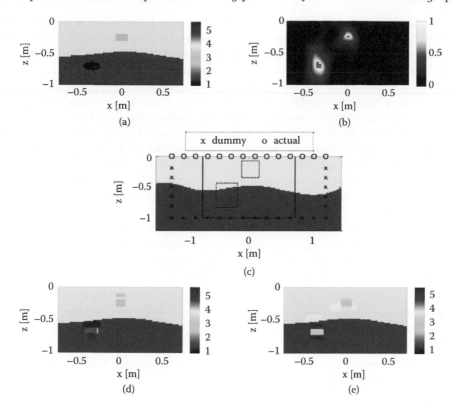

FIGURE 8.5
Imaging results with data gathered under surface GPR configuration in stratified media.
(a) Permittivity distribution of the reference profile. (b) LSM support indicator with
superimposed the pivot points and (c) sketch of the fictitious measurements and the contour
plot (dashed line) of the imaging domain Ω_I considered in the CS reconstruction approach.
Permittivity of the retrieved profile by means of (d) the VE-CS and (e) the BA-CS approach.

virtual arrays, $1\,m$ long, are placed perpendicular to the actual array and for each of them
5 measurement points are considered.

The achieved reconstructions are shown in Figure 8.5d and e. As it can be seen the
proposed method again allows to discriminate the different nature of the two targets, as
well as their position and dimension, while BA-CS only allows to fairly localize the target,
without providing quantitative results, especially for the deepest target. Note that the
retrieved conductivity is not shown since in both cases it is negligible with respect to the
permittivity values.

8.7   Conclusion

In this chapter, we have presented a strategy for quantitative microwave tomographic
imaging of nonweak scatterers. The reconstruction strategy is based only on linear
processings and relies on three main concepts. The first one is concerned with the adoption
of a solution framework, the Virtual Experiments, as well as a new effective total field
approximation, which allow to enlarge the class of retrievable profiles with respect to the
usual BA. The second contribution is concerned with the adoption of an original strategy to

counteract the unavoidable imaging deterioration of tomographic imaging in aspect-limited measurement configurations. The last contribution is essential to ensure quantitative results and relies on the exploitation of the CS theory. In particular, this strategy is able to obtain nearly optimal reconstruction of nonweak targets with piecewise constant profiles.

The overall imaging strategy is made up of two linear steps. The first one is concerned with the solution of a "design equation" which recast the original scattering experiments into a set of "virtual" equivalent ones. The second one tackles the solution of a linearized inverse scattering problem by means of a CS-inspired approach. Interestingly, the first step concerned with the design equation, requires a very low computational burden, so that the overall imaging strategy entails a very low computational effort which is fully comparable with the commonly adopted linearized approaches. On the other hand, this approach is able to retrieve a full quantitative characterization of the surveyed area without resorting to nonlinear optimization.

The numerical analysis concerned with full wave simulated data shows a good performance of the method in achieving satisfactory reconstruction of buried targets in lossy soils. As a result, this may pave the way to completely new reliable and fast strategies to perform 2D and 3D tomographic imaging in GPR surveys.

References

[1] D. J. Daniels. *Ground Penetrating Radar*, 2nd edition. London, UK: The Institution of Electrical Engineers, 2004.

[2] R. Persico. *Introduction to Ground Penetrating Radar: Inverse Scattering and Data Processing*. New York: Wiley, 2014.

[3] F. Soldovieri and L. Crocco. Electromagnetic tomography. In: *Subsurface Sensing*, Edited by A. S. Turk, K. A. Hocaoglu, and A. A. Vertiy. New York: Wiley, 2011, pp. 228–254.

[4] L. Crocco and F. Soldovieri. Nonlinear inversion algorithms. In: *Subsurface Sensing*, Edited by A. S. Turk, K. A. Hocaoglu, and A. A. Vertiy. New York: Wiley, 2011, pp. 365–376.

[5] M. Pastorino. *Microwave Imaging*. Wiley Online Library, 2010.

[6] I. Catapano, A. Randazzo, E. Slob, and R. Solimene. GPR imaging via qualitative and quantitative approaches. In: *Civil Engineering Apllications of Ground Penetrting Radar*, Edited by A. Benedetto and L. Pajewski. Berlin: Springer, 2015, pp. 239–280.

[7] D. Colton and R. Kress. *Inverse Acoustic and Electromagnetic Scattering Theory*. Berlin, Germany: Springer-Verlag, 1992.

[8] M. Bertero and P. Boccacci. *Introduction to Inverse Problems in Imaging*. Bristol, UK: Institute of Physics, 1998.

[9] G. Leone and F. Soldovieri. Analysis of the distorted Born approximation for subsurface reconstruction: Truncation and uncertainties effects. *IEEE Trans. Geosci. Remote Sens.* 41, 1 (2003), pp. 66–74.

[10] L. Crocco, I. Catapano, L. Di Donato, and T. Isernia. The Linear Sampling Method as a way for quantitative inverse scattering. *IEEE Trans. Antennas Propag.* 4, 60 (2012), pp. 1844–1853.

[11] M. Bevacqua, L. Crocco, L. Di Donato, and T. Isernia. Microwave imaging of non-weak targets via compressive sensing and virtual experiments. *IEEE Antennas and Wireless Propag. Lett.* 14 (2015), pp. 1035–1038.

[12] L. Di Donato, M. Bevacqua, L. Crocco, and T. Isernia. Inverse scattering via virtual experiments and contrast source regularization. *IEEE Trans. Antennas Propag.* 63, 4 (2015), pp. 1669–1677.

[13] M. Bevacqua, L. Crocco, L. Di Donato, and T. Isernia. An algebraic solution method for nonlinear inverse scattering. *IEEE Trans. Antennas Propag.* 63, 2 (2015), pp. 601–610.

[14] L. Di Donato and L. Crocco. Model based quantitative cross-borehole GPR imaging via virtual experiments. *IEEE Trans. Geosci. Remote Sens.* 53, 8 (2015), pp. 4178–4185.

[15] M. Bevacqua, L. Crocco, L. Di Donato, T. Isernia, and R. Palmeri. Exploiting sparsity and field conditioning in subsurface microwave imaging of nonweak buried targets. *Radio Sci.* 51, 4 (2016), pp. 301–310.

[16] L. Di Donato, R. Palmeri, G. Sorbello, T. Isernia, and L. Crocco. A new linear distorted-wave inversion method for microwave imaging via virtual experiments. *IEEE Trans. Microw. Therory Tech.* 64, 8 (2016), pp. 2478–2488.

[17] D. Donoho. Compressed sensing. *IEEE Trans. Inf. Theory* 52, 4 (2006), pp. 1289–1306.

[18] R. G. Baraniuk. Compressive sampling. *IEEE Signal Process. Mag.* 24, 4 (2007), pp. 118–124.

[19] M. Ambrosanio and V. Pascazio. A compressive-sensing-based approach for the detection and characterization of buried objects. *IEEE J. Sel. Top. Appl. Earth Observ. Remote Sens.* 8, 7 (2015), pp. 3386–3395.

[20] J. H. Bradford. Measuring water content heterogeneity using multifold GPR with reflection tomography. *Vadose Zone J.* 7 (2008), pp. 184–193.

[21] H. Gerhards, U. Wollschlager, Q. Yu, P. Schiwek, X. Pan, and K. Roth. Continuous and simultaneous measurement of reflector depth and average soil-water content with multichannel ground penetrating radar. *Geophysics* 73 (2008), J15–J23.

[22] R. G. Francese, E. Finzi, and G. Morelli. 3-D high-resolution multi-channel radar investigation of a Roman village in Northern Italy. *J. Appl. Geophys.* 67, 1 (2009), pp. 44–51.

[23] O. M. Bucci and T. Isernia. Electromagnetic inverse scattering: Retrievable information and measurement strategies. *Radio Sci.* 32 (1997), pp. 2123–2138.

[24] M. Slaney, A. C. Kak, and L. E. Larsen. Limitations of imaging with first-order diffraction tomography. *IEEE Trans. Microw. Theory Tech.* 32 (1984), pp. 860–874.

[25] W. C. Chew. *Waves Anf Fields in Inhomogeneous Media.* Piscataway, NJ: The Institute of Electrical and Electronics Engineers, 1995.

[26] Q. H. Liu, Z. Q. Zhang, and X. M. Xu. The hybrid extended Born approximation and CG-FFT method for electromagnetic induction problems. *IEEE Trans. Geosci. Remote Sens.* 39, 2 (2001), pp. 347–355.

[27] G. Leone, R. Persico, and R. Solimene. A quadratic model for electromagnetic subsurface prospecting. *Int. J. Electron. Commun.* 57, 1 (2003), pp. 33–46.

[28] T. J. Cui, W. C. Chew, A. A. Aydiner, and S. Chen. Inverse scattering of two-dimensional dielectric objects buried in a lossy earth using the distorted Born iterative method. *IEEE Trans. Geosci. Remote Sens.* 39, 2 (2001), pp. 339–346.

[29] F. Li, Q. H. Liu, and L. P. Song. Three-dimensional reconstruction of objects buried in layered media using Born and distorted Born iterative methods. *IEEE Trans. Geosci. Remote Sens. Lett.* 1, 2 (2004), pp. 107–111.

[30] F. Cakoni and D. Colton. *Qualitative Methods in Inverse Scattering Theory.* Berlin, Germany: Springer-Verlag, 2006.

[31] D. Colton, H. Haddar, and M. Piana. The linear sampling method in inverse electromagnetic scattering theory. *Inverse Probl.* 19 (2003), pp. 105–137.

[32] I. Catapano, L. Crocco, and T. Isernia. On simple methods for shape reconstruction of unknown scatterers. *IEEE Trans. Antennas Propag.* 55 (2007), pp. 1431–1436.

[33] L. Crocco, L. Di Donato, I. Catapano, and T. Isernia. An improved simple method for imaging the shape of complex targets. *IEEE Trans. Antennas Propag.* 2 (2013), pp. 843–851.

[34] L. Di Donato, R. Palmeri, G. Sorbello, T. Isernia, and L. Crocco. Assessing the capabilities of a new linear inversion method for quantitative microwave imaging. *Int. J. Antennas Propag.* 2015 (2015), ID:403760.

[35] R. Tibshirani. Regression shrinkage and selection via the LASSO. *J. Roy. Stat. Soc. B. Methodol.* (1996), pp. 267–288.

[36] E. J. Candes, M. Wakin, and S. Boyd. Enhancing sparsity by reweighted l_1 minimization. *J. Fourier Anal. Appl.* 14, 5 (2008), pp. 877–905.

[37] I. Catapano, L. Crocco, M. D' Urso, and T. Isernia. 3D microwave imaging via preliminary support reconstruction: Testing on the Fresnel 2008 database. *Inverse Probl.* 25, 2 (2009), pp. 1–23.

[38] I. Catapano and L. Crocco. An imaging method for concealed targets. *IEEE Trans. Geosci. Remote Sens.* 47, 5 (2009), pp. 1301–1309.

[39] I. Catapano, L. Crocco, and T. Isernia. Improved sampling method for shape reconstruction of 3D buried target. *IEEE Trans. Geosci. Remote Sens.* 46 (2008), pp. 3265–3273.

[40] CVX Research, Inc., *MATLAB software for disciplined convex programming, 2.0.*, Available at http:// cvxr.com/cvx, note [accessed 2011 April].

[41] R. L. Jesch and R. H. McLaughlin. Dielectric measurements of oil shale as functions of temperature and frequency. *IEEE Trans. Geosci. Remote Sens.* 39, 12 (2001), pp. 2713–2721.

[42] Y. Chen and M. Oristaglio. A modeling study of borehole radar for oil-field applications. *Geophysics* 67, 5 (2002), 1486–1494.

[43] A. Peter Annan. GPR methods for hydrogeological studies. *Water Science and Technology Library.* Vol. 50. Dordrecht, the Netherlands: Springer, 2005, pp. 185–213.

9

Seismic Source Monitoring with Compressive Sensing

Ismael Vera Rodriguez and Mauricio D. Sacchi

CONTENTS

9.1 Introduction

Compressive sensing (CS) has been a field of active research during the last decade [6–8]. CS provides formal recovery conditions for signals that have been compressed via random sampling. One of the main assumptions made by CS techniques is that signals can be represented in terms of a sparse superposition of basis functions. Popular applications of CS in seismic exploration include the reconstruction of seismic signals and studies pertaining to acquisition design [23,24,29,30,35]. Other applications of CS methodologies to seismic exploration include seismic source separation in simultaneous source acquisition [32,37], compressed wavefield propagation [31], and sign-bit seismic imaging [46].

This chapter explores the application of CS to the ubiquitous problem of spatiotemporal seismic source localization and the inversion of the seismic moment tensor [2]. These are prevailing problems in applied geophysics and an important component of seismic monitoring in fields ranging from uniaxial and triaxial compression experiments at laboratory scale, hydraulic injections (e.g., CO_2 sequestration, injection of wastewater or cuttings, natural gas storage, and hydraulic fracturing in geothermal and oil and gas applications), mining, and earthquake and volcano monitoring. Particular attention is paid to the problem of seismic moment tensor representation in terms of a sparse superposition of Green functions [48,49]. CS reduces the number of data samples required to detect, locate, and invert for the moment tensor of seismic events. Therefore, the greatest advantages obtained from the application of CS are observed in seismic monitoring using arrays with a large number of recording channels. Microseismic monitoring of hydraulic fractures from surface arrays of receivers is an example of a natural candidate for the incorporation of CS.

The fast imaging of the location of seismic events and the inversion of the seismic moment tensor with compressed data [47] can lead to a continuous monitoring workflow for the live characterization of systems of fractures activated during hydraulic fracturing.

It is interesting to mention that in our work basis functions are physical Green functions as opposed to analytical, predefined functions (e.g., Fourier and wavelet basis). A similar line of research has been utilized for compressive sensing radar [18,22]. The latter adopts Green functions for waves propagating in homogeneous media for data representation. Seismic problems, as will be shown, present the added complexity of having Green functions for inhomogeneous media that must be numerically computed.

9.2   The seismic source monitoring problem

A seismic source is the result of a transient imbalance of stresses in an elastic medium (note that some active seismic sources can also take place in acoustic media, e.g., fluids; we will focus our description to elastic media such as rocks), where from the total energy budget spent in the process, only a fraction is released in the form of seismic waves. Seismic sources have finite dimensions, but equivalent point-source approximations are frequently used to simplify their analysis [2]. An equivalent point-source approximation reproduces the pattern of seismic energy radiated by the actual, finite source. A widely used equivalent point-source model is based on the seismic moment tensor. The seismic moment tensor is composed of nine force couples, although it is most commonly used in its symmetric form with only six independent elements. A symmetric tensor represents internal sources with zero net torque and zero net force. This model is normally employed to study natural quakes [e.g., 12,15,50], microseismicity observed in industrial activities such as hydraulic injections and mining [e.g., 21,36,40], and acoustic emissions recorded during laboratory experiments [e.g., 1,28,33]. Nonsymmetric tensors represent unbalanced, external sources. The following description concentrates on symmetric moment tensor representations.

The displacement field at a position $\mathbf{x}$ produced by a point source located at $\boldsymbol{\xi}$ is estimated in terms of its source mechanism (moment tensor) using the expression [2, equation 3.23]:

$$u_n(\mathbf{x}, t) = M_{pq}(t) * G_{np,q}(\mathbf{x}, t; \boldsymbol{\xi}, 0), \tag{9.1}$$

where M_{pq} is the seismic moment tensor of the source and G_{np} are elastodynamic Green functions describing the wave propagation between $\boldsymbol{\xi}$ and $\mathbf{x}$. The Einstein summation notation applies to Equation 9.1, with the comma representing derivative with respect to the coordinate with index q. The subindex n denotes the direction of the coordinate system in which the displacement is being measured. In this description, we will use a coordinate system aligned with the geographical directions, with x_1 pointing east, x_2 pointing north, and x_3 being positive upward (ENU system). Where convenient, we will also refer to the coordinate directions as $x_1 \to x$, $x_2 \to y$, and $x_3 \to z$. The time variation of the moment tensor can be factored out by assuming that the history of displacements at the source position is the same for all the force couples in the tensor, that is,

$$M_{pq}(t) = M_{pq}s(t), \tag{9.2}$$

where $s(t)$ is called the source signature or source time function. The source signature can be extracted from the observations of the seismic event [38,42] or assumed using an analytical function. It changes in shape depending on whether we observe it close to the source (near-field) or farther than a few wavelengths away (far-field). Many practical applications of source monitoring have receivers deployed in the far-field and the source signature is represented with a theoretical model. A common choice of source model is the Brune pulse, which provides the spectrum of a circular shear dislocation [5]. Source models are attractive because they are parameterized in terms of variables that are linked to properties of the seismic source. However, if the objective of the monitoring is simply to detect and locate seismic events, the source signature can be approximated with simpler models that reproduce the wave shape and frequency spectrum of direct arrivals (Figure 9.1).

Dislocations account for the greatest proportion of the seismic source types investigated in geophysics. A dislocation is formed at the contact between two planes and is geometrically defined by the strike and dip of the contact surface (Figure 9.2a). When the dislocation is activated, two more angles are necessary to describe the motion of one of the planes with respect to the other. One angle is called the rake and the other is the angle α between the direction of displacement and the surface of the dislocation (Figure 9.2b). If the dislocation is activated in a way that releases seismic waves, then observations of its seismic radiation pattern can be used to invert for the four angles describing the source geometry and style

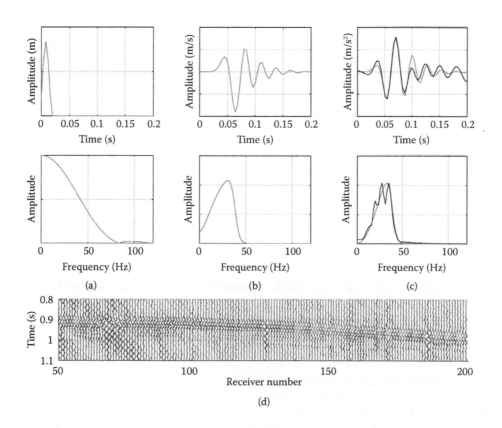

FIGURE 9.1
Example of source time function modeled using a Hanning window. (a) and (b) The displacement and velocity responses in time (top) and frequency (bottom) of the synthetic source time function, respectively. (c) The acceleration response of the synthetic source time function (gray) compared with a compressional direct arrival (black) measured in one geophone-accelerometer deployed on the earth's surface. (d) A comparison between the observations of a seismic event recorded at 150 geophone-accelerometers deployed on the earth's surface (black) and the forward modeling of a moment tensor solution obtained using the synthetic source time function (gray) from (c).

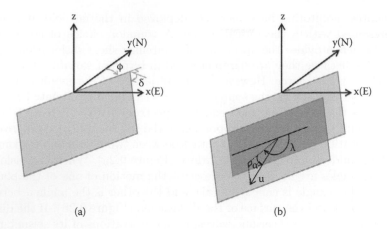

(a)   (b)

FIGURE 9.2
Angles describing the activation of a general dislocation. (a) Geometrical description with strike $\phi \in [0, 360)°$ measured in a clockwise direction from north, and dip $\delta \in [0, 90]°$ measured from a horizontal plane toward the dislocation plane at a 90° angle clockwise from the direction of strike. (b) Description of the style of failure with rake $\lambda \in (-180, 180]°$ measured from a horizontal line in the direction of strike toward the projection of the displacement vector **u** over the dislocation plane (positive in a counterclockwise direction), and $\alpha \in [-90, 90]°$ measured from the dislocation plane toward the displacement vector (positive for opening dislocations).

of activation. Real dislocations present contact surfaces with asperities and that are not perfectly planar; however, assuming that the source can be represented with a point-source equivalent also assumes that these deviations from the ideal planar dislocation can be neglected. The relationship between the source angles and the moment tensor is obtained from [44, equation 9.10]

$$M_{pq} = c_{pqkl} D_{kl}, \tag{9.3}$$

where c_{pqlk} contains the elastic parameters in the source region and D_{kl} is called the source tensor or potency tensor [10,44]. Using matrix-vector notation, the potency tensor is given by [44, equation 9.3]

$$\mathbf{D} = \frac{uS}{2}(\mathbf{n v} + \mathbf{v n}). \tag{9.4}$$

Variables **n** and **v** are unit vectors pointing in the direction normal to the dislocation plane and in the direction of displacement, respectively. The scalars u and S are the magnitude of the displacement and the surface area of the dislocation, respectively. Manipulating Equations 9.3 and 9.4, it can be demonstrated that for shearing dislocations (i.e., pure double-couples) in isotropic media, the source angles can be estimated from the moment tensor without knowledge of the elastic properties in the source region. In other cases, the elastic properties must be factored out from the moment tensor. The biaxial decomposition from [10] provides the means to achieve this.

Returning to Equation 9.1 and substituting Equation 9.2, the source displacement field can be expressed as

$$u_n(\mathbf{x}, t) = M_{pq}[s(t) * G_{np,q}(\mathbf{x}, t; \boldsymbol{\xi}, 0)] = M_{pq} g_{npq}(\mathbf{x}, t; \boldsymbol{\xi}, 0), \tag{9.5}$$

where the lowercase g_{npq} is used to represent the derivative of the Green functions (hence, the comma is dropped) band-limited with the source time function. Using matrix-vector notation, Equation 9.5 becomes

$$\mathbf{u} = \mathbf{Gm},\qquad(9.6)$$

where $\mathbf{u} = [\mathbf{u}_x, \mathbf{u}_y, \mathbf{u}_z]^T$ is an $N \times 1$ vector of the observations of the displacement field, $\mathbf{G}$ is an $N \times 6$ matrix formed with the band-limited derivatives of the Green functions, and $\mathbf{m} = [m_{xx}, m_{xy}, m_{xz}, m_{yy}, m_{yz}, m_{zz}]^T$ is a 6×1 vector of the independent moment tensor coefficients. Inverting Equation 9.6 is equivalent to taking the source displacement field, removing the wave propagation and source signature effects, and finding the moment tensor that reproduces the pattern of seismic radiation of the source. Clearly, if the radiation pattern is not sampled at a sufficient number of positions around the source (focal sphere), then the inversion becomes ill-posed because there will be multiple moment tensor solutions that can fit the limited number of observation points mapping the radiation pattern [14,43,45]. Inversion constraints can be used to stabilize the inversion; however, care must be taken on the source type that the chosen constraint may impose. Common inversion constraints consist of enforcing the dipolar components of the moment tensor to add to zero (deviatoric solution), enforcing the determinant of the tensor to be zero (pure double-couple solution), or applying damping to the normal equations [25]. Alternatively, additional observation points can also be deployed to map the focal sphere, letting the data themselves produce a unique full moment tensor solution. Incorporating new observation points, Equation 9.6 becomes

$$\begin{bmatrix} \mathbf{u}_1 \\ \mathbf{u}_2 \\ \vdots \\ \mathbf{u}_{N_r} \end{bmatrix} = \begin{bmatrix} \mathbf{G}_1 \\ \mathbf{G}_2 \\ \vdots \\ \mathbf{G}_{N_r} \end{bmatrix} \mathbf{m},\qquad(9.7)$$

where the subindex specifies the receiver number. The vector of observations has now dimensions of $(NN_r) \times 1$ and the matrix with the band-limited derivatives of the Green functions is $(NN_r) \times 6$. The variable N_r is the total number of observation points or seismic receivers in the monitoring network. Equations 9.6 and 9.7 assume that we know the location of the seismic event and, therefore, that we can directly model the wave propagation effects between the source and the receivers. In seismic monitoring, however, the source location is also unknown. A practical way to incorporate source location into the solution consists in solving Equation 9.7 at different trial positions [26] and selecting the position that optimizes a target metric, for example, the misfit between the observations and their forward modeling with the moment tensor solutions at each trial point. Equation 9.7 incorporating multiple trial positions is written as

$$\begin{bmatrix} \mathbf{u}_1 \\ \mathbf{u}_2 \\ \vdots \\ \mathbf{u}_{N_r} \end{bmatrix} = \begin{bmatrix} \mathbf{G}_1[1] & \mathbf{G}_1[2] & \cdots & \mathbf{G}_1[N_n] \\ \mathbf{G}_2[1] & \mathbf{G}_2[2] & \cdots & \mathbf{G}_2[N_n] \\ \vdots & \vdots & \ddots & \vdots \\ \mathbf{G}_{N_r}[1] & \mathbf{G}_{N_r}[2] & \cdots & \mathbf{G}_{N_r}[N_n] \end{bmatrix} \begin{bmatrix} \mathbf{m}[1] \\ \mathbf{m}[2] \\ \vdots \\ \mathbf{m}[N_n] \end{bmatrix},\qquad(9.8)$$

where the variable $\mathbf{G}_i[l]$ represents the band-limited derivatives of the Green functions computed between the trial position l and the receiver i. The moment tensor solution at trial position l is given by $\mathbf{m}[l]$, and the total number of trial positions is N_n. A final source parameter that must be determined in seismic monitoring is the time at which the seismic event takes place, or origin time. In this case, it is normally safe to assume that the Green functions are time invariant: in other words, that the medium of propagation between the trial positions and the receivers does not change with time or that any changes are negligible. In some applications (e.g., mining or hydraulic fracture monitoring), the medium of propagation can change during the monitoring period. When these changes compromise the accuracy of the estimated source parameters, updating the Green functions along time is necessary. If invariance of the Green functions is a reasonable assumption, then the final system of equations for source monitoring is

$$
\begin{bmatrix} \mathbf{u}_{1(j)} \\ \mathbf{u}_{2(j)} \\ \vdots \\ \mathbf{u}_{N_r(j)} \end{bmatrix} = \begin{bmatrix} \mathbf{G}_1[1] & \mathbf{G}_1[2] & \cdots & \mathbf{G}_1[N_n] \\ \mathbf{G}_2[1] & \mathbf{G}_2[2] & \cdots & \mathbf{G}_2[N_n] \\ \vdots & \vdots & \ddots & \vdots \\ \mathbf{G}_{N_r}[1] & \mathbf{G}_{N_r}[2] & \cdots & \mathbf{G}_{N_r}[N_n] \end{bmatrix} \begin{bmatrix} \mathbf{m}_{(j)}[1] \\ \mathbf{m}_{(j)}[2] \\ \vdots \\ \mathbf{m}_{(j)}[N_n] \end{bmatrix}. \tag{9.9}
$$

The subindex j in parenthesis denotes the time sample at which the observations start. In a continuous monitoring system, sets of observations are continuously fed into an algorithm that solves Equation 9.9. The first sample of each trace being fed is indicated by the index j. The output from the algorithm for each j index are the moment tensor solutions $\mathbf{m}_{(j)}[l]$. For a time interval defined by $j \in [j_{t=t_0}, j_{t=t_f}]$, the moment tensor solutions can be compared in terms of a target metric (e.g., misfit), and the solution that optimizes the target metric along the j and l dimensions can then be identified as a seismic source with origin time j, location l, and source mechanism $\mathbf{m}_{(j)}[l]$. This procedure will always detect one source within each interval $[j_{t=t_0}, j_{t=t_f}]$. A threshold value for the target metric can therefore be imposed to discriminate between real and false seismic sources. Using a threshold can also remove the limitation that only one source is identified within each time interval.

9.3   Source monitoring as a sparse representation problem

The least-squares solution to Equation 9.9 is given by

$$
\mathbf{m}_m = \left[\left(\mathbf{G}_m^H \mathbf{G}_m \right)^{-1} \mathbf{G}_m^H \right] \mathbf{u}_m, \tag{9.10}
$$

where the subindex m has been incorporated to refer to the multichannel, multinode monitoring network described in Equation 9.9, and the sliding in time (subindex j) is considered implicit. The superindex H represents the Hermitian or conjugate-transpose operator. The operation between square brackets is invariant along time and needs to be estimated only once per array of receivers to perform continuous monitoring. Unfortunately, the dimensions of many practical monitoring networks prevent the estimation of the inverse in Equation 9.10. Consider, for example, the monitoring network described in [41]. The network includes 7 three-component, broadband seismic stations. The length of every band-limited Green function is 120 points and the number of nodes (i.e., trial positions) in the monitoring grid is 6875. For this configuration, matrix $\mathbf{G}_m^H \mathbf{G}_m$ has dimensions of $41,250 \times 41,250$. Assuming double-precision, floating-point format, the size in memory of such matrix lies on the order of 13.6 gigabytes. This is a variable size that requires significant computer resources to find its inverse. In this case, the size of $\mathbf{G}_m^H \mathbf{G}_m$ is determined by the number of nodes in the grid and is independent of the number of monitoring stations. However, the efficiency in the estimation of the product $\mathbf{G}_m^H \mathbf{G}_m$ itself is influenced by the number of receivers. Note that these limitations only exist with least-squares inversion using full-waveform Green functions. Alternative methodologies overcome these limitations with the use of ray tracing attributes stored in compact form, although these methodologies are generally targeted to resolve only event locations [9,13,19]. In these cases, the source mechanism is normally determined in a subsequent step with an independent algorithm.

A computationally more efficient, yet approximate, solution to Equation 9.9 is given by the adjoint [11,27]

$$\hat{\mathbf{m}}_m = \mathbf{G}_m^H \mathbf{u}_m. \tag{9.11}$$

The adjoint or time-reversal imaging provides a robust imaging solution able to collapse seismic energy around its point of origin. However, because Green functions are not orthogonal, the resulting moment tensor coefficients are only roughly approximated. This is the same reason why seismic migration in exploration seismology resorts to least-squares solutions to estimate true-amplitude images.

The properties of the solution vector $\mathbf{m}_m$ can also be taken into consideration to implement an alternative approach to solve Equation 9.9. Because $\mathbf{m}_m$ is formed by the concatenation of moment tensors of virtual sources at each node position, it presents a regular structure in blocks of six coefficients (i.e., $\mathbf{m}_{(j)}[l]$ in Equation 9.9 is the lth block of $\mathbf{m}_m$ within the processing window that starts at the jth time sample). On the other hand, in many scenarios of continuous source monitoring, the processing window can be reduced to time intervals where the observations of only one seismic event are recorded at a time. Under this assumption, the solution vector $\mathbf{m}_m$ contains only one block of six nonzero coefficients every time a seismic event is observed. In these circumstances, $\mathbf{m}_m$ can be described as block 1-sparse [49]. In general, the level of block sparsity k of a vector is defined as [16]

$$k = \|\mathbf{m}_m\|_{2,0} = \sum_l I(\|\mathbf{m}_{(j)}[l]\|_2), \tag{9.12}$$

where

$$I(\|\mathbf{m}_{(j)}[l]\|_2) = \begin{cases} 1 & if \quad \|\mathbf{m}_{(j)}[l]\|_2 > 0. \\ 0 & otherwise \end{cases} \tag{9.13}$$

By exploiting the sparsity of the solution, two new objective functions can be defined to solve Equation 9.9, these are,

$$\textbf{minimize } \|\mathbf{m}_m\|_{2,0} \quad \text{s.t.} \quad \|\mathbf{u}_m - \mathbf{G}_m\mathbf{m}_m\|_2 < \varepsilon \tag{9.14}$$

and

$$\textbf{minimize } \|\mathbf{m}_m\|_{2,1} \quad \text{s.t.} \quad \|\mathbf{u}_m - \mathbf{G}_m\mathbf{m}_m\|_2 < \varepsilon, \tag{9.15}$$

where ε is a scalar that controls the level of fitting in case of noise in the observations and/or when only non-exact solutions exist. The mixed norm $\|\mathbf{m}_m\|_{2,1}$ is given by

$$\|\mathbf{m}_m\|_{2,1} = \sum_l \|\mathbf{m}_{(j)}[l]\|_2. \tag{9.16}$$

The observations $\mathbf{u}_m$ are a linear combination of the columns of $\mathbf{G}_m$; therefore, the problem of solving Equations 9.14 and 9.15 can be worded as finding the linear combination that uses the smallest number of elements from $\mathbf{G}_m$ to approximate $\mathbf{u}_m$ within an ε margin in a least-squares sense. Intuitively, the more "alike" are the columns of $\mathbf{G}_m$, the harder it is to identify a unique linear combination of its elements to accomplish the optimizations in Equation 9.14 and/or Equation 9.15. In the best-case scenario, the columns of $\mathbf{G}_m$ are mutually orthogonal but this is a property not found in our Green functions. Consider, for example, the block orthogonal matching pursuit (BOMP) algorithm [17]. The convergence to $\mathbf{m}_m$ using BOMP is guaranteed if [17]

$$kd < \frac{1}{2}\left(\frac{1}{\mu_B} + d - (d-1)\frac{v}{\mu_B}\right), \tag{9.17}$$

where d is the number of coefficients in each block. The variable ν is called the subcoherence of $\mathbf{G}_m$ and is computed as

$$\nu = \max_l \left(\max_{i,j \neq i} |\mathbf{g}_i^T \mathbf{g}_j| \right), \quad \mathbf{g}_i, \mathbf{g}_j \in \mathbf{G}[l], \tag{9.18}$$

where $\mathbf{g}_i$ is the ith column of $\mathbf{G}[l]$ and the notation without the subindex in $\mathbf{G}[l]$ refers to a complete column-block of $\mathbf{G}_m$; in other words

$$\mathbf{G}[l] = \begin{bmatrix} \mathbf{G}_1[l] \\ \mathbf{G}_2[l] \\ \vdots \\ \mathbf{G}_{N_r}[l] \end{bmatrix}.$$

The variable μ_B is called the block-coherence of $\mathbf{G}_m$ and is defined as

$$\mu_B = \max_{l,r \neq l} \left(\frac{1}{d} \rho(\mathbf{M}[l,r]) \right), \tag{9.19}$$

where $\rho(\cdot)$ is the spectral norm [39] of the matrix $\mathbf{M}[l,r]$ given by

$$\mathbf{M}[l,r] = \mathbf{G}^H[l]\mathbf{G}[r]. \tag{9.20}$$

Condition 9.17 can be rewritten in the block 1-sparse source monitoring problem as

$$\mu_B < \frac{1 - 5\nu}{6}. \tag{9.21}$$

The subcoherence gives an indication of how "alike" are the columns within a block $\mathbf{G}[l]$, while the block-coherence can be seen as an indicator of how "alike" are different blocks $\mathbf{G}[l]$. The fulfillment of Condition 9.21 requires $\nu < 0.2$. Because the functions $\mathbf{g}_i$ within a block $\mathbf{G}[l]$ are formed by wave arrivals occurring at the same time (only amplitudes and polarities can be different), it is reasonable to expect the subcoherence of $\mathbf{G}_m$ to be generally large. Note, however, that Condition 9.17 refers to a worst-case scenario, which leaves a gray area where BOMP can successfully find correct solutions.

9.3.1   Synthetic example

Consider the example displayed in Figure 9.3. The monitoring network consists of a two-dimensional (2D) line of 41 vertical-component receivers deployed at zero depth. The separation between receivers is constant at 50 m. The velocity model consists of three homogeneous, isotropic, horizontal layers, with the middle layer generating high velocity contrasts with the upper and lower layers. The focal coverage offered by the array reduces the number of elements of the moment tensor that can be resolved to m_{xx}, m_{xz}, and m_{zz} [see, for example, 43]. For processing with the sparse solver, the blocks $\mathbf{G}[l]$ are normalized by their spectral norm. This normalization reduces the intrinsic bias toward solutions that lie closer to the receivers, which display higher amplitudes in the band-limited Green functions. The normalization by column, necessary to bound the values of ν and μ_B, is not applied in this case.

The observations of a seismic source located at an offset of -660 m and a depth of -3300 m display mainly the arrivals of compressional (P-) waves because of the relatively small aperture of the array with respect to the depth of the monitored area (Figure 9.4a, top). The least-squares solution to recover the source mechanism and location provides

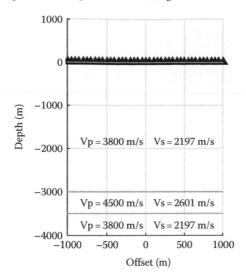

FIGURE 9.3

Elastic, isotropic model with a high-velocity layer. Black triangles are receiver positions and the gray horizontal lines denote the contacts between layers with different seismic velocities.

a perfect answer in the noise-free scenario. In terms of location, similar performance is observed in the case of the adjoint and the sparse solver (Figure 9.4a). Only the adjoint presents a small amount of energy smearing around the source location, but this effect does not compromise the identification of the right source coordinates. As expected, the adjoint solution is also not able to recover the correct coefficients of the moment tensor (Table 9.1).

Between the three solvers, the least-squares solution is the most sensitive to noise in the input data. The adjoint and the sparse solver remain practically unchanged in their distribution of nonzero coefficients in the solution vector $\mathbf{m}_m$ (Figure 9.4b). Although the least-squares solution is still able to provide the correct source coordinates in this example, this sensitivity to noise is what makes least squares less robust than the adjoint for imaging purposes. The source mechanism is reasonably well recovered by the least-squares and sparse solvers but not by the adjoint. In this case, the approximation provided by the sparse solver is better than the least-squares solution. The reason is that the sparse solver inverts the observations using only the band-limited Green functions of the identified source location. The least-squares solver in Equation 9.10, on the other hand, attempts to resolve the source mechanism using the band-limited Green functions from all the nodes in the grid at the same time. This smears part of the solution outside the actual source position, reducing thereby its accuracy.

It is interesting to note that BOMP performs a time reversal or adjoint at every iteration. The contraction of the adjoint solution (by taking the ℓ_2-norm of the moment tensors at each grid node) is what BOMP then uses to select the next block that will become part of the sparse solution. For instance, the first iteration of BOMP in this example produces the contracted-adjoint solution displayed in Figure 9.4. From this solution, BOMP takes only the block or source location with the largest moment tensor in terms of its ℓ_2-norm. At this stage, BOMP has used the most robust attribute from the time-reversal solver, which is the imaging of the source location. In the next step, BOMP uses the band-limited Green functions of the selected location to invert for the source mechanism. From this perspective, we can see that BOMP uses the best of the least-squares and adjoint solvers: robust imaging to estimate source locations and inversion to obtain an accurate source mechanism.

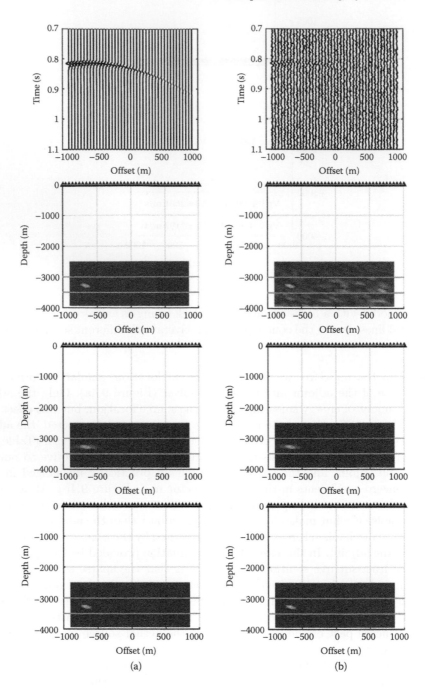

FIGURE 9.4

From top to bottom: synthetic arrivals for a source located at -660 and -3300 m using the example model in Figure 9.3; ℓ_2-norm of moment tensor solutions obtained with least-squares (Equation 9.10); ℓ_2-norm of moment tensor solutions obtained with the adjoint (Equation 9.11); ℓ_2-norm of moment tensor solutions obtained with BOMP. Column (a) shows results with noise-free synthetics. Column (b) shows results with Gaussian noise added. The noise is zero mean with standard deviation equal to $\frac{1}{3}$ of the maximum absolute amplitude in the synthetics.

TABLE 9.1

Comparison of Moment Tensor Solutions Using Different Solvers

Moment Tensor Coefficient	Input	Least Squares		Adjoint		BOMP	
		Noise-Free	Noise	Noise-Free	Noise	Noise-Free	Noise
m_{xx}	0.78	0.78	0.70	-0.08	-0.09	0.78	0.75
m_{xz}	0.78	0.78	0.83	0.06	0.10	0.78	0.80
m_{zz}	0.39	0.39	0.39	1.41	1.40	0.39	0.39

Note: Moment tensors are normalized to a seismic moment of 1.

Condition 9.21 is not fulfilled in this example. This non compliance means that there is no 100% guarantee that the correct solution will always be found by BOMP, but it does not imply that a correct solution could not be estimated in a number of cases. The results presented in this example are evidence of that. In these circumstances, it is crucial to develop an understanding of the uncertainty in the recovered source parameters. The source location is a function of travel time and direction of arrival in some limited aperture monitoring arrays. It depends mainly on the velocity model of the medium. The source mechanism estimation in this description uses as input the source location and the velocity model. Therefore, an accurate location imaging is crucial to recover a reliable source mechanism. The location of the source in our parameterization of the source monitoring problem is given by the rth index of the block of band-limited Green functions $\mathbf{G}[r]$ selected to participate in the forward modeling of the observations. The identification of the correct block r to be part of the solution is influenced by how "alike" are different blocks in terms of the metric being used to make a choice. Consider, for example, the time reversal step followed by BOMP at iteration number it, that is,

$$\hat{\mathbf{m}}_{it} = \mathbf{G}_m^H \mathbf{u}_m = \mathbf{G}_m^H(\mathbf{G}[r]\mathbf{m}[r]). \tag{9.22}$$

The block with the rth index in Equation 9.22 contains the correct band-limited Green functions that expand the observations $\mathbf{u}_m$. The product $\mathbf{G}_m^H \mathbf{G}[r]$ can be seen as a comparison between each of the blocks in $\mathbf{G}_m^H$ with the block $\mathbf{G}[r]$ in terms of their trace-to-trace correlation. The comparisons are equally weighed by the vector $\mathbf{m}[r]$ and maximized at $\mathbf{G}^H[l]\mathbf{G}[r], \forall l = r$. For this way of comparing between band-limited Green functions from different blocks, the spectral norm of the cross product between blocks (see Equations 9.19 and 9.20) provides an indication of the uncertainty in location imaging for BOMP.

Figure 9.5 shows examples of the analysis of the spectral norm of cross products for the synthetic example. Each image provides an indication of the uncertainty in the estimation of the source location at different positions within the monitoring grid. As expected, the higher spectral norm values occur closer to the node position under analysis. This is consistent with the uncertainties observed when identifying the maximum of an adjoint imaging solution. It is interesting to extract from this analysis the shape and size of the uncertainty region for a given threshold in the spectral norm values. In this case, the higher values distribute preferentially in the horizontal directions. This suggests higher uncertainty in the estimation of the offset coordinate of the source. This uncertainty distribution is imprinted by the limitations of the coverage provided by the geometry of the array of receivers. The size of the uncertainty region under consideration could also be used to produce an estimation of the expected errors in location.

The analysis of the spectral norm of cross products provides an initial estimate of the expected uncertainties in the estimation of source locations using BOMP. Once a location has been selected, the uncertainties in the estimation of the source mechanism can be obtained through standard tools in least-squares inversion (e.g., condition number and

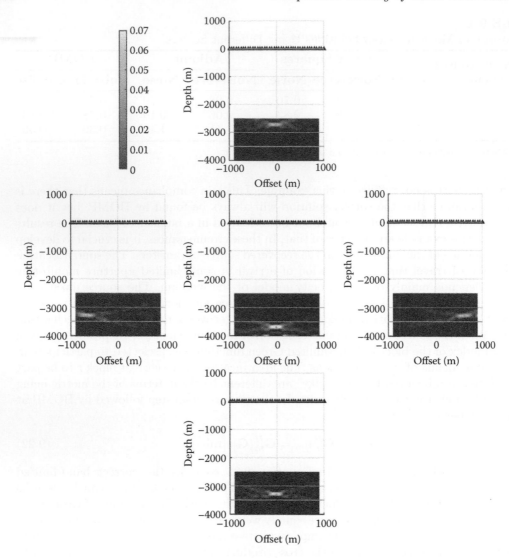

FIGURE 9.5
Normalized spectral norm of the products $\mathbf{G}^{H}[l]\mathbf{G}[r]$ (see Equation 9.19), where r is a fixed block/node and l ranges over all the blocks/nodes in the grid. The dark dot inside the bright amplitude anomaly in each image denotes the position of the fixed block/node r in each case.

resolution matrix). The analysis of spectral norms is interesting because it provides a tool for network design, where the coverage of the receivers can be optimized to control the shape of the location-uncertainty distributions and reduce their size. Note, however, that the analysis of the spectral norms is limited in its scope. Other common elements influencing the accuracy of estimated source parameters are noise in the observations, errors in the velocity model, incomplete account of the physics of the problem, and interpolation of locations within the monitoring grid. These elements impact the outcome of the time reversal step in BOMP and are not accounted for in the analysis of spectral norms. A more complete picture of resolvability and uncertainty for a given monitoring setting can be obtained by numerical simulation of the solution to multiple sources with random locations and source mechanisms introducing perturbations into all the elements mentioned above.

9.4 Incorporation of compressive sensing for seismic source monitoring

The sparsity of the vector of model coefficients $\mathbf{m}_m$ makes ℓ_0 and ℓ_1 solvers ideal to estimate inverse solutions in seismic source monitoring. However, the size of the dictionaries $\mathbf{G}_m$ in practical applications can quickly make unacceptable the computational cost of the solver. CS provides an alternative to reduce turnaround time by reducing the dimensionality of the problem in one direction. CS is incorporated through the introduction of an encoding matrix $\mathbf{\Phi}$, that is [48]

$$\mathbf{\Phi}\mathbf{u}_m = \mathbf{\Phi}\mathbf{G}_m\mathbf{m}_m. \tag{9.23}$$

Using the notation above, matrix $\mathbf{\Phi}$ is designed to have a larger number of columns than rows. Hence, its product with the vector of observations, on the left-hand side, and with the dictionary, on the right-hand side, effectively reduces the number of rows in the system of equations. This strategy works well, for example, to increase the number of receivers and/or to increase the modeling time in the Green functions while minimizing the impact in the response time of the monitoring network. On the other hand, there is no compression effect over the dimension that is controlled by the size of the monitoring grid (columns of $\mathbf{G}_m$), limiting the benefits of the compression over the number of grid nodes.

The product between the encoding matrix and the dictionary $\mathbf{G}_m$ must also result in an incoherent compressed dictionary to guarantee that a sparse solver will converge to unique solutions. Probably the simplest way to design a matrix $\mathbf{\Phi}$ is by drawing independent, identically distributed samples from a Gaussian distribution [7]. Other suitable random ensembles have also been described in the literature [3,34]. For the examples presented in this chapter, we will use $\mathbf{\Phi}$ with Gaussian entries.

The incorporation of CS in Equation 9.23 requires knowledge of all the observations recorded in the array of receivers. This is disadvantageous because it implies that the recordings must be transmitted, fully sampled, to a central facility where the compression can be implemented. The main benefit and original aim of CS, however, is to record signals with sampling requirements below the limit suggested by the Nyquist–Shannon theorem. For this purpose, the compression step in Equation 9.23 can be moved a step back to the receiver locations, that is,

$$
\begin{bmatrix}
\mathbf{\Phi}\mathbf{u}_{1(j)} \\
\mathbf{\Phi}\mathbf{u}_{2(j)} \\
\vdots \\
\mathbf{\Phi}\mathbf{u}_{N_r(j)}
\end{bmatrix}
=
\begin{bmatrix}
\mathbf{\Phi}\mathbf{G}_1[1] & \mathbf{\Phi}\mathbf{G}_1[2] & \cdots & \mathbf{\Phi}\mathbf{G}_1[N_n] \\
\mathbf{\Phi}\mathbf{G}_2[1] & \mathbf{\Phi}\mathbf{G}_2[2] & \cdots & \mathbf{\Phi}\mathbf{G}_2[N_n] \\
\vdots & \vdots & \ddots & \vdots \\
\mathbf{\Phi}\mathbf{G}_{N_r}[1] & \mathbf{\Phi}\mathbf{G}_{N_r}[2] & \cdots & \mathbf{\Phi}\mathbf{G}_{N_r}[N_n]
\end{bmatrix}
\begin{bmatrix}
\mathbf{m}_{(j)}[1] \\
\mathbf{m}_{(j)}[2] \\
\vdots \\
\mathbf{m}_{(j)}[N_n]
\end{bmatrix}, \tag{9.24}
$$

where the elements $\mathbf{G}_i[l]$ are normalized by the spectral norm of the block $\mathbf{G}[l]$. Equation 9.24 leads to the idea of the design of a new type of seismic receiver that can incorporate the encoding of the signal in time windows. The encoded recordings could then be sent to a centralized processing facility at a reduced transmission cost. The encoding matrix does not need to be the same at each receiver, but it can simplify its manipulation to have a fixed encoder for the system. In theory, by using Equation 9.24, the need to expand the recorded signals into a fully sampled version is no longer necessary because the representation vector $\mathbf{m}_m$ already provides the information that we want about the seismic source. Notwithstanding that, because the encoding of the signal has no dependency on the compressed dictionary, the dictionary $\mathbf{G}_m$ can be swapped with a more suitable set of basis functions to decode a fully sampled version of the recordings.

The incorporation of CS into the source monitoring problem is straightforward and requires only the introduction of the encoding matrix. Many of the challenging aspects of CS source monitoring are found in its practical implementation. In the next section, we use a real data set to illustrate some of these challenges.

9.4.1   Real data example

The data in this example correspond to a hydraulic fracturing treatment. Hydraulic fractures are regularly performed in geothermal and oil and gas operations to increase rock permeability in the vicinity of a target well. Other operations involving fluid injections include wastewater disposal and CO_2 sequestration. During hydraulic fracturing, microseismic events are produced through the activation of preexisting and/or new planes of weakness in the rock. The microseismicity produced by the hydraulic treatment is often recorded by monitoring arrays of receivers to produce interpretations of the fracture geometry, complexity, and local state of stresses. These attributes provided by the microseismic data are useful to diagnose the containment of the fracture within the target rock formation, the extent of the stimulated volume, and even to get an idea of the increase in permeability as a result of the treatment. Multiple research groups around the world are continuously developing innovative ways to extract new information from the microseismic data to better characterize and even control hydraulic injections. Hydraulic fracture monitoring is normally performed with arrays of receivers deployed on the surface of the earth or down in wells located nearby the well under treatment. Surface monitoring is an ideal candidate for a CS-based monitoring system because of the large number of receivers deployed in the arrays and the large volumes of data recorded.

The monitoring array in this example consists of 1082 vertical component receivers spread over 10 surface lines (Figure 9.6a). The target well is approximately vertical down to the target rock formation, where it turns horizontal over a distance on the order of 1 km (see Figure 9.6). The wellbore is isolated from the surrounding rock formations with a series of steel casings. The purpose of the casing is to ensure that the fluids flowing from the hydraulic fracture toward the well travel directly to the surface without leaking back into a different rock formation. At the position along the well where the hydraulic treatment is desired, a set of perforation shots are fired to pierce the casing. The perforations open communication paths between the well and the target formation and also act as initiation points for the hydraulic fracture. In this case, the casing was perforated in three contiguous locations separated by 30 m intervals at a vertical depth of −2000 m (see Figure 9.6).

The velocity model in the medium between the well and the receivers is homogeneous with vertical transverse isotropy (VTI). The model was calibrated through an inverse optimization that minimizes the difference between observed and theoretical travel times for a number of microseismic events with known location. The height of the receivers was also changed to a common datum (see Figure 9.6b). By fixing the velocity model, a set of residual corrections for the receivers were calculated to account for the change in datum and other heterogeneities not considered within the homogeneous model. The calibration of the velocity model and estimation of residual corrections are described in [4].

The monitored volume is discretized with a regular grid rotated 50° clockwise from north (see Figure 9.6a). Surface monitoring arrays present a known limitation in the poor constraining of the depth of microseismic events. Increasing the aperture of the array (maximum horizontal distance from the $x - y$ position of the event) improves the depth constrain. Unfortunately, because the effective seismic path length between the event and the farther receivers also increases, there is a maximum aperture that can effectively contribute to constraining the location of the event. On the plus side, this low sensitivity to the depth of the event can be used to our advantage to improve the efficiency of the monitoring algorithm.

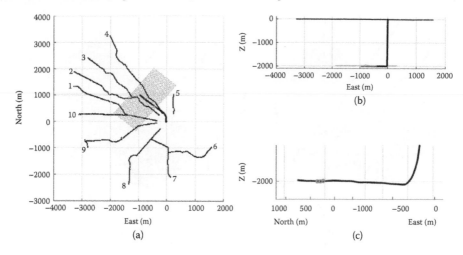

FIGURE 9.6

(a) Top view of monitoring geometry for the real data example. Black dots are receiver positions (crooked lines with numbers identifying lines of receivers). Gray dots are grid nodes. The black solid line starting at the origin of the coordinate system is the well trajectory, where gray dots along this line denote the position of perforations. (b) Side view from south. All the receivers are repositioned to the same datum in depth (0 m) and the monitoring grid has a constant depth (-2000 m). (c) Side view zoomed in around the depth of the perforation shots and from an angle approximately normal to the well trajectory.

Instead of designing a three-dimensional grid to estimate the hypocenter ($x - y - z$ location) of the microseismic events, we can use a 2D grid to estimate only epicenters ($x - y$ location). Subsequently, origin time, depth, and moment tensor can be simultaneously optimized only for those detections that we identify as "most likely" microseismic events. Using this strategy, the grid is designed on a horizontal plane at a depth of -2000 m (see Figure 9.6a and b). The number of nodes in the grid is 21 and 45 in the short and long directions, respectively. The separation between nodes is 50 m, constant in both orthogonal directions. Final locations are selected in BOMP through interpolation of the time-reversal images into a finer grid.

As has been mentioned before, the advantage of CS in source monitoring lies in the efficient handling of full-waveform information. For simplicity, in this example, we compute Green functions using ray tracing, but the traces are treated as full waveforms. Both direct P- and Sv- (vertically polarized shear) waves are included in the modeling. Considering the longest Sv arrival time between all possible combinations of grid node and receiver, the total duration of the Green functions is set at 2.896 s. The sampling rate is 0.002 s, which is common in this type of application, and the source time functions for P- and Sv-waves are modeled from clean arrivals extracted from a large-magnitude event detected in the monitoring array. This combination of variables results in a number on the order of 6 million Green functions that are required to set up the full-waveform monitoring system. Considering the sampling rate, duration of each Green function, and assuming double-precision, floating-point format, the size in memory of the dictionary $\mathbf{G}_m$ lies on the order of 71 gigabytes.

The first logical question for the implementation of CS into the monitoring system is: "how much compression can be applied without compromising the accuracy of the results?" The analysis of the spectral norm of cross products provides an initial idea of the answer. Figure 9.7 presents an example for a source located at the central perforation in this case study. The compression levels are 0.1%, 1%, and 10%, which correspond to sizes of 94 megabytes, 660 megabytes, and 6.9 gigabytes, respectively, for the compressed dictionaries.

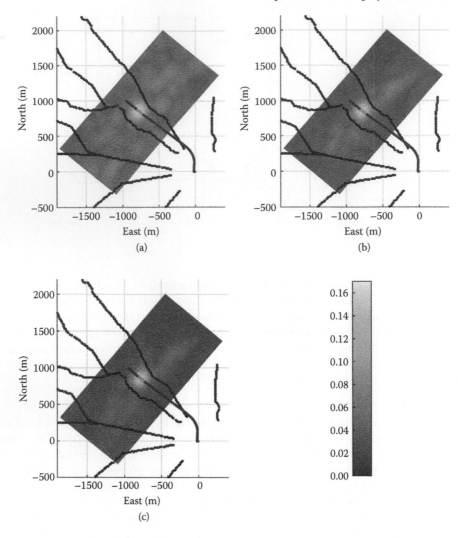

FIGURE 9.7
Analysis of the spectral norm of cross products for a microseismic event located at the central perforation in this example (dots along the well trajectory, which is depicted as a black solid line starting at the origin of the coordinate system). The compression levels are (a) 0.1%, (b) 1%, and (c) 10%, where 100% corresponds to the fully sampled case. Black dots are receiver positions (crooked lines).

Compression levels in this chapter are reported as the percentage given by the ratio of the number of samples left after compression to the original number of samples. The effect of the compression is more obvious at the 0.1% level (Figure 9.7a), where the difference between the maximum spectral norm at the position of the event and the rest of the grid is smaller than in the other two cases (Figure 9.7b and c). The contrast in spectral norm at the position of the event must be maximized to guarantee uniqueness in the solution. There is not much change in the spectral norm images between the 1% and 10% compression levels, which suggests the possibility to apply compression down to the 1% level in this monitoring system. The size of the amplitude anomalies in the spectral norm images also provides an idea of the location uncertainty. In this case, the most significant change in spectral norm happens around a radius of 30 m from the event location under analysis. This can be considered an indication of the horizontal location uncertainty.

As mentioned before, the analysis of the spectral norm of cross products does not provide information on the impact of other important factors in the resolvability of the problem. For this purpose, we must resort to numerical modeling with synthetics. Two important elements influencing the accuracy of the solutions are noise and modeling errors in the Green functions. Noise can be introduced by environmental sources and as part of the construction of the recording instrument (e.g., electronic noise and resonance). Furthermore, environmental noise conditions can change from one receiver position to another and over time. Figure 9.8 shows an instance of the variation of noise conditions across the monitoring array in this example. The root mean square (rms) amplitude for a 5-min interval of noise recording is generally stable with localized channels displaying higher noise-amplitude conditions. A group of receivers at the end of line 9 was switched off during this interval of time; hence, they display an rms amplitude of zero in the plot. We will use this 5-min interval of noise records to extract 100 realizations of noise to model its effect on the CS monitoring system. The advantage of using real noise time series is not only to account for its spatial variation, but also to have access to a sample of the mix of coherent and incoherent noise that the monitoring system will face in reality. Normalizing the synthetic observations per channel helps equalize the spatial variation of noise conditions at the cost of distorting the moment tensor solutions. In the following, we normalize the input observations by their rms per channel for location purposes. Then, the normalization is dropped to estimate final moment tensor solutions.

Modeling errors in the Green functions are normally the result of an inaccurate knowledge of the velocity structure in the medium of propagation. These inaccuracies can be in the form of a simplified velocity model and limited information about attenuation and heterogeneities in the near surface. Additional modeling errors can be the result of limitations in our capabilities to simulate the physics of the wave propagation due to the complexity of the medium. We will not analyze here the effects of the above mentioned

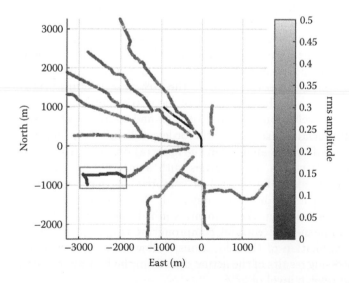

FIGURE 9.8
Root mean square (rms) amplitude at each monitoring channel for an interval of 5 min of environmental noise records. The black solid line at the center of the plot is the well trajectory, with the perforations depicted as gray dots. The channels enclosed by the rectangle were switched off during the time interval considered in this plot.

factors in the CS system, but will study the impact of the most fundamental form of inaccuracy in the Green functions, which is when the event location does not coincide with a node of the grid. Note that this type of inaccuracy is always present, even if the velocity model was perfectly known and we could model the complete physics of the problem.

The source location in this test is selected arbitrarily. The only condition is that it should be in the region where events from the actual hydraulic fracture are expected, and also it must lie approximately at the center of the nearest four grid nodes. The selected source mechanism corresponds to a dislocation with its plane striking 40° from north and dipping 80° from the horizontal. The dislocation is activated with a rake of −85°, opening at an angle α of 10°. The selected source geometry is approximately aligned with the regional stresses, while the style of activation corresponds to an extensive stress regime. All these parameters reflect general observations of source mechanisms consistent with hydraulic fracturing. We analyze three scenarios of signal-to-noise ratio (SNR, with SNR = $20\log_{10}(\frac{rms_{signal}}{rms_{noise}})$), which were selected to reflect plausible conditions that the CS system could face. Both the source–receiver distance and the radiation pattern associated with the source mechanism impact the amount of signal that reaches each recording channel. In this case, it is possible to appreciate larger SNR conditions in the northern section of the array as compared to the southern channels (Figure 9.9a). The average SNR across the array increases by about 15 dB from one scenario to the next. The most common SNR conditions expected during the actual monitoring correspond to the first scenario (Figure 9.9 top).

Figure 9.10 shows the results of the processing of one realization of noisy synthetics in the three SNR scenarios. The compression levels are 1%, 10%, and 20%. The 1% compression level is unable to resolve the correct source location in the scenario with poorest SNR, while the 10% and 20% compression levels provide more accurate locations. As the SNR increases, the location results in the 1% compression level improve. As the scenario with poorest SNR within these three cases is expected to be the most common during the actual monitoring, a compression level no lower than 10% would be necessary. Note that in other applications of source monitoring (e.g., earthquake monitoring), higher SNR could be expected and then, compression levels down to the order of 1% could be successfully used. Furthermore, recent publications using sign-bit CS have shown promising results in pushing the compression level down in scenarios of poorer SNR [46]. Examining the location errors in the complete set of 100 realizations of noise recordings confirms the assertions above (Figure 9.11). Furthermore, at the poorest SNR scenario, we can observe that the 20% compression level still provides a visible improvement over the 10% case (Figure 9.11a). The location error is mostly concentrated on the estimation of the north coordinate. This distribution of error is influenced by our selection of source location and source mechanism and cannot be generalized. The total location error is on the order of 40–50 m, which is consistent with the uncertainty suggested by the analysis of the spectral norm of cross products.

Compared to spectral norms, the analysis with synthetics and real noise recordings provides a more realistic idea of the level of compression that can be applied in a CS monitoring system. Even if we did not analyze all the possible variables that influence the existence of a unique solution, the elements studied above suggest that a compression level of 20% can be implemented without compromising the detection of weak events and the accuracy in the estimation of their locations. Following a conservative standpoint, we proceed to show processing results of the actual data from the hydraulic fracturing treatment implementing a compression level of 25%.

We compare the CS location results against two processing alternatives in a fully sampled domain (Figure 9.12). The first alternative has only a noise-whitening filter applied to the data as preprocessing. The second alternative incorporates in addition the cross-correlation of the data with the recordings of a high SNR event, followed by the formation of subgroups of traces using nonlinear stacking techniques [20]. In comparison, the CS system in our example has

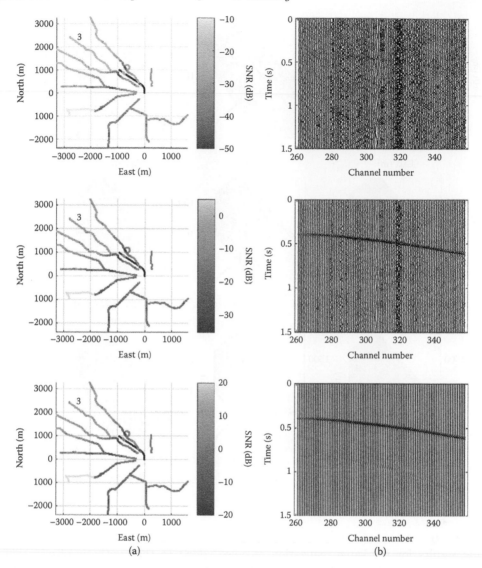

FIGURE 9.9

(a) Signal-to-noise ratio (SNR) conditions for different scenarios of signal magnitude and one realization of real noise. The black solid line at the center of the plots is the well trajectory, with the perforations depicted as gray dots. The circle around (−650 and 1100 m) denotes the event's location. (b) Corresponding seismic traces for channels along one of the lines with higher SNR (line 3, depicted on the left column plots with the line number next to it). The time series of noise is constant in all cases, while the magnitude of the signal increases from top to bottom.

the noise whitening followed by a band-pass filter applied before encoding. The CS system in this example has two advantages over the simpler processing alternative. The first advantage is that station statics are corrected in the Green functions. As the two methods rely on the stacking of channels, station statics can make a difference in the detection of smaller magnitude events and also influence the accuracy of event locations. The second advantage is the incorporation of S-wave information, which further constrains the estimation of locations. S-waves, however, are highly attenuated in the near-surface and in many cases cannot be used.

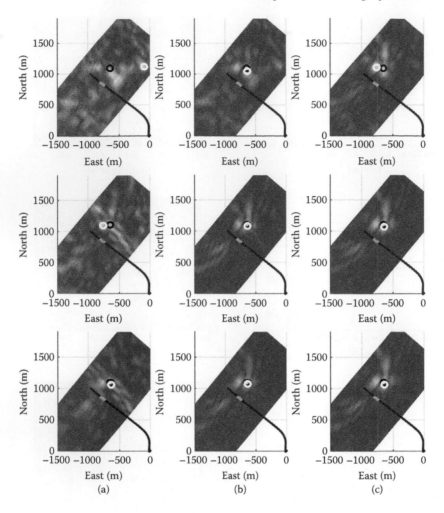

FIGURE 9.10

Processing results for the same realization of noise added to the synthetic observations while imposing different SNR and compression level. The SNR increases from (a) to (c) (see Figure 9.9). From top to bottom, the compression levels are 1%, 10%, and 20%, respectively. The black solid line is the well trajectory, with the perforations depicted as gray dots. The black circle is the true source location. The white circle is the source location estimated with BOMP using the compressed observations.

In this comparison, the simpler processing chain produced 149 locations potentially related to the hydraulic fracture (Figure 9.12b). The CS system, on the other hand, output 195 (Figure 9.12a). The locations from the CS system also display a better delineation of the structures stimulated by the treatment, where at least three branches extending from the well can be distinguished. This better clustering of locations and the larger number of detections are mostly attributed to the incorporation of static corrections.

The largest number of detections (376) from the three cases and best relative locations are obtained with the more sophisticated preprocessing chain (Figure 9.12c). The cross-correlation with the reference event relaxes the dependency of the detection and location

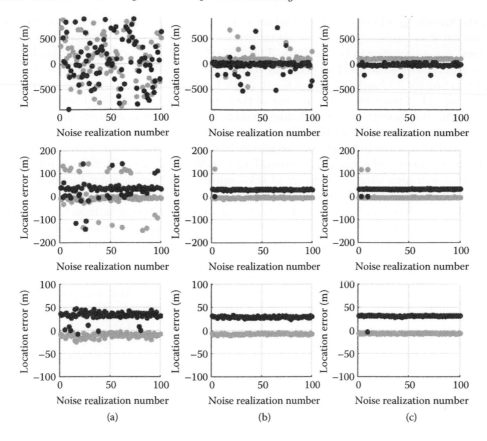

FIGURE 9.11

Location errors of the east (gray points) and north (black points) coordinates estimated with BOMP and the compressed observations. Each point in a plot corresponds to a different realization of noise recordings added to the synthetic signal while imposing different SNR and compression level. The SNR increases from (a) to (c) (see Figure 9.9). From top to bottom, the compression levels are 1%, 10%, and 20%, respectively.

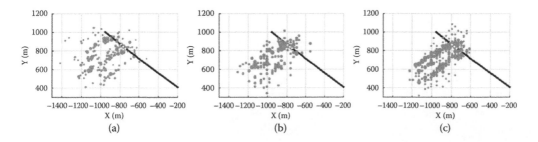

FIGURE 9.12

Comparison of location results obtained with CS (a), P-only with noise whitening in fully sampled domain (b), and P-only with state-of-the-art data conditioning in fully sampled domain (c).

steps on the knowledge of the velocity structure of the medium. In other words, it accounts to some extent for changes in waveform signature across the array, static corrections, heterogeneities, and anisotropy in the medium. This powerful strategy increases the number of detections and improves the location accuracy relative to the position of the reference event. The requirements for its implementation are that the location and source mechanism of the events to be detected must be similar to that of the reference event. The nonlinear stacking techniques result in a more powerful SNR enhancement than their linear equivalents, although there is a stronger sensitivity to the alignment of the signals from different traces. The cross-correlated traces remove most of the effects that affect signal alignment, improving thereby the performance of nonlinear stacking. Unfortunately, the identification of reference events requires scanning over the data set at least until a good candidate is found. This is an important barrier for its implementation in continuous monitoring. The statics in the CS system, on the other hand, are an average for the region under monitoring and can be obtained from previous treatments or even from perforation shots if the signals are of good quality.

Waveform fitting moment tensor inversion is highly sensitive to the alignment between band-limited Green functions and observations. Ideally, every event should be treated manually to ensure the reliability of moment tensor solutions. However, the requirement of fast information to make decisions drives the need for robust automatic estimation of source mechanisms. Figures 9.13 and 9.14 present a comparison between automatic moment

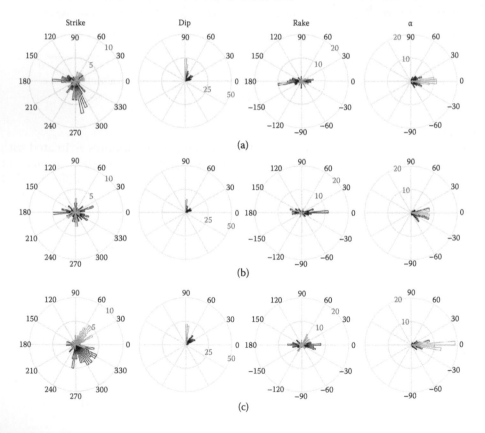

FIGURE 9.13
Dislocation geometries and style of activation interpreted from the biaxial decomposition of moment tensor solutions in a fully sampled domain (a), using CS without local static corrections (b) and CS with local static corrections (c). Each color represents one family of possible solutions.

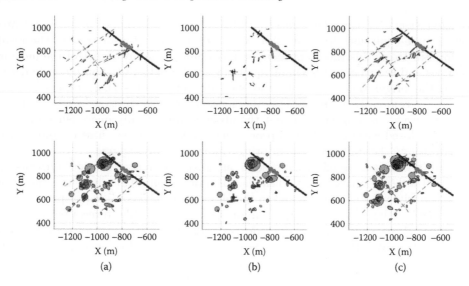

FIGURE 9.14
Dislocation planes (dark discs) interpreted from the biaxial decomposition of moment tensor solutions in a fully sampled domain (a), using CS without local static corrections (b) and CS with local static corrections (c). Top and bottom plots correspond to the gray and black families in Figure 9.13, respectively. Gray dashed lines are a possible interpretation of main fracture planes activated by the hydraulic fracturing treatment. The diameter of the discs is an indicator of the relative size of the events.

tensor solutions estimated with P arrivals in a fully sampled domain and incorporating local static corrections, against CS solutions with and without local static corrections. The static corrections are obtained from the analysis of a high SNR event; therefore, they can be incorporated only after all or at least part of the data set has been scanned. The CS moment tensor solutions obtained without local statics roughly reproduce the style of activation and dip estimated without compression (Figure 9.13a and b). Incorporating local statics improves the estimated dislocation azimuths (Figure 9.13c). If we consider only the solutions from the larger events, similar interpretations of fracture activation can be attained using the moment tensor solutions in the fully sampled domain and the CS solutions with statics (Figure 9.14 red dashed lines). In this example, a family of near-vertical dislocations aligned with the regional stresses can be tracked to delineate at least three different main fracture planes. A second family of subhorizontal fractures (dip $\sim 30°$) is also identified. It is interesting to observe a lineament of dislocations joining different fractures at an angle of approximately 90°. This lineament is consistent with an ancient state of stresses in the area, which suggests that preexisting planes of weakness were reactivated by the hydraulic fracturing treatment.

9.5 Conclusions

Continuous monitoring of seismic activity is a problem of interest at scales ranging from laboratory experiments (e.g., compression tests), industrial applications (e.g., mining, geothermal, oil and gas, and CO_2 sequestration), and hazard identification systems (e.g., earthquake and tsunami warning systems, volcano monitoring, and nuclear test monitoring).

In most of these scenarios, the source mechanism of the seismic events can be represented with a point-source equivalent in terms of a symmetric moment tensor (i.e., as the result of internal forces). Using this representation, and assuming time invariance of the medium between sources and receivers, a system of equations can be defined to continuously detect and estimate the origin time, location, and moment tensor of seismic events as they take place. Furthermore, in many of these applications, the separation in time between different seismic events is long enough that the observations of only one seismic event at a time can be enclosed within predefined processing windows. Under these circumstances, the solution vector to the continuous monitoring system is always sparse.

Solving the continuous monitoring system as a sparse representation presents an advantage over the least-squares solution in the form of enhanced resolution in the event location, particularly when noise is present in the observations. The adjoint displays similar robustness in the imaging of the event location; however, it is unable to recover the correct source mechanism because Green functions do not possess the property of orthogonality. Sparse solvers such as BOMP overcome this limitation by inverting for the amplitudes of the coefficients in the solution vector, which in the source monitoring problem means inverting for the moment tensor only at the position of the detected event.

Practical source monitoring scenarios can consist of large monitoring grids accompanied by a large number of receivers in the monitoring array. Handling of full-waveform processing in these scenarios is computationally expensive and can increase the response time of the monitoring system to unreasonable levels. CS is a novel alternative for the efficient handling of full-waveform information. An attractive property of the CS system presented in this chapter is that fully sampled data are not required to estimate the information of interest about the seismic sources. Because a sparse parameterization of the observations can be stated in terms of a dictionary of band-limited Green functions, the seismic source properties can be retrieved directly from the compressed samples.

Dictionaries of band-limited Green functions are adaptive (i.e. they change with each monitoring setting) and highly coherent. Numerical modeling is crucial to assess whether CS can be implemented in a particular monitoring system and to determine the level of compression that can be achieved without compromising detection and accuracy of the estimated source parameters. The practical effect of compression is observed as an increase in the minimum SNR at which an event can be detected and reliably located. In other words, the cost of compression is the nondetection of the lower SNR seismic events. The smaller the compression level, the higher the threshold SNR at which an event can be detected. In general, for a given level of compression, the robustness of the source parameters decreases from origin time detection (most reliable), then location, and, finally, moment tensor. Current lines of research are focused on decreasing the floor of the SNR at which seismic events can be reliably characterized while reducing the compression level in the system. Furthermore, the manufacturing of a seismic receiver that incorporates the encoding of the observations is still a pending matter.

Acknowledgments

We would like to thank Schlumberger for granting permission for the publication of this work. Special thanks to Aline Gendrin for sharing her processing results in the real data example, and also to Phil Christie and Gwenola Michaud whose feedback significantly improved the presentation of this manuscript.

References

[1] E. Aker, D. Kuhn, V. Vavrycuk, M. Soldal, and V. Oye. Experimental investigation of acoustic emissions and their moment tensors in rock failure. *International Journal of Rock Mechanics & Mining Sciences*, 70:286–295, 2014.

[2] K. Aki and P. Richards. *Quantitative Seismology*. University Science Books, Sausalito, CA, 2009.

[3] R. Baraniuk, M. Davenport, R. DeVore, and M. Wakin. A simple proof of the restricted isometry property for random matrices. *Constructive Approximation*, 28(3):253–263, 2008.

[4] T. Becker. Moment tensor inversion using combined surface and downhole hydraulic fracture monitoring. MSc thesis, ETH Zurich, 2015.

[5] J. Brune. Tectonic stress and the spectra of seismic shear waves from earthquakes. *Journal of Geophysical Research*, 75:4997–5009, 1970.

[6] E. Candes, J. Romberg, and T. Tao. Robust uncertainty principles: Exact signal reconstruction from highly incomplete frequency information. *IEEE Transactions on Information Theory*, 52:489–509, 2006.

[7] E. Candes and T. Tao. Near-optimal signal recovery from random projections: Universal encoding strategies? *IEEE Transactions on Information Theory*, 52:5406–5425, 2006.

[8] E. Candes and M. Wakin. An introduction to compressive sampling. *IEEE Signal Processing Magazine*, 25:21–30, 2008.

[9] K. Chambers, J. Kendall, S. Brandsberg-Dahl, and J. Rueda. Testing the ability of surface arrays to monitor microseismic activity. *Geophysical Prospecting*, 58:821–830, 2010.

[10] C. Chapman and W. Leaney. A new moment-tensor decomposition for seismic events in anisotropic media. *Geophysical Journal International*, 188:343–370, 2012.

[11] J. Claerbout. *Basic Earth Imaging*. Stanford University, Stanford, CA, 2010.

[12] D. Dreger and D. Helmberger. Determination of source parameters at regional distances with single station or sparse network data. *Journal of Geophysical Research*, 98: 8107–8125, 1993.

[13] J. Drew, R. White, F. Tilmann, and J. Tarasewicz. Coalescence microseismic mapping. *Geophysical Journal International*, 195:1773–1785, 2013.

[14] H. Dufumier and L. Rivera. On the resolution of the isotropic component in moment tensor inversion. *Geophysical Journal International*, 131:595–606, 1997.

[15] A. Dziewonski, T. Chou, and J. Woodhouse. Determination of earthquake source parameters from waveform data for studies of global and regional seismicity. *Journal of Geophysical Research*, 86:2825–2852, 1981.

[16] Y. C. Eldar and H. Bolcskei. Block-sparsity: Coherence and efficient recovery. In *IEEE International Conference on Acoustics, Speech, and Signal Processing*, pages 2885–2888, 2009.

[17] Y. C. Eldar, P. Kuppinger, and H. Bolcskei. Block-sparse signals: Uncertainty relations and efficient recovery. *IEEE Transactions on Signal Processing*, 58(6):3042–3054, 2010.

[18] J. H. Ender. On compressive sensing applied to radar. *Signal Processing*, 90(5): 1402–1414, 2010. Special Section on Statistical Signal and Array Processing.

[19] D. Gajewski, D. Anikiev, B. Kashtan, E. Tessmer, and C. Vanelle. Localization of seismic events by diffraction stacking. In *77th Annual International Meeting, SEG, Expanded Abstracts*, 2007.

[20] A. Gendrin, A. Ozbek, T. Probert, I. Bradford, and J. LeCalvez. Toward and optimized data conditioning for surface-acquired microseismic data. In *78th EAGE Conference and Exhibition, Expanded Abstracts*, 2016.

[21] M. Godano, E. Gaucher, T. Bardainne, M. Regnier, A. Deschamps, and M. Valette. Assessment of focal mechanisms of microseismic events computed from two three-component receivers: Application to the Arkema-Vauvert field (France). *Geophysical Prospecting*, 58:775–790, 2010.

[22] M. Herman and T. Strohmer. Compressed sensing radar. In *2008 IEEE Radar Conference*, pages 1–6, IEEE, 2008.

[23] F. Herrmann. Randomized sampling and sparsity: Getting more information from fewer samples. *Geophysics*, 75:WB173–WB187, 2010.

[24] F. Herrmann, M. Friedlander, and O. Yilmaz. Fighting the curse of dimensionality: Compressive sensing in exploration seismology. *IEEE Signal Processing Magazine*, 29(3):88–100, 2012.

[25] M. Jost and R. Herrmann. A student's guide to and review of moment tensors. *Seismological Research Letters*, 60:37–57, 1989.

[26] H. Kawakatsu. On the realtime monitoring of the long-period seismic wavefield. *Bulletin of the Earthquake Research Institute*, 73:267–274, 1998.

[27] H. Kawakatsu and J. Montagner. Time-reversal seismic-source imaging and moment tensor inversion. *Geophysical Journal International*, 175:686–688, 2008.

[28] R. Kranz, T. Satoh, O. Nishizawa, K. Kusunose, M. Takahashi, K. Masuda, and A. Hirata. Laboratory study of fluid pressure diffusion in rock using acoustic emissions. *Journal of Geophysical Research*, 95:21593–21607, 1990.

[29] C. Li, C. C. Mosher, and S. T. Kaplan. Interpolated compressive sensing for seismic data reconstruction. In *SEG Technical Program Expanded Abstracts 2012*, pages 1–6, 2012.

[30] C. Li, C. C. Mosher, L. C. Morley, Y. Ji, and J. D. Brewer. Joint source deblending and reconstruction for seismic data. In *SEG Technical Program Expanded Abstracts 2013*, pages 82–87, 2013.

[31] T. Lin and F. Herrmann. Compressed wavefield extrapolation. *Geophysics*, 72(5):SM77–SM93, 2007.

[32] H. Mansour, H. Wason, T. T. Lin, and F. J. Herrmann. Randomized marine acquisition with compressive sampling matrices. *Geophysical Prospecting*, 60(4):648–662, 2012.

[33] G. Manthei. Characterization of acoustic emission sources in a rock salt specimen under triaxial compression. *Bulletin of the Seismological Society of America*, 95:1674–1700, 2005.

[34] S. Mendelson, A. Pajor, and N. Tomczak-Jaegermann. Uniform uncertainty principles for Bernoulli and subgaussian ensembles. *Constructive Approximation*, 28:277–289, 2008.

[35] C. C. Mosher, C. Li, L. C. Morley, F. D. Janiszewski, Y. Ji, and J. Brewer. Non-uniform optimal sampling for simultaneous source survey design. In *SEG Technical Program Expanded Abstracts 2014*, pages 105–109, 2014.

[36] R. Nolen-Hoeksema and L. Ruff. Moment tensor inversion of microseisms from the b-sand propped hydrofracture, m-site, Colorado. *Tectonophysics*, 336:163–181, 2001.

[37] M. D. Sacchi. Sparse inversion of the Radon coefficients in the presence of erratic noise with application to simultaneous seismic source processing. In *2014 IEEE International Conference on Acoustics, Speech and Signal Processing (ICASSP)*, pages 2386–2389, May 2014.

[38] S. A. Sipkin. Estimation of earthquake source parameters by the inversion of waveform data: Synthetic waveforms. *Physics of the Earth and Planetary Interiors*, 30(2–3): 242–259, Salt Lake City, Utah, 1982.

[39] G. Strang. *Linear Algebra and Its Applications*. Thomson Brooks/Cole, 2006.

[40] C. Trifu and V. Shumila. Reliability of seismic moment tensor inversions for induced microseismicity at Kidd mine, Ontario. *Pure and Applied Geophysics*, 159:145–164, 2002.

[41] H. Tsuruoka, H. Kawakatsu, and T. Urabe. Grid mt (grid-based real-time determination of moment tensors) monitoring the long-period seismic wavefield. *Physics of the Earth and Planetary Interiors*, 175:8–16, 2009.

[42] D. Vasco. Deriving source-time functions using principal component analysis. *Bulletin of the Seismological Society of America*, 79:711–730, 1989.

[43] V. Vavrycuk. On the retrieval of moment tensors from borehole data. *Geophysical Prospecting*, 55:381–391, 2007.

[44] V. Vavrycuk. Tensile earthquakes: Theory, modeling and inversion. *Journal of Geophysical Research*, 116: B12320, 2011. doi:10.1029/2011JB008770.

[45] I. Vera Rodriguez, Y. Gu, and M. Sacchi. Resolution of seismic-moment tensor inversions from a single array of receivers. *Bulletin of the Seismological Society of America*, 101(6):2634–2642, 2011.

[46] I. Vera Rodriguez and N. Kazemi. Compressive sensing imaging of microseismic events constrained by the sign-bit. *Geophysics*, 81:1–10, 2016.

[47] I. Vera Rodriguez and M. Sacchi. Microseismic source imaging in a compressed domain. *Geophysical Journal International*, 198:1186–1198, 2014.

[48] I. Vera Rodriguez, M. Sacchi, and Y. Gu. A compressive sensing framework for seismic source parameter estimation. *Geophysical Journal International*, 191:1226–1236, 2012.

[49] I. Vera Rodriguez, M. Sacchi, and Y. Gu. Simultaneous recovery of origin time, hypocentre location and seismic moment tensor using sparse representation theory. *Geophysical Journal International*, 188:1188–1202, 2012.

[50] F. Walter, J. Clinton, N. Deichmann, D. Dreger, S. Minson, and M. Funk. Moment tensor inversions of icequakes on Gornergletscher, Switzerland. *Bulletin of the Seismological Society of America*, 99(2A):852–870, 2009.

10

Seismic Data Regularization and Imaging Based on Compressive Sensing and Sparse Optimization

Yanfei Wang and Jingjie Cao

CONTENTS

10.1 Introduction

Seismic exploration method is a crucial method to explore oil/gas, coal, and other resources in the underground. Geophones that are arranged at the earth's surface or in wells can record vibrations of the earth. Data recorded by geophones can be used to extract velocity information from the underground medium after a series of processing flow. In order to reconstruct the structure of the earth correctly, seismic acquisition should satisfy the Nyquist/Shannon sampling theorem not only in the time domain but also in the space domain. Generally, the sampling process in the time domain can fulfill the sampling theorem. However, sampling in the space domain often violates the sampling theorem due to the influence of obstacles, rivers, bad traces, noise, acquisition aperture, topography, and acquisition costs. Seismic data that violate the Nyquist/Shannon sampling theorem may bring harmful aliases and deteriorate the results of migration (Liu and Sacchi 2004), multiple elimination (Naghizadeh 2009), denoising (Soubaras 2004), and amplitude versus offset (AVO) analysis (Liu 2004; Sacchi and Liu 2005; Naghizadeh and Sacchi 2010). In order to remove the influences of subsampled data, the seismic regularization/interpolation/reconstruction/restoration technique is a key step to provide reliable data from the subsampled data. Thus, seismic interpolation is a crucial research direction in exploration seismology (Wang et al. 2011).

Many interpolation methods have been proposed in the past decades (Mostafa and Sacchi 2007), and these methods can be classified into wave equation–based methods and signal processing methods (Cao et al. 2015). The wave equation–based methods utilize the physical properties of wave propagation to reconstruct seismic data. An integral with a continuous operator is often used to obtain the complete wave field (Ronen 1987; Stolt 2002). This kind of method calls for some information on velocity distribution in the interior of the earth and is computationally demanding. Signal processing–based methods play an important role in providing reliable wave field information (Spitz 1991; Duijndam et al. 1999; Liu 2004; Liu and Sacchi 2004; Zwartjes and Gisolf 2006, Herrmann and Hennenfent 2008; Naghizadeh and Sacchi 2010) and can be further divided into four kinds.

The first signal-processing method is the predictive error filtering method. Based on this method, seismic data can be obtained by finding a convolution filter that predicts the data in such a way that the error is white noise. The F–X predictive error filtering method (Spitz 1991; Porsani 1999) is the first predictive error filtering method that can reconstruct spatially aliased, regularly subsampled data. The low frequency data components are utilized to recover the high frequency data. A modified predictive error filtering method was proposed by Soubaras (2004). Gulunay (2003) proposed a similar method in the F–K domain for band-limited signals using the Fourier transform to predict the complete wave field. The T–X domain predictive error filtering method was introduced by Claerbout (1992), Crawley (2000), and Fomel (2002); the dip in a sliding time window should be an estimate at first, followed by interpolating along the dip direction in each time window. However, this kind of method is based on the assumption of linear events, and the interpolation results will deteriorate for cross events.

The second signal processing method is the sparse transform–based method combined with a regularization strategy. For this method, seismic interpolation is treated as an inverse problem, and seismic data are assumed to be sparse in some transformed domain. For seismic processing, the Fourier transform (Sacchi and Ulrych 1996; Sacchi et al. 1998; Duijndam et al. 1999; Liu 2004; Xu et al. 2005; Zwartjes and Sacchi 2007) and the linear Radon transform (Trad et al. 2002) are commonly used sparse transforms. Methods based on the Fourier transform can be applied to seismic signals with spatially band-limited properties; the sparse linear Radon transform can transform line events into several points in the transformed domain. Satisfactory results based on them can be obtained under the assumption of linear events. However, they all operate on a single scale and the decomposition into multiresolution elements is not used. For curve events, the parabolic Radon transform (Darche 1990) and the curvelet transform (Herrmann and Hennenfent 2008) are more suitable than the two abovementioned transforms. In the last decade, multiscale methods have gained much interest, especially the curvelet transform. The curvelet transform, which was proposed by Candes and Donoho (2000), is a multiscale, multidirectional, and sparse representation of curve-like signals. The curvelets are localized not only in the spatial domain and the frequency domain but also in the angular orientation. A new directional parameter provides an additional angular geometric property with a high degree of orientation that identifies the directional singularities (Candes and Donoho 2004). As a multiscale, multidirectional, anisotropic tight frame, it is strictly localized in the Fourier domain. Furthermore, it provides an optimal representation of objects that have discontinuities along edges (Candes and Donoho 2000; Starck et al. 2002; Candes and Donoho 2004). The curvelet transform can represent curved events sparsely and avoid the assumption of linear events for seismic interpolation (Herrmann and Hennenfent 2008). The curvelet transform was introduced to seismic interpolation by Hennenfent and

Herrmann (2008) at first and was proved to be the sparsest transform for seismic data compared with the Fourier transform and the wavelet transform.

Recently, the matrix/tensor completion method was introduced to seismic interpolation (Kreimer and Sacchi 2012, 2013; Yang et al. 2012). This method can be seen as an extension of the compressed sensing theory into two-dimensional or high-dimensional data. The complete seismic data can be assumed to be a low-rank matrix, and subsampled data may increase the trace of a matrix. Based on matrix completion theory, seismic data can be treated as a matrix for two-dimensional data or a tensor for three-dimensional data, and singular value decomposition of matrix should be used to estimate the trace of a matrix. Numerical experiments using this method show robust results in two-dimensional and high-dimensional data (Kreimer and Sacchi 2012, 2013).

The seismic interpolation problem can be deemed as a compressive sensing problem, because seismic acquisition is the sampling processing of seismic data. Based on compressive sensing theory, the sparse transform–based method and the matrix/tensor completion method change the seismic interpolation problem into sparse optimization problems. This chapter focuses on these two interpolation methods. Basic compressive sensing theory is introduced in Section 10.2, then the mathematical models of seismic interpolation based on these methods are illustrated in detail in Section 10.3. Following that, we discuss some sparse optimization methods that can solve the induced sparse optimization models efficiently. In Section 10.5, synthetic and experimental tests based on some proposed models and methods are given to show the efficiency of compressive sensing theory and sparse optimization. Finally, some concluding remarks are given.

10.2 Compressive sensing

As the foundation of digital signal and image processing, the Nyquist/Shannon sampling theorem was widely used in many areas such as remote sensing (Wang et al. 2009), medical image processing (Lustig et al. 2007), and geophysics (Herrmann and Hennenfent 2008; Cao et al. 2011), to name a few. Based on this theorem, a signal must be sampled at a rate at least twice its highest frequency in order to be represented without error. However, the sampling process and storage are time-consuming because of the large amount of data sampled. Because of the restriction of sampling time and sampling numbers in applications, the sampled data usually violate this theorem. The recently emerged compressive sensing theory demonstrates that complete and accurate signals can be reconstructed from incomplete measurements if the signals are sparse in some domains and the sampling methods fulfill certain conditions (Candes 2006; Donoho 2006).

If the sampling process is linear, then it can be written as follows:

$$b = \Phi f, \tag{10.1}$$

where $f \in R^N$ is the original signal, $b \in R^M$ is the sampled data, and $\Phi \in R^{M \times N}$ is the sampling matrix. If the sampled data is incomplete, that is, $M < N$, then Equation 10.1 is underdetermined. This indicates that solution of the problem 10.1 is discrete ill-posed since there exist infinite solutions that satisfy Equation 10.1.

However, compressive sensing theory has proven that, as an inverse problem, the original data can be reconstructed from the subsampled data under certain conditions (Donoho 2006). This is a special case of the general regularization scheme in inversion community. The relationship between compressive sensing and the Tikhonov regularization was addressed by Wang et al. (2012). First, the original signal f must be sparse or compressible under a certain transform; that is, f can be expressed as $f = \Psi s$, where s has a few nonzero entries. If this condition is fulfilled, Equation 10.1 can be changed into the following:

$$b = As, \tag{10.2}$$

where $A = \Phi\Psi$. Another condition of compressive sensing is that A must satisfy the restricted isometry property (RIP) (Candes and Tao 2004; Candes and Tao 2005; Candes et al. 2006). If Φ is the Gaussian random matrix, the partial Fourier matrix, the uniform spherical matrix, the binary random matrix, the partial Hadamard matrix, or the Toeplitz matrix, then A satisfies the RIP. If $K < C \cdot M / \log(N/M)$, the sparse solution can be solved, where C is a universal constant and K is the number of nonzero elements of s (Candes et al. 2006).

If the above two conditions are satisfied simultaneously, s can be found by solving Equation 10.2 with the constraint that s be sparse or compressible, which is actually a combinatorial optimization problem:

$$\min \|s\|_0, \text{ s.t. } As = b, \tag{10.3}$$

where $\| \cdot \|_0$ denotes the number of nonzero entries of vectors. Theoretically, solving this problem requires exhaustive searches over all subsets of columns of A (Candes and Tao 2005; Candes 2006).

Based on the above discussion, compressive sensing theory includes four ingredients: the sampling methods, the sparse transforms, the optimization models, and the solving methods. The sampling methods will vary from problem to problem, whereas the other three parts are generic for application problems. The Fourier transform is the most commonly used transform in signal processing and seismic processing. The discrete cosine transform and wavelet transform are two frequently used transforms for image processing and magnetic resonance imaging (Lustig et al. 2007). In the following sections, we focus on sparse inversion models and solution methods.

10.3   Seismic data regularization

According to compressive sensing theory, seismic acquisition can be denoted as follows:

$$\Phi d + n = d_{obs}, \tag{10.5}$$

where Φ is the sampling matrix, d is the complete wave field data, n is the additive random noise, and d_{obs} is the observed data. Because the sampling is incomplete, there are infinitely many solutions theoretically, but we can utilize some prior information to find the solutions with physical meaning. Sparsity of the seismic data in a transformed domain is commonly used because seismic events can be sparsely expressed by some transforms. If the original data d can be sparsely expressed by some transform—that is, $s = \Psi^T d$ is sparse or compressible, Ψ^T is a sparse orthogonal transform or tight frame—Equation 10.5 can be transformed to the following:

$$As + n = d_{obs}, \tag{10.6}$$

where $A = \Phi\Psi$. Various methods have been developed to find sparse solutions of Equation 10.6, such as greedy algorithms (Mallat and Zhang 1993), convex optimization methods (Chen et al. 1998; Beck and Teboulle 2009; van den Berg and Friedlander 2009), and nonconvex optimization methods (Mohimani et al. 2009). Greedy algorithms such as matching pursuit (MP) are computationally simple and quite effective. However, because the algorithm is greedy, there are situations where the basic algorithm does not result in effective sparse solutions. Convex optimization methods are the most commonly used, because they are theoretically rigorous and suitable for large-scale computation (Chen et al. 1998; Cao et al. 2012). The most commonly used convex optimization is the basis pursuit model:

$$\min \|s\|_1, \text{ s.t. } \|As - d_{\text{obs}}\|_2 \leq \sigma, \tag{10.7}$$

where $\| \cdot \|_1$ is a measure of sparsity and σ is the energy of noise. If $\sigma = 0$, Equation 10.7 can be transformed into a linear optimization and solved by interior point methods (Chen et al. 1998; Candes and Tao 2005). However, the objective function of Equation 10.7 is nondifferentiable; it cannot be solved by general purpose methods such as the conjugate gradient method or Newton-type methods directly. Because there are sufficient algorithms to solve the unconstrained form of Equation 10.7, researchers prefer to solve the following problem:

$$\min \|As - d_{obs}\|_2^2 + \lambda \|s\|_1, \tag{10.8}$$

where λ is the regularization parameter, which should be adjusted carefully.

In order to overcome the nondifferentiability of the L_1 norm, the L_1 norm can be replaced by its smooth approximations, which can be called the *smoothL₁* *method*. Thus, Equation 10.7 can be changed into the following:

$$\min F(s), \text{ s.t. } As = d_{obs}, \tag{10.9}$$

where $F(s)$ is a smooth approximation of the L_1 norm, which can be called the *smoothL₁ norm function*. SmoothL₁ norm functions are generally separable, that is, $F(s) = \sum_{i=1}^{N} f(s_i)$, so we need only to analyze the 1D function $f(s_i)$, where s_i is a scalar. The first function we discuss is the following:

$$f_\varepsilon(s_i) = \sqrt{s_i^2 + \varepsilon}. \tag{10.10}$$

As a continuous, convex, and differential function (Wang et al. 2011), $f_\varepsilon(s_i)$ approximates to $|s_i|$ quite well when ε is very small. However, it is not zero exactly at the original point. This may affect the sparsity of solutions. The second function is as follows:

$$f_\theta(s_i) = \frac{1}{\theta}[\ln(1 + e^{-\theta s_i}) + \ln(1 + e^{\theta s_i})], \tag{10.11}$$

which approximates to $|s_i|$ closely when θ is large enough. This function is convex and differential (Chen and Mangasarian 1996); it is also nonzero at the original point.

Another familiar function is the Huber function:

$$f_{Huber}(s_i) = \begin{cases} s_i^2/2a, & |s| \leq a \\ |s_i| - a/2, & |s| > a \end{cases}, \tag{10.12}$$

which is smooth everywhere and approaches to $|s_i|$ extremely well when a turns to zero (Bube and Nemeth 2007). The Huber function is a hybrid of the L_1 norm and the L_2 norm; it behaves like the L_2 norm for small a and like the L_1 norm for large a. The smooth transition from L_2 norm to L_1 norm behavior is controlled by a. Huber functions are not new in geophysical inverse problems; Sacchi (1997) used the Huber functions for sparse deconvolution to get a sparse reflectivity series, so it is also a measure of sparsity of solutions. The three functions mentioned are not only separable but also convex and smooth. The Cauchy function can be also taken as the sparsity measure, but it is nonconvex in the whole domain.

The above smoothL$_1$ norm functions have some common points. A superparameter exists in each function to control the approximation to $|s_i|$, so the superparameter should be chosen carefully. In addition, they are all differential and separable.

Convex optimization methods can be seen as the convex relaxation of the zero norm because there exist numerous methods to solve the convex optimization model. Additionally, some nonconvex relaxation of the zero norm can also be used to build inversion models. A classic model is the L$_p$ norm regularization model (Chartrand 2007):

$$\min \|s\|_p, \text{ s.t. } \|As - d_{obs}\|_2 \leq \sigma, 0 < p < 1, \tag{10.13}$$

where $\|s\|_p$ denotes the L$_p$ norm of s. Similar to the L$_1$ norm, the L$_p$ norm is also nondifferentiable, so some smooth approximations of the L$_p$ norm can also be chosen to build inversion models. This kind of strategy can be called the *smoothL$_0$* method.

If the smooth approximation of the L$_0$ norm is taken as the objective function, then the quality of solutions will be better than the L$_1$ norm–based models (Chartrand 2007). For the one-dimensional case, $f_\kappa(s_i) = 1 - \exp(\frac{-s_i^2}{2\kappa^2})$ is a smooth L$_0$ function, where κ is a superparameter. This function has the following properties: (1) $f_\kappa(s_i)$ is continuously differentiable and (2) when κ turns to zero, $f_\kappa(s_i)$ turns to the L$_0$ norm, that is,

$$\lim_{\kappa \to 0} f_\kappa(s_i) = \begin{cases} 0, & s_i = 0 \\ 1, & s_i \neq 0. \end{cases} \tag{10.14}$$

Thus, we can build a new optimization model, based on which Equation 10.9 can be transformed to the following:

$$\min \sum_{i=1}^{N} f_\kappa(s_i), \text{ s.t. } As = d_{obs}. \tag{10.15}$$

Besides the abovementioned function, there are also some functions that can be used to build the nonconvex model, such as

$$f_\kappa(s_i) = \begin{cases} 0, & |s_i| \leq \kappa \\ 1 - (s_i/\kappa)^2, & |s_i| \geq \kappa \end{cases} \tag{10.16}$$

and

$$f_\kappa(s_i) = 1 - \kappa^2/(s_i^2 + \kappa^2). \tag{10.17}$$

Generally, the unconstrained optimization model of Equation 10.15,

$$\min \|As - d_{obs}\|_2^2 + \lambda \sum_{i=1}^{N} f_\kappa(s_i) \tag{10.18}$$

is easier to solve.

In summary, sparse inversion models of seismic interpolation can be written in a unified framework as the following model:

$$\min \frac{1}{2}\|As - d_{obs}\|_2^2 + \lambda R(s), \tag{10.19}$$

where λ is the regularization parameter to balance the fitting error and sparsity of s, and $R(s)$ is an operator that represents the sparsity of the solution. $R(s)$ can be $\|s\|_1$, $\|s\|_p$, $(0 < p < 1)$, smoothL$_1$ functions, or nonconvex smooth functions of s. For different chosen values of $R(s)$, special methods should be utilized to solve Equation 10.19. For example, if $R(s) = \|s\|_1$, the iterative soft thresholding (IST) method (Daubechies et al. 2004) is a practical strategy to solve Equation 10.19, and the iterative hard thresholding (IHT) method (Blumensath and Davies 2008) can be used when $R(s) = \|s\|_0$. More details about the solving method are discussed in Section 10.4.

In addition to the abovementioned sparse transform–based methods, the matrix completion method is a novel method for seismic interpolation. In a certain sense, the matrix completion problem can be regarded as applying compressed sensing on matrix variables. The complete two-dimensional seismic data in a time–space window can be arranged as a low-rank block Toeplitz/Hankel matrix (Oropeza and Sacchi 2011); however, the subsampled data will violate this condition. Based on this condition, the following matrix optimization problem can be built:

$$\min \|M_d\|_{nu}, \text{ s.t. } M_R \cdot M_d = M_{d_{obs}}, \tag{10.20}$$

where M_R is a block Toeplitz/Hankel matrix that comes from the sampling matrix R, M_d is a block Toeplitz/Hankel matrix that comes from the complete data d, $M_{d_{obs}}$ is a block Toeplitz/Hankel matrix that comes from the sampled data d_{obs}, and $\|M_d\|_{nu}$ is the nuclear norm of matrix M_d, which indicates the one norm of the singular values of M_d. In fact, the nuclear norm is a convex relaxation of the trace of a matrix. If M_d is obtained by solving Equation 10.20, then transforming it into the time–space domain will yield the reconstructed data.

Similar to the smoothL$_0$ strategy, we can also build a nonconvex optimization model of matrix completion:

$$\min F(M_d), \text{ s.t. } M_R \cdot M_d = M_{d_{obs}}, \tag{10.21}$$

where $F(M_d)$ is a nonconvex smooth function of the singular values of M_d. If the singular values of M_d are $s_1, s_2, \cdots s_K$, where K is the number of singular values or rank, then $F(M_d) = \sum_{i=1}^{K} f(s_i)$, where $f(s_i) = 1 - \exp^{(-s_i^2/2\theta^2)}$, or $f(s_i) = 1 - \frac{\theta^2}{s_i^2 + \theta^2}$. There are some additional nonconvex functions that can be used to build a nonconvex optimization model.

10.4 Sparse optimization methods

Since the emergence of compressive sensing theory, various sparse optimization methods have continually been developed at a rapid pace because high-quality algorithms improve the quality of compressive sensing results and accelerate the speed of computation. This section discusses some classical methods and newly developed methods of transform for sparse transform–based methods and matrix completion–based methods.

10.4.1 Quick review of classical methods

Finding a sparse solution for underdetermined problems has been studied in many areas; the commonly used methods are based on L$_1$ norm optimization. These methods find a sparse solution by solving Equations 10.7 or 10.8. Equation 10.7 can be changed into linear programming and then solved by interior point methods (Chen et al. 1998; Wang et al. 2009). Although the underlying problem is a linear program, conventional algorithms such as interior point methods suffer from poor scalability for large-scale real-world problems. The Lasso problem,

$$\min \|As - d_{obs}\|_2^2, \text{ s.t. } \|s\|_1 \leq \delta \tag{10.22}$$

is another model closely connected to Equations 10.7 and 10.8. With appropriate parameter choices of σ, λ, and δ, the solutions for Equations 10.7, 10.8, and 10.22 coincide, and these problems are in some sense equivalent.

However, computational challenges arise from the following reasons. First, real-world applications are invariably large scale. For example, there are more than a million variables in an image reconstruction problem. Second, real-time or near-real-time processing is required in some applications. Consequently, algorithms requiring matrix decomposition or factorization are not practical. Moreover, the sparsity of the solutions presents a unique opportunity for achieving relatively fast convergence with a first-order method (Daubechies et al. 2004). These features make the development of efficient optimization algorithms for compressive sensing applications an interesting research area. In light of the high interest in finding more efficient algorithms to solve these problems, many new algorithms have been proposed. It is impossible to summarize all existing algorithms in the literature; we provide a comprehensive review of some representative methods. Examples of such algorithms include shrinkage-based algorithms (Figueiredo and Nowak 2003; Combettes and Pesquet 2007; Wright et al. 2009; Osher et al. 2010), interior point algorithms such as L1_Ls (Kim et al. 2007), the spectral projected gradient (SPGL1) method (van den Berg and Friedlander 2008) for the least absolute shrinkage and selection operator (LASSO) problem, Nesterov's algorithm (NESTA) (Becker et al. 2011) for the basis pursuit denoising problem, the gradient projection for sparse reconstruction (GPSR) (Figueiredo et al. 2007), augmented Lagrange multiplier methods (Rockafellar 1976), homotopy (Osborne et al. 2000), and so on.

Chen et al. (1998) solved Equation 10.7 with $\sigma = 0$ by first reformulating it as a "perturbed linear program" and then applying a standard primal-dual interior point method. The linear equations are solved by iterative methods such as Least-squares QR factorization (LSQR) or conjugate gradient method. Another interior point method is the L1_Ls method (Kim et al. 2007), which is based on changing 8 into a quadratic programming problem solved by the preconditioned conjugate gradient method. Equation 10.22 can also be solved by reconsidering it as second-order cone programming and then applying a log-barrier method (Wang et al. 2009).

Equation 10.8 can be transformed into a non-negative constraint quadratic program by splitting the variable s into its positive and negative parts; GPSR is a gradient projection method for the transformed quadratic program (Figueiredo et al. 2007). The computational complexity and convergence of GPSR is difficult to estimate precisely. Another issue is that the formulated quadratic program doubles the dimension of the equations from Equation 10.6.

The IST method is a kind of shrinkage-based algorithm for solving Equation 10.8 (Figueiredo and Nowak 2003). The original shrinkage-based algorithms may be sensitive to the initial values and are not always stable. In addition, much iteration is required for convergence.

Recently, a spectral gradient-projection method was developed for solving Equation 10.22 (van den Berg and Friedlander 2008). This method relies on root-finding of the parameter δ by solving the nonlinear, convex, monotone equation $\|As(\delta) - d_{obs}\|_2 = \sigma$. The root-finding method is the famous discrepancy principle in the regularization theory for ill-posed problems (Wang and Xiao 2001).

To be fair, there exists no overall winner that achieves the best performance in terms of both speed and accuracy for all applications. Yang et al. (2010) compared some classical algorithms through some simple examples and face recognition examples; the interior point methods for the L_1-norm minimization problem suffer from poor scalability for large-scale real-world problems. Without concerns about speed and data noise, the success rates of the interior point method is the highest among gradient projection (Figueiredo et al. 2007), homotopy (Osborne et al. 2000), the iterative shrinkage thresholding method (Daubechies et al. 2004), the proximal gradient method (Beck and Teboulle 2009), and augmented Lagrange multiplier methods (Rockafellar 1976). A reweighted L_1 norm regularization method was proposed by Candes et al. (2008), which outperforms L_1 minimization in the sense that substantially fewer measurements are needed for exact recovery.

Methods based on nonconvex objects were studied by Mohimani et al. (2009) and Gasso et al. (2009), where the L_0 norm is replaced by nonconvex smooth functions. The initial values should be carefully chosen to prevent local optimal solutions.

Alternatively, heuristic greedy algorithms have been developed to get sparse solutions for Equation 10.6. Greedy algorithms work when the data satisfy certain conditions, such as the RIP (Candes and Tao 2004). These algorithms include MP (Mallat and Zhang 1993), orthogonal matching pursuit (OMP) (Davis et al. 1997), least angle regression (Efron et al. 2004), compressive sampling matching pursuit (CoSaMP) (Needell and Tropp 2008), subspace pursuit (SP) (Dai and Olgica 2008), stagewise OMP (StOMP) (Donoho et al. 2012), and many other variants. MP (Mallat and Zhang 1993) and OMP (Davis et al. 1997) were derived from the sparse approximation problem over redundant dictionaries. The vector d_{obs} is expressed as a linear combination of a few columns of A, where the active set of columns to be used is built in a greedy fashion. At each iteration, a new column is added to the active set. OMP includes an extra orthogonal step so that it is known to perform better than the standard MP. OMP is fundamental, extremely simple, and cheap, but the theoretical guarantees are not very strong and practical performance varies. StOMP adds one or more indices at each iteration. Rather than being a monotonically growing index set as in OMP and StOMP, at each iteration of CoSaMP and SP the active set is the union of the indices of the K most significant components of the current point and the K most significant components of the gradient. Empirically, these greedy algorithms work better when the solution is very sparse but will deviate from the solution for Equation 10.6 when the number of nonzero elements of solution increases. In other words, the greedy methods do not come with strong theoretical guarantees for global convergence. However, when the RIP is good, greedy methods can succeed in recovering sparse signals. Blumensath and Davies (2008) proved that the IHT method, which is also a greedy method, can be used for zero-norm regularized least-squares problems. Wen et al. (2010) proposed a hybrid method combing shrinkage, subspace optimization, and continuation strategies for sparse optimization.

In fact, greedy methods are active-set methods; the homotopy method belonging to this kind of method was originally designed for solving noisy overdetermined L_1-penalized least-squares problems (Osborne et al. 2000). It was also used to solve underdetermined problems by Donoho and Tsaig (2006). As a modification of the homotopy method, the least angle regression method (Efron et al. 2004) was investigated for solving Equation 10.22. When the solution is sufficiently sparse, both of these methods are more rapid than general-purpose linear programming methods even for high-dimensional cases. If the solution is not rigorously sparse, the homotopy method may fail to find the correct solution in certain scenarios.

By using the active set as a constraint, the iterative support detection (ISD) method is proposed to address failed reconstructions of L_1 minimization due to insufficient measurements (Wang and Yin 2010). Different from greedy methods such as OMP, the index set in ISD is not necessarily nested or increasing, and all the solution components are updated at the same time.

As a big-data, large-scale computing problem, seismic data interpolation requires efficient methods to reduce the increasing computational cost. The iterative-reweighed least-squares method was introduced to solve the seismic interpolation problem (Sacchi and Ulrych 1996). Abma and Kabir (2006) introduced the projection onto convex sets (POCS) method to irregular seismic interpolation because the seismic interpolation problem can be seen as an image inpainting problem. The IST method (Daubechies et al. 2004) was introduced by Herrmann and Hennenfent (2008) for subsampled seismic interpolation using the curvelet transform as a sparse transform. The SPGL1 method for Equation 10.7 (van den Berg and Friedlander 2008) shows robust numerical examples of seismic interpolation. Cao et al. (2012) proposed a projected gradient method for a nonconvex optimization model of wave field reconstruction.

10.4.2 Gradient projection method for smoothL$_1$ norm optimization

In this subsection, we give a simple and efficient algorithm for seismic interpolation (Cao et al. 2015). The sparse transform can be orthogonal transform or tight frame. The inversion model of seismic interpolation is the smoothL$_1$ model:

$$\min F(s) = \sum_{i=1}^{N} f_{Huber}(s_i), \quad \text{s.t. } As = d_{obs} \tag{10.23}$$

Because A is an underdetermined matrix in its discrete form, $S = \{s | As = d_{obs}\}$ is a convex set; thus Equation 10.23 can be solved by a gradient projection algorithm, which can be designated as follows:

Algorithm 10.1

Step 1.   Give the maximum iteration number L, the parameter $a = 0.0001$, which controls the approximation degree of the Huber function to the L$_1$ norm, $k = 0$, and the initial solution s^0.

Step 2.   Solve the gradient $\nabla F(s^k)$. If the stopping criterion is satisfied, go to Step 4; otherwise, give a trial iteration $s_{k+1}^{pre} = s^k - \mu \nabla F(s^k)$, where μ is the step length (which can be solved by the back-tracing method).

Step 3.   Update the iteration point: $s^{k+1} = s_{k+1}^{pre} - P_S(s_{k+1}^{pre})$ ($P_S(\bullet)$ is the projection onto $S = \{s | As = d_{obs}\}$); let $s^k = s^{k+1}$, $k = k + 1$, and return to Step 2.

Step 4.   Give the final solution: $s^{final} = s^k$.

The initial solution is set as $s = A^* d^{obs}$, where A^* is the Hermitian transpose of A. The main computations of this algorithm are focused on Step 3 theoretically, because the projection in Step 3, $P_S(s_{k+1}^{pre}) = A^*(AA^*)^{-1}(As_{k+1}^{pre} - d_{obs})$, contains a forward transform, an inverse transform, and a large-scale matrix inversion. However, if the forward transform is an orthogonal transform or a tight frame, the inversion of AA^* can be omitted; thus the computational efficiency of this algorithm will be enhanced significantly. In this case, the projection can be simplified to $P_S(s_{k+1}^{pre}) = A^*(As_{k+1}^{pre} - d_{obs})$.

10.4.3 Trust region method for sparse inversion

Because seismic reconstruction is computationally demanding, fast and robust methods are required to improve the computational efficiency (Trad 2009). However, most of the abovementioned methods belong to line-searching methods; the trust region method, which can provide globally convergent solutions for nonlinear problems, has gotten less attention in sparse optimization (Yuan 1993). For the trust region method, a quadratic approximation of the objective function at the current iteration point is built as a trust region subproblem, and then a descent direction is obtained by solving the subproblem (Yuan 1993). Generally, traditional trust region methods use the L$_2$ norm as the trust region. In this subsection, a novel L_1 norm trust region method is proposed, in which a novel L_1 norm trust region subproblem (TRSL1) is addressed and a rapid projected gradient method for the subproblem is proposed to improve the computational efficiency significantly. The TRSL1 method is a novel research avenue for seismic reconstruction.

The aim of our work is to get a sparse solution for Equation 10.6 and in the meantime to reach the minimum value of the following least-squares problem:

$$\min J(s) = \|As - d_{obs}\|_2^2 \tag{10.24}$$

If each descent direction of Equation 10.24 is sparse, then sparse solutions will be found quickly. In order to obtain sparse descent directions, the trust region subproblem using the L_1 norm as the trust region constraint is a feasible choice. The main merit of the L_1 trust region method is that the sparsity of the descent directions makes it possible to obtain sparse solutions quickly. In order to find the sparse descent directions of Equation 10.24, its L_1 norm trust region subproblem should be solved:

$$\min \varphi(x) = \frac{1}{2} x^T H_k x + g_k^T x,$$
$$\text{s.t. } \|x\|_1 \leq \Delta_k, \tag{10.25}$$

where H_k and g_k are the Hessian matrix and the gradient of $J(s)$ at the kth iterative point, respectively; x is the descent direction at the current point; and Δ_k is the trust region radius.

In order to introduce the idea of the trust region strategy, a classical algorithm framework for the trust region method for solving nonlinear problems is given as follows (Wang 2007):

Algorithm 10.2

Step 1. Choose $0 < \tau_3 < \tau_4 < 1 < \tau_1$, $0 < \tau_0 < \tau_2 < 1$, give the initial feasible solution s_0 and the initial trust region step $\Delta_0 > 0$, and let $k = 0$.

Step 2. If the stopping condition is satisfied, then stop iteration; otherwise, solve Equation 10.25 to get the descent direction x_k.

Step 3. Calculate the ratio r_k. Update the iterative point:

$$s_{k+1} = \begin{cases} s_k, & r_k \leq \tau_0 \\ s_k + x_k, & \text{otherwise} \end{cases}$$

Update the trust region radius:

$$\Delta_{k+1} \in \begin{cases} [\tau_3 \|x_k\|, \tau_4 \Delta_k], & r_k \leq \tau_2 \\ [\Delta_k, \tau_1 \Delta_k], & \text{otherwise} \end{cases}$$

Step 4. Solve the gradient g_{k+1} at s_{k+1}, and let $k = k + 1$; go to Step 2.

This algorithm will not stop until the residual reaches a threshold or the maximum iteration number is reached. The ratio $r_k = \frac{J(s_k) - J(s_k + x_k)}{\phi_k(0) - \phi_k(x_k)}$ controls the updating of s_k and Δ_k. The initial trust region step Δ_0 is chosen empirically. Interested readers can refer to Wang (2007) for detailed information about this method.

The efficiency of the trust region method hinges on solving Equation 10.25 efficiently. Birgin et al. (2000) proposed a robust projected gradient method for constrained optimization; here, we modify it to solve Equation 10.25. By defining $\varphi(x) = \frac{1}{2}x^T H_k x + g_k^T x$ and its gradient as $\nabla\varphi(x)$, the projected gradient method is as follows:

Algorithm 10.3

Step 1.   Give the initial direction x, the trust region step Δ_k, the minimal step length $\alpha_{\min}$, the maximal step length $\alpha_{\max}$, the initial step length $\alpha_0 \in [\alpha_{\min}, \alpha_{\max}]$, the sufficient descent parameter $\gamma \in (0,1)$, and the maximum number of iterations allowed by back tracing M. Calculate the initial projection $x_0 = P_{S_k}(x)$ (where $P_{S_k}(\bullet)$ is the projection of x onto $S_k = \{x | \|x\|_1 \leq \Delta_k\}$), the gradient of the objective function $\nabla\varphi(x_0)$, and let $l = 0$.

Step 2.   If the stopping criterion is satisfied, go to Step 8; otherwise, go to Step 3.

Step 3.   Calculate the step length using the back tracing method:

   Step 3.1. $\alpha = \alpha_l$.

   Step 3.2.   Calculate the projection $\bar{x} = P_{\Delta_k}[x_l - \alpha\nabla\varphi(x_l)]$.

   Step 3.3.   If $\varphi(\bar{x}) \leq \max_{j \in [0,\min\{k, M-1\}]} \varphi(x_{l-j}) + \gamma(\bar{x} - x_l)^T \nabla\varphi(x_l)$, go to Step 4; otherwise, $\alpha = \alpha/2$ and go to Step 3.2.

Step 4.   Update the iterative point $x_{l+1} = \bar{x}$ and calculate the new gradient g_{l+1}.

Step 5.   Calculate $\Delta x = x_{l+1} - x_l$, $\Delta g = \nabla\varphi(x_{l+1}) - \nabla\varphi(x_l)$.

Step 6.   Update the step length:

   If $\Delta x^T \Delta g \leq 0$, $\alpha_{l+1} = \alpha_{\max}$.

   Otherwise, $\alpha_{l+1} = \min\left\{\alpha_{\max}, \max\left\{\alpha_{\min}, (\Delta x^T \Delta x)/(\Delta x^T \Delta g)\right\}\right\}$.

Step 7. $l = l + 1$ and go to Step 3.

Step 8.   Let $x_{\Delta_k} = x_l$.

In Algorithm 10.3, projections onto $S_k = \{x | \|x\|_1 \leq \Delta_k\}$ can be calculated according to van den Berg and Friedlander (2008). For more choice of parameters in this algorithm, refer to van den Berg and Friedlander (2008).

In this subsection, the gradient and Hessian matrix can be calculated exactly, which will save computational effort and improve the accuracy of solutions. In addition, the projected gradient method for solving subproblems is very efficient. These merits make the TRSL1 method suitable for large-scale computation.

10.4.4   Operator splitting method

The alternating direction method is a kind of operator splitting method, which has received more attention in recent years (see He et al., 2000, and Wang et al., 2012, and references therein). For the general equality-constrained convex optimization problem

$$\min f(x),$$
$$\text{s.t. } Ax = d, \tag{10.26}$$

where $x \in \mathbb{R}^N$, $A \in \mathbb{R}^{M \times N}$, $f : \mathbb{R}^N \to \mathbb{R}$ is convex and separable, the splitting operator method refers to the following problem:

$$\min f_1(x) + f_2(y),$$
$$\text{s.t. } Ax + By = c, \tag{10.27}$$

where $y \in \mathbb{R}^p$, $B \in \mathbb{R}^{M \times p}$, and $c \in \mathbb{R}^M$; f_1 and f_2 are two convex functions. It is clear that the original input signal x is split into two parts, called x and y here, with the objective function separation meeting this splitting.

Using the method of multipliers, we form the augmented Lagrangian function:

$$L^a(x, y, \lambda) = f_1(x) + f_2(y) + \lambda^T(Ax + By - c) + \frac{1}{2}\nu\|Ax + By - c\|_{l_2}^2, \qquad (10.28)$$

where λ is the Lagrangian multiplier and $\nu > 0$ is a preassigned parameter. This formulation is clearly a regularized form of the Lagrangian problem:

$$\min f_1(x) + f_2(y) + \frac{1}{2}\nu\|Ax + By - c\|_{l_2}^2, \qquad (10.29)$$
$$\text{s.t. } Ax + By - c = 0,$$

where $\nu > 0$ serves as a regularization parameter. With the splitting form, the original problem (Equation 10.26) can be solved by alternating directions:

$$x_{k+1} = \text{argmin}_x L^\nu(x, y_k, \lambda_k),$$
$$y_{k+1} = \text{argmin}_y L^\nu(x_{k+1}, y, \lambda_k), \qquad (10.30)$$
$$\lambda_{k+1} = \lambda_k + \nu(Ax_{k+1} + By_{k+1} - c).$$

Referring to our minimization problem,

$$f(a) = \|As - d_{obs}\|_{l_2}^2 + \nu\|s\|_{l_1} \to \min, \qquad (10.31)$$

the problem can be written in splitting form

$$\min f_1(s) + f_2(y), \qquad (10.32)$$
$$\text{s.t. } s - y = 0,$$

where $f_1(s) = \|As - d_{obs}\|_{l_2}^2$ and $f_2(y) = \nu\|y\|_{l_1}$. Using the alternating directions method of multipliers mentioned above, Equation 10.31 can be solved by alternating directions:

$$s_{k+1} = \text{argmin}_s \left(f_1(s) + \frac{1}{2}\nu\|s - (y_k - u_k)\|_{l_2}^2 \right),$$
$$y_{k+1} = S_C(s_{k+1} + u_k), \qquad (10.33)$$
$$u_{k+1} = u_k + (s_{k+1} - y_{k+1}),$$

where $u = \frac{1}{\nu}\lambda$, λ is the Lagrangian multiplier for Equation 10.32, and S_C serving as a proximal operator is a projection operator that projects some iteration points onto C. The definition of $S_c(s)$ is as follows: $S_c(s) = (s - c)_+ - (-s - c)_+, c \in C$, and $(\cdot)_+ = \max(\cdot, 0)$, which provides some soft thresholding to s. A detailed algorithm for the alternating directions method of multipliers for seismic data regularization was given by Wang et al. (2012).

10.4.5 Sparse optimization methods for matrix completion–based seismic interpolation

As for solving the nuclear norm optimization problem (Equation 10.20), the singular value soft thresholding algorithm is an efficient method, with the following iteration form in each step:

$$M_d^{k+1} = S_\lambda(M_d^k - \alpha^k M_R \cdot (r^k)) \qquad (10.26)$$

with $r^k = M_R \cdot M_d^k - M_{d_{obs}}$, and $S_\lambda(Z) = U\text{diag}(\bar{\sigma})V^T, \bar{\sigma} = S_\lambda(\sigma(Z))$. $S_\lambda(\cdot)$ is the soft thresholding operation.

A method similar to the POCS method can be used for solving Equation 10.21 (Abma and Kabir 2006). During each iteration, assuming that the kth iterative solution is M_d^k, and its singular value decomposition is $M_d^k = U_k S_k V_k$, where the diagonal elements of S_k are $s_1^k, s_2^k, \cdots s_N^k$, then update these singular values as $\tilde{s}_i^k = s_i^k - \lambda\nabla f(s_i^k), i = 1, 2, \cdots, N$; thus M_d^k is updated to the new matrix $\tilde{M}_d^k = U_k \tilde{S}_k V_k$, projection of this matrix $\tilde{M}_d^k$ onto the convex set $\{M|M_R \cdot M_d = M_{d_{obs}}\}$ will yield the $k+$ 1th iteration solution M_d^{k+1}. After many iterations, the matrix completion will be done.

Oropeza and Sacchi (2011) introduced the multichannel singular spectrum analysis (MSSA) method to seismic interpolation, where seismic traces are reconstructed as block Toeplitz/Hankel matrices, and simultaneous interpolation and denoising can be realized by reducing the rank of the Toeplitz/Hankel matrices. The rank number in the MSSA method plays the role of regularization. In order to get stable solutions, it needs to be reduced gradually over iterations. Therefore, the idea of the MSSA is similar to that of the thresholding methods.

Because this operation of singular value decomposition-based matrix completion is computationally demanding, some random singular value decomposition (SVD) methods and SVD-free methods were proposed to improve the computational efficiency of the matrix completion method (Ma et al. 2011; Aravkin et al. 2014); these methods are more suitable for large-scale seismic interpolations. Gao et al. (2015) introduced a parallel matrix factorization algorithm of tensor for 5D seismic data reconstruction and denoising, which is an SVD-free method (Xu et al. 2015).

10.5   Applications

We only test the smoothL$_1$ method and the TRSL1 method in this section and make a comparison of the two methods with well-developed methods from the literature—FISTA (fast iterative shrinkage-thresholding algorithm), SPGL1, and IST with continuation (ISTc).

10.5.1   Experimental examples of the smoothL$_1$ method

The performance of the smoothL$_1$ method in Subsection 10.4.2 was evaluated on two real data experiments. To demonstrate the efficiency of the proposed method, comparisons were conducted with the FISTA method (Beck and Teboulle 2009) and the SPGL1 method (van den Berg and Friedlander 2008).

The shot data contained 115 traces with a receiver interval of 12.5 m; there are 600 time samples per trace with 2 ms as the time sampling interval. Incomplete acquisition was simulated by randomly sampling 69 traces. The original data are given in Figure 10.1a, and the sampled data are given in Figure 10.1b. We conducted experiments using the smoothL$_1$, FISTA (Beck and Teboulle 2009), and SPGL1 methods (van den Berg and Friedlander 2008). Some parameters in these algorithms are listed below: the max iteration number was 15 for the smoothL$_1$ method, 20 for the FISTA method, and 30 for the SPGL1 method. Based on the above parameters, these methods can get comparable results with much different CPU times. The CPU time, signal-to-noise ratio (SNR), and relative error are listed in Table 10.1, where SNR is defined as $SNR = 10\log_{10}\frac{\|d_{orig}\|_2^2}{\|d_{orig}-d_{rest}\|_2^2}$ (d_{orig} means the original data and d_{rest} is the restored data) and the relative error is defined as $\frac{\|d_{orig}-d_{rest}\|_2}{\|d_{orig}\|_2}$. The interpolation results using the smoothL$_1$ method are given in Figure 10.2a and the difference

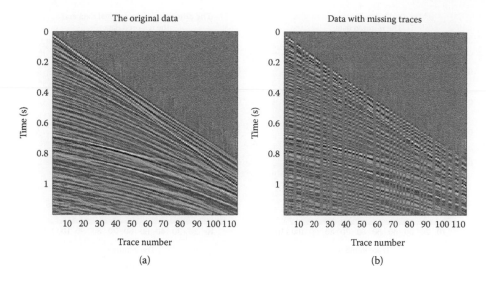

FIGURE 10.1

(a) The original data; (b) the sampled data.

TABLE 10.1

Comparison of the smoothL$_1$, FISTA, and SPGL1
Methods for Shot Data

	smoothL$_1$	FISTA	SPGL1
CPU time (sec)	56	73	156
SNR (db)	10.4975	9.8556	9.9523
Relative error	0.2986	0.3215	0.3180

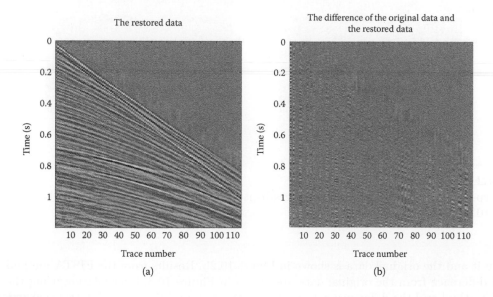

FIGURE 10.2

(a) Interpolation of the smoothL$_1$ method; (b) difference between Figure 10.2a and the
original data.

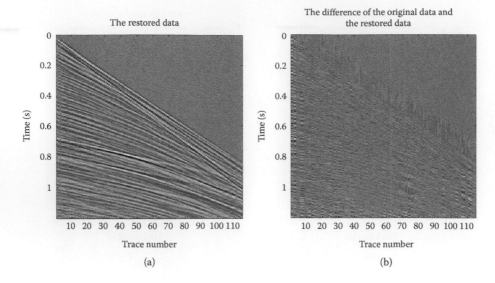

FIGURE 10.3

(a) Interpolation of the FISTA method; (b) difference between Figure 10.3a and the original data.

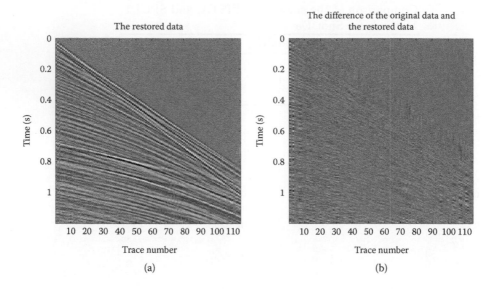

FIGURE 10.4

(a) Interpolation of the spectral projected gradient (SPGL1) method; (b) difference between Figure 10.4a and the original data.

between it and the original data is shown in Figure 10.2b. Results from the FISTA method and its difference from the original data are shown in Figure 10.3; interpolation using the SPGL1 method and the difference between the restoration and the original data are shown in Figure 10.4. Table 10.1 lists the SNR, relative error, and CPU time for each method; we can conclude that the smoothL$_1$ method is faster than the FISTA method and about 1/3 the CPU time of the SPGL1 method.

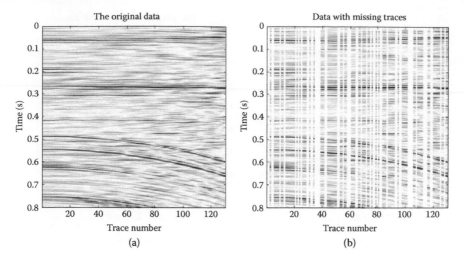

FIGURE 10.5

(a) Original data; (b) sampled data.

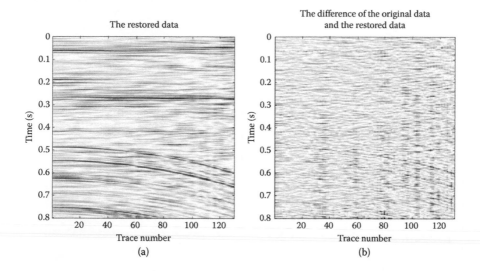

FIGURE 10.6

Interpolation of the smoothL$_1$ method; (b) difference between Figure 10.6a and the original data.

We further examine the efficiency of the smoothL$_1$ method with stacked data. Figure 10.5a gives stacked data consisting of 130 traces with a trace interval of 25 m and 401 time samples per trace with 2 ms as the time interval. The subsampled gather is shown in Figure 10.5b, with 40% of the original traces randomly deleted. The maximum number of iterations for the smoothL$_1$ method is 20. The interpolation of the smoothL$_1$ method and the difference between the interpolation and the original data are displayed in Figure 10.6. The interpolation results using the FISTA method with a maximum of 30 iterations and its difference from the original data are shown in Figure 10.7. Interpolation based on the SPGL1 method with a maximum of 50 iterations and its difference from the original data are shown in Figure 10.8. The CPU time, SNR, and relative error of these methods are shown in Table 10.2. These results demonstrate that, when the interpolation results are

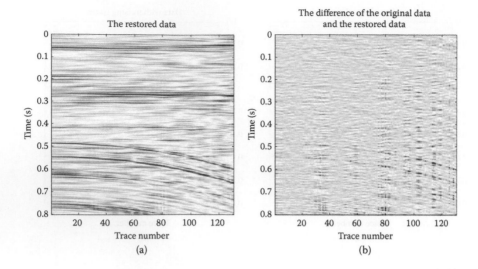

FIGURE 10.7
(a) Interpolation of the FISTA method; (b) difference between Figure 10.7a and the original data.

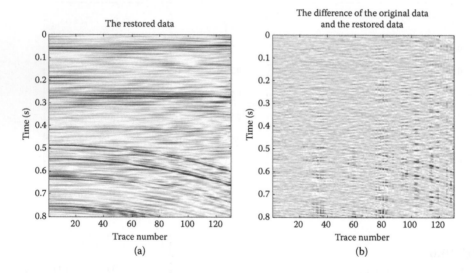

FIGURE 10.8
(a) Interpolation of the SPGL1 method; (b) difference between Figure 10.8a and the original data.

TABLE 10.2

Comparison of the smoothL$_1$, FISTA, and SPGL1 Methods for Stacked Data

	SmoothL$_1$	FISTA	SPGL1
CPU time (sec)	56	80	163
SNR (db)	22.1805	22.7094	22.9518
Relative error	0.0778	0.0732	0.0712

almost the same, the smoothL$_1$ method is faster than the FISTA method and about 1/3 the CPU time of the SPGL1 method. Thus, the proposed method is efficient and can reduce the amount of computation significantly.

10.5.2 Numerical and experimental examples of the TRSL1 method

The performance of the TRSL1 method is evaluated on synthetic as well as real data in this subsection. The goal of the synthetic data experiment was to demonstrate its potential for large-scale computation; the field data example verifies its ability to restore field data.

We considered seismic data with six layers, modeled with a 15 m receiver interval, 2 ms sampling interval, and a source function given by a Ricker wavelet with a central frequency of 25 Hz. The dataset contained 256 traces with 256 time samples in each trace. Incomplete acquisition was simulated by randomly extracting 160 traces from 256 traces. The original data and their F-K spectrum are shown in Figure 10.9; the sampled data and their F-K spectrum are displayed in Figure 10.10.

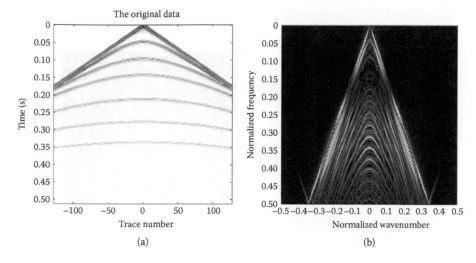

FIGURE 10.9

(a) Original data and (b) its F-K spectrum.

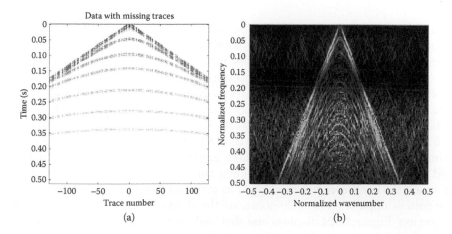

FIGURE 10.10

(a) Random subsampled data and (b) its F-K spectrum.

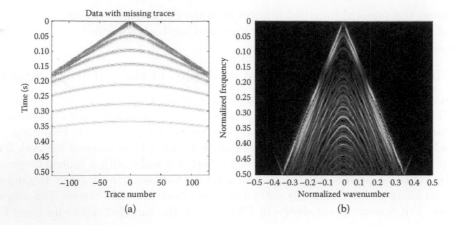

FIGURE 10.11
(a) Restoration of the TRSL1 method and (b) its F-K spectrum.

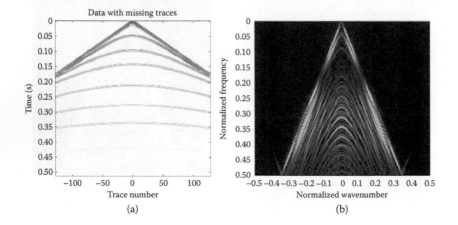

FIGURE 10.12
(a) Restoration of the SPGL1 method and (b) its F-K spectrum.

The SPGL1 method is robust and can yield high precision results; ISTc is present in Herrmann and Hennenfent (2008). Both methods are well-known in seismic processing. Thus, the proposed TRSL1 method contrasts with the SPGL1 method and the ISTc method. In order to get comparable results, the number of inner and outer loops of the TRSL1 method is five and three, respectively; restoration of the TRSL1 method is shown in Figure 10.11. The maximum number of iterations for the SPGL1 method is 70; restoration with the SPGL1 method is shown in Figure 10.12. For the ISTc method, there are 5 inner loops and 15 outer loops; the restoration of the ISTc method is shown in Figure 10.13. The CPU time, relative error, and SNR of these methods are given in Table 10.3. It can be concluded from Table 10.3 that, when yielding comparable reconstruction, the TRSL1 method can reach the same computation speed as excellent line-searching methods.

A marine data example is given to demonstrate the ability of the TRSL1 method for field data in this section. Figure 10.14 displays one shot and their F-K spectrum. It contains 200 traces with the first 0.6 seconds of data. The sampling rate is 2 ms with a receiver spacing of 15 m. The decimated data with 75 traces randomly removed and their F-K spectrum are shown in Figure 10.15. For the TRSL1 method, there are five inner loops and three

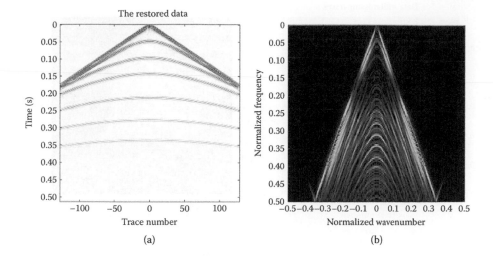

FIGURE 10.13
(a) Restoration of the ISTc method and (b) its F-K spectrum.

TABLE 10.3
Comparison of Synthetic Data Restoration with
TRSL1, SPGL1, and ISTc

Methods	CPU (sec)	Relative Error	SNR
TRSL1	337	0.2874	10.8309
SPGL1	268	0.2818	11.0010
ISTc	306	0.2794	11.0765

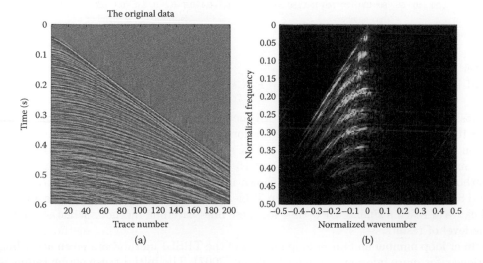

FIGURE 10.14
(a) The original data and (b) its F-K spectrum.

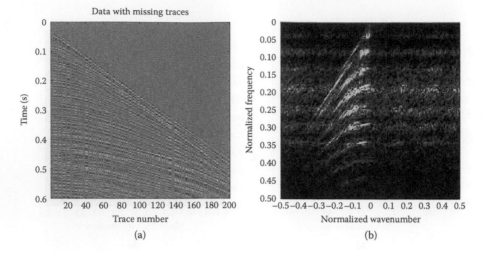

FIGURE 10.15

(a) The sampled data and (b) its F-K spectrum.

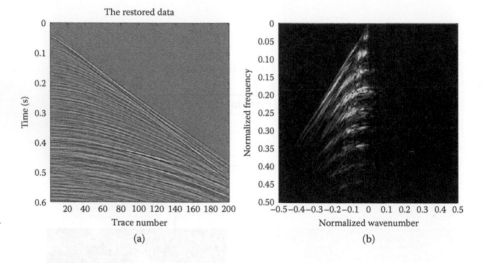

FIGURE 10.16

The TRSL1 method: (a) restoration and (b) F-K spectrum.

outer loops; the restored data with the TRSL1 method and their F-K spectrum are shown in Figure 10.16. The maximum number of iterations of the SPGL1 method is set at 70; the restoration with the SPGL1 method and its F-K spectrum are given in Figure 10.17. For the ISTc method, there are 5 inner loops and 15 outer loops; the restored data with the ISTc method and their F-K spectrum are shown in Figure 10.18. The CPU time, relative error, and SNR of these methods are listed in Table 10.4. Based on these results, the TRSL1 method can still reach a similar CPU time for these line-searching methods while obtaining the same level of results.

The inner loop number and outer loop number of the TRSL1 method are given according to experience; for more information, refer to Wang (2007). The initial trust region radius is related to the efficiency of this method, and thus it must be carefully chosen. Generally, it should be increased with data scale and amplitude of the data.

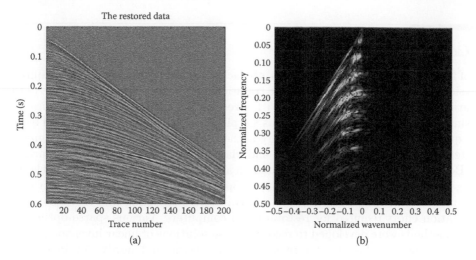

FIGURE 10.17
The SPGL1 method: (a) restoration and (b) F-K spectrum.

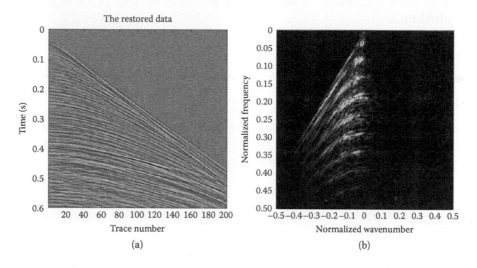

FIGURE 10.18
The ISTc method: (a) restoration and (b) F-K spectrum.

TABLE 10.4
Comparison of Field Data Restoration with the
TRSL1, SPGL1, and ISTc Methods

Methods	CPU (sec)	Relative Error	SNR
TRSL1	293	0.4026	7.9018
SPGL1	308	0.4176	7.5849
ISTc	286	0.4071	7.8050

10.6 Conclusion

Because of the influence of bad traces, rivers, obstacles, and other factors, seismic acquisition in geophysical exploration often violates the traditional Nyquist/Shannon theorem. Seismic regularization is a commonly used method for recovering complete data for subsampled data. Seismic regularization can be treated as a typical compressive sensing problem because seismic acquisition in exploration seismology is the sampling process of seismic signals. In this chapter, we gave a basic background and classification of the seismic regularization problem. Because this problem can be treated as a compressive sensing problem, the basis of compressive sensing theory was illustrated. Sparse optimization constitutes a crucial part of compressive sensing theory, because it transforms sampling and storage cost into the computation effort required for restoration in computers. Various sparse optimization models and methods were developed to recover sparse solutions of linear inversion problems from different aspects. This chapter reviewed some classic and newly developed sparse optimization methods from the point of view of compressive sensing theory. Numerical experiments of seismic regularization based on some methods proved the contribution of compressive sensing theory and sparse optimization to solving this special geophysics problem. As an ongoing research direction, excellent sparse optimization methods should be developed to make compressive sensing theory practical for this problem and other geophysical problems.

Acknowledgments

This work was supported by the National Natural Science Foundation of China under grant numbers 41325016, 91630202 and 41674114 and by the Strategic Priority Research Program of the Chinese Academy of Science (grant number XDB10020100).

References

Abma, R., and Kabir, N., 2006, 3D interpolation of irregular data with a POCS algorithm, *Geophysics*, 71(6): E91–E97.

Aravkin, A., Kumar, R., Mansour, H., Recht, B., and Herrmann, F. J., 2014, Fast methods for denoising matrix completion formulations, with applications to robust seismic data interpolation, *SIAM Journal on Scientific Computing*, 36(5): S237–S266.

Beck, A., and Teboulle, M., 2009, A fast iterative shrinkage-thresholding algorithm for linear inverse problems, *SIAM Journal on Imaging Sciences*, 2(1): 183–202.

Becker, S., Bobin, J., and Candes, E., 2011, NESTA: A fast and accurate first-order method for sparse recovery, *SIAM Journal on Imaging Sciences*, 4(1): 1–39.

Birgin, E., Martinez, J., and Raydan, M., 2000, Nonmonotone spectral projected gradient methods on convex sets, *SIAM Journal on Optimization*, 10(4): 1196–1211.

Blumensath, T., and Davies, M., 2008, Iterative thresholding for sparse approximations, *Journal of Fourier Analysis and Applications*, 14(5–6): 629–654.

Bube, K., and Nemeth, T., 2007, Fast line searches for the robust solution of linear systems in the hybrid and Huber norms, *Geophysics*, 72(2): A13–A17.

Candes, E., 2006, Compressive sampling, *Proceedings of International Congress of Mathematicians, European Mathematical Society Publishing House*, vol. 3, Madrid, Spain, pp. 1433–1452.

Candes, E., and Donoho, D., 2000, *Curvelets: A Surprisingly Effective Nonadaptive Representation for Objects with Edges*, Stanford University, California, Department of Statistics, Stanford, CA.

Candes, E., and Donoho, D., 2004, New tight frames of curvelets and optimal representations of objects with piecewise singularities, *Communications on Pure and Applied Mathematics*, 57(2): 219–266.

Candes, E., Romberg, J., and Tao, T., 2006, Robust uncertainty principles: Exact signal reconstruction from highly incomplete frequency information, *IEEE Transactions on Information Theory*, 52(2): 489–509.

Candes, E., and Tao, T., 2004, Near optimal signal recovery from random projections: Universal encoding strategies? *IEEE Transactions on Information Theory*, 52(12): 5406–5425.

Candes, E., and Tao, T., 2005, Decoding by linear programming, *IEEE Transactions on Information Theory*, 51(12): 4203–4215.

Candes, E., Wakin, M., and Boyd, S., 2008, Enhancing sparsity by reweighted l1 minimization, *Journal of Fourier Analysis and Applications*, 14(5–6): 877–905.

Cao, J., Wang, Y., and Wang, B., 2015, Accelerating seismic interpolation with a gradient projection method based on tight frame property of curvelet, *Exploration Geophysics*, 46(3): 253–260.

Cao, J., Wang, Y., and Yang, C., 2012, Seismic data restoration based on compressive sensing using regularization and zero-norm sparse optimization, *Chinese Journal of Geophysics*, 55(2): 239–251.

Cao, J., Wang, Y., Zhao, J., and Yang, C., 2011, A review on restoration of seismic wavefields based on regularization and compressive sensing, *Inverse Problems in Science and Engineering*, 19(5): 679–704.

Chartrand, R., 2007, Exact reconstructions of sparse signals via nonconvex minimization, *IEEE Signal Processing Letters*, 14(10): 707–710.

Chen, C., and Mangasarian, O., 1996, A class of smoothing functions for non-linear and mixed complementarity problems, *Computational Optimization and Applications*, 5(2): 97–138.

Chen, S., Donoho, D., and Saunders, M., 1998, Atomic decomposition by basis pursuit, *SIAM Journal on Scientific Computing*, 20(1): 33–61.

Claerbout, J., *Earth Soundings Analysis: Processing Versus Inversion*, Blackwell Science, Boston, MA, 1992.

Combettes, P., and Pesquet, J., 2007, Proximal thresholding algorithm for minimization over orthonormal bases, *SIAM Journal on Optimization*, 18(4): 1351–1376.

Crawley, S., *Seismic Trace Interpolation with Non-Stationary Predictionary-Error Filters*, PhD thesis, Stanford University, CA, 2000.

Dai, W., and Olgica, M., 2008, *Subspace Pursuit for Compressive Sensing: Closing the Gap between Performance and Complexity*, Technical Report, Department of Electrical and Computer Engineering, University of Illinois at Urbana-Champaign, Urbana, IL.

Darche, G., 1990, Spatial interpolation using fast parabolic transform, *60th Annual International Society Exploration Geophysicist*, pp. 1647–1650.

Daubechies, I., Defrise, M., and Mol, C. D., 2004, An iterative thresholding algorithm for linear inverse problems with a sparsity constraint, *Communications on Pure and Applied Mathematics*, 57(11): 1413–1457.

Davis, G., Mallat, S., and Avellaneda, M., 1997, Adaptive greedy approximation, *Constructive Approximation*, 13(1): 57–98.

Donoho, D., 2006, Compressed sensing, *IEEE Transactions on Information Theory*, 52(4): 1289–1306.

Donoho, D., and Tsaig, Y., *Fast Solution of L1-Norm Minimization Problems When the Solution may be Sparse*, Technical Report, Stanford University, Stanford, CA, 2006.

Donoho, D., Tsaig, Y., Drori, I., and Starck, J., 2012, Sparse solution of underdetermined systems of linear equations by stagewise orthogonal matching pursuit, *IEEE Transactions on Information Theory*, 58(2): 1094–1121.

Duijndam, A., Schonewille, M., and Hindriks, C., 1999, Reconstruction of band-limited signals, irregularly sampled along one spatial direction, *Geophysics*, 64(2): 524–538.

Efron, B., Hastie, T., Johnstone, I., and Tibshirani, R., 2004, Least angle regression, *The Annals of Statistics*, 32(2): 407–499.

Figueiredo, M., and Nowak, R., 2003, An EM algorithm for wavelet-based image restoration, *IEEE Transactions on Image Processing*, 12(8): 906–916.

Figueiredo, M., Nowak, R., and Wright, S., 2007, Gradient projection for sparse reconstruction: Application to compressed sensing and other inverse problems, *IEEE Journal of Selected Topics in Signal Processing*, 1(4): 586–597.

Fomel, S., 2002, Application of plane-wave destruction filters, *Geophysics*, 67(6): 1946–1960.

Gao, J., Stanton, A., and Sacchi, M., 2015, Parallel matrix factorization algorithm and its application to 5D seismic reconstruction and denoising, *Geophysics*, 80(6): V173–V187.

Gasso, G., Rakotomamonjy, A., and Canu, S., 2009, Recovering sparse signals with a certain family of non-convex penalties and DC programming, *IEEE Transactions on Signal Processing*, 57(12): 4686–4698.

Gulunay, N., 2003, Seismic interpolation in the Fourier transform domain, *Geophysics*, 68(1): 355–369.

Hale, E., Yin, W., and Zhang, Y., 2008, Fixed-point continuation for L1-minimization: Methodology and convergence, *SIAM Journal on Optimization*, 19(3): 1107–1130.

He, B. S., Yang, H., and Wang, S. L., 2000, Alternating direction method with self-adaptive penalty parameters for monotone variational inequalities, *Journal of Optimization Theory and Applications*, 106: 337–356.

Hennenfent, G., and Herrmann, F., 2008, Simply denoise: Wavefield reconstruction via jittered undersampling, *Geophysics*, 73(3): V19–V28.

Herrmann, F., and Hennenfent, G., 2008, Non-parametric seismic data recovery with curvelet frames, *Geophysical Journal International*, 173(1): 233–248.

Kim, S., Koh, K., Lustig, M., Boyd, S., and Gorinevsky, D., 2007, An interior-point method for large-scale L1-regularized least squares, *IEEE Journal of Selected Topics in Signal Processing*, 1(4): 606–617.

Kreimer, N., and Sacchi, M., 2012, A tensor higher-order singular value decomposition for prestack seismic data noise reduction and interpolation, *Geophysics*, 77(3): V113–V122.

Kreimer, N., and Sacchi, M., 2013, Tensor completion based on nuclear norm minimization for 5D seismic data reconstruction, *Geophysics*, 78(6): V273–V284.

Larner, K., Gibson, B., and Rothman, D., 1981, Trace interpolation and the design of seismic surveys, *Geophysics*, 46(9): 407–409.

Liu, B., *Multi-Dimensional Reconstruction of Seismic Data*, PhD thesis, University of Alberta, Alberta, Canada, 2004.

Liu, B., and Sacchi, M., 2004, Minimum weighted norm interpolation of seismic records, *Geophysics*, 69(6): 1560–1568.

Lustig, M., Donoho, D., and Pauly, J., 2007, Sparse MRI: The application of compressed sensing for rapid MR imaging, *Magnetic Resonance in Medicine*, 58(6): 1182–1195.

Ma, S., Goldfarb, D., and Chen, L., 2011, Fixed point and Bregman iterative methods for matrix rank minimization, *Mathematical Programming*, 128(1–2): 321–353.

Mallat, S., and Zhang, Z., 1993, Matching pursuits with time-frequency dictionaries, *IEEE Transactions on Signal Processing*, 41(12): 3397–3415.

Mohimani, H., Babaie-Zadeh, M., and Jutten, C., 2009, A fast approach for overcomplete sparse decomposition based on smoothed L0 norm, *IEEE Transactions on Signal Processing*, 57(1): 289–301.

Mostafa, N., and Sacchi, D., 2007, Multistep autoregressive reconstruction of seismic records, *Geophysics*, 72(6): V111–V118.

Naghizadeh, M., *Parametric Reconstruction of Multidimensional Seismic Records*, PhD thesis, University of Alberta, 2009.

Naghizadeh, M., and Sacchi, M., 2010, Beyond alias hierarchical scale curvelet interpolation of regularly and irregularly sampled seismic data, *Geophysics*, 75(6): WB189–WB202.

Needell, D., and Tropp, J., 2008, COSAMP: Iterative signal recovery from incomplete and inaccurate samples, *Applied and Computational Harmonic Analysis*, 26(3): 301–321.

Oropeza, V., and Sacchi, M., 2011, Simultaneous seismic data de-noising and reconstruction via multichannel singular spectrum analysis, *Geophysics*, 76(3): V25–V32.

Osborne, M., Presnell, B., and Turlach, B., 2000, A new approach to variable selection in least squares problems, *IMA Journal of Numerical Analysis*, 20(3): 389–403.

Osher, S., Mao, Y., Dong, B., and Yin, W., 2010, Fast linearized Bregman iteration for compressive sensing and sparse denoising, *Communications in Mathematical Sciences*, 8(1): 93–111.

Porsani, M., 1999, Seismic trace interpolation using half-step prediction filters, *Geophysics*, 64(5): 1461–1467.

Rockafellar, R., 1976, Augmented Lagrangians and applications of the proximal point algorithm in convex programming, *Mathematics of Operations Research*, 1(2): 97–116.

Ronen, J., 1987, Wave-equation trace interpolation, *Geophysics*, 52(7): 973–984.

Sacchi, M., 1997, Re-weighting strategies in seismic deconvolution, *Geophysical Journal International*, 129(3): 651–656.

Sacchi, M., and Liu, B., 2005, Minimum weighted norm wavefield reconstruction for AVA imaging, *Geophysical Prospecting*, 53(6): 787–801.

Sacchi, M., and Ulrych, T., 1996, Estimation of the discrete Fourier transform, a linear inversion approach, *Geophysics*, 61(4): 1128–1136.

Sacchi, M., Ulrych, T., and Walker, C., 1998, Interpolation and extrapolation using a high-resolution discrete Fourier transform, *IEEE Transactions on Signal Processing*, 46(1): 31–38.

Soubaras, R., 2004, Spatial interpolation of aliased seismic data, *74th Annual International Society of Exploration Geophysicist, Expanded Abstract*, pp. 1167–1170.

Spitz, S., 1991, Seismic trace interpolation in the F-X domain, *Geophysics*, 56(6): 785–794.

Starck, J., Candes, E., and Donoho, D., 2002, The curvelet transform for image denoising, *IEEE Transactions on Image Processing*, 11(6): 670–684.

Stolt, R., 2002, Seismic data mapping and reconstruction, *Geophysics*, 67(3): 890–908.

Trad, D., 2009, Five-dimensional interpolation: Recovering from acquisition constraints, *Geophysics*, 74(6): V123–V132.

Trad, D., Ulrych, T., and Sacchi, M., 2002, Accurate interpolation with high-resolution time-variant Radon transforms, *Geophysics*, 67(2): 644–656.

van den Berg, E., and Friedlander, M., 2008, Probing the Pareto frontier for basis pursuit solutions, *SIAM Journal on Scientific Computing*, 31(2): 890–912.

Wang, Y., *Computational Methods for Inverse Problems and their Applications*, Higher Education Press, Beijing, 2007.

Wang, Y., Cao, J., and Yang, C., 2011, Recovery of seismic wavefields based on compressive sensing by an L1-norm constrained trust region method and the piecewise random sub-sampling, *Geophysical Journal International*, 187(1): 199–213.

Wang, Y., Ma, S., Yang, H., Wang, J., and Li, X., 2009, On the effective inversion by imposing *a priori* information for retrieval of land surface parameters, *Science in China Series D: Earth Sciences*, 52(4): 540–549.

Wang, Y., and Xiao, T., 2001, Fast realization algorithms for determining regularization parameters in linear inverse problems, *Inverse Problems*, 17(2): 281–291.

Wang, Y., Yagola, A. G., and Yang, C. C., editors, 2012, *Computational Methods for Applied Inverse Problems*, published in Series: Inverse and Ill-Posed Problems Series 56, Walter de Gruyter, ISBN 978-3-11-025905-6.

Wang, Y., Yang, C. C., and Cao J. J., 2012, On Tikhonov regularization and compressive sensing for seismic signal processing, *Mathematical Models and Methods in Applied Sciences*, 22(2): 1150008.

Wang, Y., and Yin, W., 2010, Sparse signal reconstruction via iterative support detection, *SIAM Journal on Imaging Sciences*, 3(3): 462–491.

Wen, Z., Yin, W., Goldfarb, D., and Zhang, Y., 2010, A fast algorithm for sparse reconstruction based on shrinkage, subspace optimization, and continuation, *SIAM Journal on Scientific Computing*, 32(4): 1832–1857.

Wright, S., Nowak, R., and Figueiredo, M., 2009, Sparse reconstruction by separable approximation, *IEEE Transactions on Signal Processing*, 57(7): 2479–2493.

Xu, S., Zhang, Y., Pham, D., and Lambare, G., 2005, Antileakage Fourier transform for seismic data regularization, *Geophysics*, 70(4): V87–V95.

Xu, Y., Hao, R., Yin, W., and Su, Z., 2015, Parallel matrix factorization for low-rank tensor completion, *Inverse Problems and Imaging*, 9(2): 601–624.

Yang, A., Ganesh, A., Zhou, Z., Sastry, S., and Ma, Y., 2010, A review of Fast L1-minimization algorithms for robust face recognition, *2010 IEEE International Conference on Image Processing*, pp. 1849–1852.

Yang, Y., Ma, J., and Osher, S., 2013, *Seismic data reconstruction via matrix completion: Inverse problem and imaging*, 7: 1379–1392.

Yuan, Y., *Numerical Methods for Nonlinear Programming*, Shanghai Science and Technology Publication, Shanghai, 1993.

Zwartjes, P., and Gisolf, A., 2006, Fourier reconstruction of marine-streamer data in four spatial coordinates, *Geophysics*, 71(6): V171–V186.

Zwartjes, P., and Sacchi, M., 2007, Fourier reconstruction of nonuniformly sampled, aliased seismic data, *Geophysics*, 72(1): V21–V32.

11

Land Use Classification with Sparse Models

Mohamed L. Mekhal , Farid Melgani, Yakoub Bazi, and Naif Alajlan

CONTENTS

11.1 Introduction

Owing to today's abundance of advantageous satellites, supplied with various acquisition sensors, information-rich data pertaining to earthly as well as atmospheric mutations can be deduced easier than for instance a decade or two ago. These data, if suitably exploited, can be used to efficiently discern useful environmental and urban manifestations that may have tangible impacts on both the social and economic shapes of a certain region.

In contrast, the way such acquired data (usually in the form of multispectral images) are processed depends upon the application being addressed. Instances include cloud removal [1], shadow removal [2], and change detection [3]. Another very important ongoing topic is remote sensing (RS) image classification [4,5], which offers inestimable aid in pinpointing the locations highlighting specific object classes of concern.

With regard to the literature, existing works normally approach the RS image classification concern from two distinct but complementary aspects, namely (1) image representation and (2) classification. In the former consideration, the ultimate aim is to adequately mitigate the gap between the low-level inherent spectral/spatial content and the high-level semantic attributes of the image in hand. As for the latter concern, the solution is to design a suitable model able to classify the respective image features. In this respect, the optimal intent is to adequately tie the two earlier components into a reliable classification paradigm. Yang and Newsam [4], for instance, took the initiative to put forth a land use (LU) image classification dataset totaling 21 object classes. They proposed three bag-of-visual-words (BOW) models to address the classification issue. The first one is a basic BOW, which proceeds by establishing a diminished set of features called a *codebook*, followed by checking the presence of its features (words) in the feature space of a given image. Based on this, a sharply reduced BOW image representation can be produced. To compensate for the spatial information loss incurred by this basic BOW model, two other variants have been suggested, namely a spatial pyramid match kernel and a spatial co-occurrence kernel. Feature extraction was performed by means of the scale-invariant feature transform (SIFT). The methods yielded reasonable classification results on the dataset mentioned above. Another alternative, called *spatial pyramid co-occurrence*, was introduced by Yang and Newsam [5], who gathered relative and absolute spatial arrangements of the visual words over a spatial partitioning of the images, which has indeed shown improvement over the basic BOW model. For the same reason, another model, termed the *pyramid of spatial relatons*, was proposed by Shizhi and Tian [6] and demonstrated important gains over previous works. A concentric circle structured multiscale BOW fed into a support vector machine classifier was presented by Zhao, Tan, and Huo [7] and scored satisfactory outcomes.

Note that thus far the BOW has been a staple component for LU image representation. Although interesting findings have been achieved, the BOW signature, among many other feature extractors, if considered alone, does not seem to provide cutting-edge classification rates. Such observation suggests a potential fusion of multiple feature extractors so as to render classification more efficient. Another element is that the color/spectral information is not included in the aforementioned BOW-based LU image classification, as suggested by the fact that the constructed codebooks are generated from an input feature space (e.g., SIFT) that is normally extracted from gray-level images. It is evident that spectral information is also a valuable asset that can serve as a discriminative factor between image classes that exhibit similar spatial behaviors (e.g., textures).

In this context, we capitalize in this chapter on the two earlier directions (gray image-based feature fusion and spectral fusion), and we show that both scenarios favor the LU image classification in terms of classification rates with respect to the literature. In dealing with that, we put forward a compressive sensing (CS)–based fusion method. CS is a theory meant to sparsely recover a generic signal/image over an ensemble of signals/images grouped into what is commonly referred to as a *dictionary*. This implies that a compact representation can be obtained depending on the size of the dictionary. Henceforth, CS is opted for in this chapter on account of its capacity to convey concise representations. The main contributions of this chapter can be confined to the following:

1. CS-based residual fusion, which has not been adopted in the general computer vision nor the pattern recognition literatures so far. We particularly show that the CS can indeed play a major role as a decision-making tool.

2. Fusion of different feature extractors to improve classification accuracy within the context of the LU problem described in [4]. Based on this, we confirm that making use of multiple features at once is more reliable than considering individual features solely.

3. Fusion of different spectral layers to raise the classification rates. In this part, we stress that spectral information is key information that can make the difference in successfully classifying RS images pertaining to a high number of heterogeneous object classes.

The remainder of this chapter is framed as follows. Section 11.2 describes the two fusion strategies mentioned above. Section 11.3 details CS theory and the adopted reconstruction technique. Section 11.4 presents experimental results and interpretations. Finally, Section 11.6 draws conclusions and briefs potential ameliorations.

11.2 Proposed fusion methodologies

The aim of this chapter, as previously stated, is to tackle the problem of LU image classification. The task of classification consists of assigning labels to fixed-size test images based on a ground truth training set. Thus, the issue is addressed from two perspectives. The first one is articulated over fusing several feature vectors extracted from the input image, whereas the second one consists in fusing feature vectors extracted from the inherent spectral layers of the input image. In both scenarios, the proposed fusion concept is the same, inspired by the CS theory that stands for its remarkable property of concisely representing a generic signal by means of a sparse signature. In what follows, we describe both fusion procedures.

11.2.1 Feature fusion scenario

Let I be a probe image to be classified. Let us assume that we have available a labeled gallery of images where those belonging to the same class are collected into a single group. Let C and N be the numbers of classes and feature types adopted for image representation, respectively. Each feature type (among the N adopted ones) is extracted from the test image and afterward converted into a vector (likewise for the gallery images, for which the vectors are computed and stored offline in the form of a matrix, called a *dictionary*). The next stage takes as input each probe feature vector and performs a CS-based reconstruction out of all the C gallery dictionaries (i.e., C reconstructions are performed). By performing a CS recovery of the probe pattern over all the dictionaries, as many residuals (scalars) as the number of classes will be generated, thereby forming a sequence of residuals denoted by $R_{ij}(i = 1 \ldots N$, $j = 1 \ldots C)$. The smallest quantity within R_{ij} over j indicates the estimated class of the probe vector associated with the ith feature. For the N features, N residual sequences will thus be produced. In order to infer a global decision among the N feature types, a residual fusion layer is applied. The pipeline of the entire framework is illustrated in Figure 11.1.

As for the fusion step, we present two strategies, detailed in the next two subsections.

11.2.1.1 Sum operator–based strategy

In the previous section, we pointed out that in the case of N features, N residual sequences are generated. The linear residual fusion strategy starts by normalizing, for each feature type, all the R_{ij} arrays so that the corresponding maximum peak will be at unity. Such a normalization step is important because the residual values pertaining to a certain kind of feature might overweigh the remaining residuals in the fusion process. Afterward, all the R_{ij} vectors are linearly summed up to form a single sequence of the same length. Finally, the class of the probe pattern corresponds to the lowest residual of the final sequence.

$$R_j = \sum_{i=1}^{N} R_{ij} \quad \text{and} \quad class = \operatorname*{argmin}_{j}(R_j) \tag{11.1}$$

11.2.1.2 Majority-based operator strategy

In this strategy, from each of the N sequences of residuals, a decision is made by choosing the class corresponding to the smallest residual of the considered sequence. Hence, as many decisions (classes) are produced as the number of sequences. The next step is to pick up the class label that is the most frequent among the N made decisions and assign it as the final class of the probe pattern. When a tie occurs, we choose as the final class that with the smallest residual among the classes in conflict.

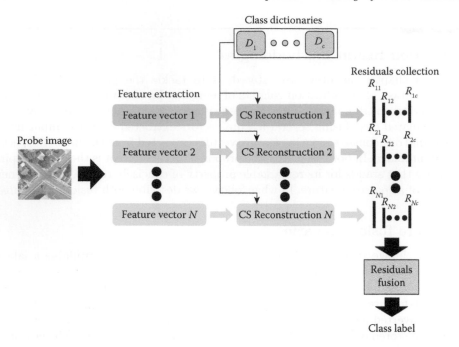

FIGURE 11.1
Block diagram of the multifeature fusion strategy.

11.2.2 Spectral fusion scenario

The described method deals with the problem of LU image classification. The goal is thus to determine a class label for the test image given a bundle of training images labeled *a priori*. Accordingly, it can be formulated as an image-matching issue given that the test image is believed to acquire the same label as the closest training image (nearest neighbor). However, the solution appears to be two-pronged. On the one hand, the images dealt with may reflect various rotation, scale, and illumination differences, meaning that comparison at a spectral level is likely to fail. Such a hurdle may be partly tractable with any appropriate feature extraction technique available in the literature, which could deal with those changes. On the other hand, a salient feature alone might be insufficient to adequately address the LU classification issue. In other words, it needs to be coupled with an efficient classification paradigm. There have been several attempts to address the LU problem described in [4–7]; however, most of the undertaken works have tended to overlook spectral information that can be of good saliency.

To this end, we present in this chapter a method that represents an image through an ensemble of CS encodings originating from features that are extracted from the available spectral channels and opportunely fused. In this regard, we take advantage of the well-known local binary pattern (LBP) for image representation [8], on account of its notable ability related to encoding and gathering gradient changes across the treated image. As for decision-making, we rely on a fusion method based on CS theory. Given an input image, the LBP feature vector is first extracted from each of the K available channels. Hence, as many LBP vectors as the number of channels are generated. The next stage is to proceed with a CS-based recovery of all the LBP vectors by making use of training dictionaries prepared *a priori* (i.e., developed by means of LBPs extracted offline from the training images). Thus, for the case of C classes, K residual sequences (i.e., one residual vector out

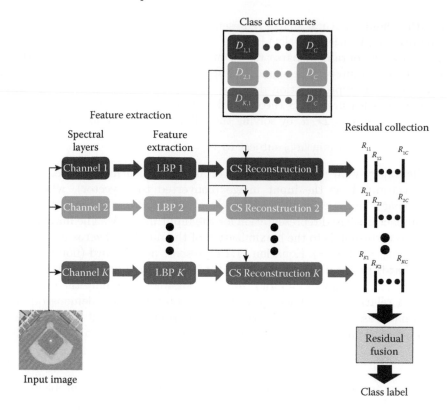

FIGURE 11.2
Pipeline of the proposed multispectral classification scheme.

of each spectral channel) of C bins each are produced from the reconstruction and then summed up in order to form a C-length final residual vector. Assuming that the smaller the residual reflected by a certain class, the more likely the test image is to be attributed to that class, we thereupon assign the label of the smallest residual to the test image. The flow of the proposed method is depicted in Figure 11.2.

11.3 Sparse representation

11.3.1 CS theory

CS, also known as *compressive sampling*, *compressed sensing*, or *sparse sampling*, was recently introduced by Donoho [9] and Candès [10]. CS theory aims at recovering an unknown sparse signal from a small set of linear projections. By exploiting this new and important result, it is possible to obtain equivalent or better representations by using less information compared with traditional methods. CS has proved to be a powerful tool for several tasks, such as acquisition, representation, regularization in inverse problem, feature extraction, and compression of high-dimensional signals, and has been applied in different research fields such as signal processing, object recognition, data mining, and bioinformatics [11]. In these fields, CS has been adopted to cope with several tasks such as face recognition [12], image super-resolution [13], segmentation [14], denoising [15], inpainting and reconstruction [16,17], and classification [18]. Note that images are a special case of signals that hold a natural sparse representation, with respect to fixed bases, also called a *dictionary* (i.e., Fourier, wavelet) [19].

CS is thus a way to obtain a sparse representation of a signal. It relies on the idea to exploit redundancy (if any) in the signals [9,10]. Usually signals like images are sparse, as they contain, in some representation domains, many coefficients close to or equal to zero. The fundamental feature of CS theory is the ability to recover with relatively few measurements defined by $V = D \cdot \alpha$, by solving the following L_0-minimization problem:

$$\min \|\alpha\|_0 \text{ subject to } V = D \cdot \alpha, \tag{11.2}$$

where D is a dictionary with a certain number of atoms (which in our case, are images converted into vectors); V is the input image (converted to a vector), which can be represented as a sparse linear combination of these atoms; and α is the set of coefficients intended as a compact CS-based representation for the input image **V**. The minimization of $\|\cdot\|_0$, the L_0 norm, corresponds to the maximization of the number of zeros in α. Following this formulation, $\|\alpha\|_0 = (l: \alpha_1 \neq)$. Equation 11.2 represents an NP-hard (nondeterministic polynomial-time hard) problem, which means that it is computationally infeasible to solve. Following the discussion of Candès and Tao [20], it is possible to simplify the evaluation of Equation 11.2 in a relatively easy linear programming solution. They demonstrated that, under some reasonable assumptions, minimizing the L_1 norm is equivalent to minimizing the L_0 norm, which is defined as $\|\alpha\|_1 = \sum_i |\alpha_i|$. Accordingly, it is possible to rewrite Equation 11.2 as follows:

$$\min \|\alpha\|_1 \quad \text{subject to} \quad V = D \cdot \alpha. \tag{11.3}$$

In the literature, there exist several algorithms for solving optimization problems similar to the one expressed in Equation 11.3—for instance, basic matching pursuit, orthogonal matching pursuit (OMP), and basis pursuit. In the following, we briefly introduce another alternative called *stagewise orthogonal matching pursuit* (StOMP) [21], which will be used in our chapter. In contrast to the basic OMP algorithm, StOMP involves many coefficients at each stage (iteration), whereas in OMP only one coefficient can be involved. Additionally, StOMP runs over a fixed number of stages, whereas OMP may take numerous iterations. Hence, StOMP is preferred in our chapter on account of its fast computation capability.

11.3.2 Stagewise orthogonal matching pursuit

The use of CS theory for image representation in our chapter is thus motivated by its capability to concisely represent a given image. For such purpose, the batch of training LU images are converted to grayscale and then into vectors. Their concatenation forms the dictionary D. Given a query image V, its compact representation α (whose dimension is reduced to the number of training images) is achieved by means of the procedure summarized below.

Step 1: Consider an initial solution $\alpha_0 = 0$, an initial residual $r_0 = V$, a stage counter s set to 1, and an index sequence denoted as $T_1, \ldots, T_s$ that contains the locations of the nonzeros in α_0.

Step 2: Compute the inner product between the current residual and the considered dictionary D:

$$C_s = D^T \cdot r_{s-1} \tag{11.4}$$

Step 3: Perform hard thresholding in order to find out the significant nonzeros in C_s by searching for the locations corresponding to the "large coordinates" J_s:

$$J_s = \{J : \alpha_s(J) > t_s \sigma_s\}, \tag{11.5}$$

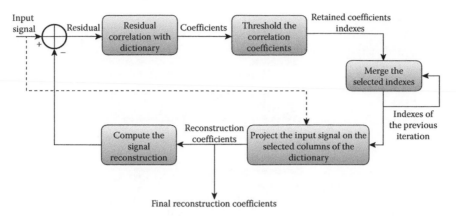

FIGURE 11.3
Basic flowchart of the stagewise orthogonal matching pursuit (StOMP) technique.

where σ_s represents a formal noise level and t_s is a threshold parameter taking values in the range $2 \leq t_s \leq 3$.

Step 4: Merge the selected coordinates J_s with the previous support.

$$T_s = T_{s-1} \cup J_s \tag{11.6}$$

Step 5: Project the vector V on the columns of D that correspond to the previously updated T_s. This yields a new approximation α_s:

$$(\alpha_S)_{T_S} = (D_{T_S}^t D_{T_S})^{-1} D_{T_S}^t V \tag{11.7}$$

Step 6: Update the residual according to $r_s = V - D \cdot \alpha_s$.

Step 7: Check whether a stopping iterative condition (e.g., $s_{max} = 10$) is met; if so, α_s is considered the final solution. Otherwise, the stage counter s is incremented and the next-stage process is repeated starting from Step 2. A detailed pipeline of the StOMP routine is illustrated in Figure 11.3.

11.4   Experimental results

To assess the presented CS fusion-based classification schemes, we exploited the UC Merced LU database, which was made available by Yang and Newsam [4,5]. The dataset was manually derived from another dataset of large aerial orthoimagery with about 30 cm of pixel resolution. It was downloaded from the United States Geological Survey (USGS) National Map of the following US regions: Birmingham, Boston, Buffalo, Columbus, Dallas, Harrisburg, Houston, Jacksonville, Las Vegas, Los Angeles, Miami, Napa, New York, Reno, San Diego, Santa Barbara, Seattle, Tampa, Tucson, and Ventura. The dataset contains 2100 images, each 256×256 pixels, categorized into 21 classes, where each class comprises 100 images.

The respective class labels are as follows: agricultural, airplane, baseball diamond, beach, buildings, chaparral, dense residential, forest, freeway, golf course, harbor, intersection, medium-density residential, mobile home park, overpass, parking lot, river, runway, sparse residential, storage tanks, and tennis court. For the sake of illustration, samples of the LU database are depicted in Figure 11.4.

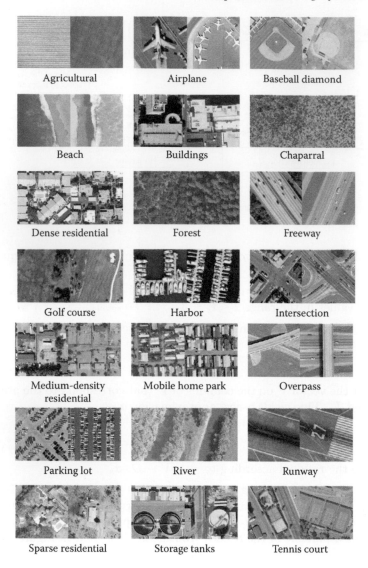

FIGURE 11.4
Two image samples per class from the land-use dataset.

In particular, the two main aspects underscoring this dataset are the high number of classes and the heterogeneity of the images, which may render the classification problem difficult.

11.4.1   Feature fusion scenario

The set of features adopted in our chapter consists of three types:

1. Histogram of oriented gradients (HOG): This kind of feature has gained a sound reputation in computer vision owing to its capacity of comprehensively describing an image through its local gradients [22,23].

2. Co-occurrence of adjacent local binary patterns (CoALBP): This is a variant to the popular LBP. CoALBP covers spatial information of all the LBPs pertaining

to a given image and takes into account their co-occurrence. CoALBP has shown richer representation and a robustness to illumination change [24].

3. Gradient local auto-correlations (GLAC): This type of feature, in addition to the gradient information, conveys the behavior of a given image surface in terms of curvatures. Thus, it has demonstrated improvement over other reference features such as HOG [25].

These three different feature types are considered in our chapter due to their sound discrimination capability as discussed in [22–25].

For the sake of consistency with reference works, we articulate the evaluation upon a fivefold cross-validation, where the database is randomly split into five parts (or folds), each conveying 20 images per class. One held-out subset is used as a probe (test), and the remaining four subsets are employed as the gallery (training images). This process is performed five times, and their average is considered the final classification accuracy, which is defined as the ratio of the correctly classified samples to the total number of test samples.

We first report the results obtained by using each type of feature individually; thus the fusion process is not involved at this point. Table 11.1 summarizes the results. The lowest rate is observed for HOG features, scoring 68.67%, followed by the GLAC pattern by yielding a raise of about 8% over the first one. The best accuracy of 80.52% is recorded for the CoALBP features. Note that the performances are different among the three features, which might be explained by the fact that the images of the same class reveal various orientations and scale changes, not to mention the total number of classes (i.e., 21 categories), which is subject to raise a large within-class and low interclass variability.

With regard to the fusion of the previous features by means of the two simple strategies described above, we detail the results in Table 11.1. It appears that the sum operator–based strategy (SOS) exhibits a better accuracy (94.33%) than the majority-based operator strategy (MOS) (87.95%). First, as compared with relying on only one kind of feature, it is noteworthy that the fusion of multiple features by means of the CS representation has led to a drastic improvement. Second, the lower accuracy of MOS compared with SOS can be explained by the fact that the former disregards partial information (i.e., the residuals) used for the individual decisions on which it relies. Exploiting all available information in the decision has emerged as the best way to proceed with the fusion as performed by SOS.

As for comparing the proposed scheme with reference state-of-the-art methods on the same dataset, a number of interesting works were taken into account [4,6,7]. They all opted for a similar fivefold-based evaluation, except for [7], where a twofold validation was adopted. We therefore ran the experiments considering this latter validation too. The comparison is provided in Table 11.2.

TABLE 11.1

Classification Accuracies Achieved by Single-Feature CS Representations and CS Fusion Strategies

	Feature Typology			Fusion	
	HOG	**CoALBP**	**GLAC**	**SOS**	**MOS**
Accuracy (%)	68.67	80.52	77.1	94.33	87.95

CoALBP, co-occurrence of adjacent local binary patterns; GLAC, gradient local auto-correlations; HOG, histogram of oriented gradients; MOS, majority-based operator strategy; SOS, sum operator-based strategy.

TABLE 11.2

Comparison with State-of-the-Art Methods

Work	Description	Accuracy (%)	Validation
Yang and Shawn [4]	Spatial BOW	81.19	Fivefold
Chen and Tian [6]	Pyramid of spatial relatons	89.1	Fivefold
Proposed method	CS multifeature fusion	**94.33**	Fivefold
Zhao, Tang, and Hou [7]	Concentric multiscale BOW	86.64	Twofold
Proposed method	CS multifeature fusion	**91.1**	Twofold

Note: The bold values represent the best accuracy.
BOW, bag of visual words; CS, compressive sensing.

As shown, the proposed method resulted in sharply superior accuracies compared with all the considered works, where the lowest performance was observed for the method in [7]. Our method exhibits two advantages over these reference methods:

1. The representation is very compact (reduced to 21, namely the number of classes), whereas the reference methods rely on the BOW, which produces representations with hundreds or thousands of bins.

2. In our case, the reconstruction residuals already represent information from which decisions can be inferred (normalized residuals can be viewed as one minus posterior probabilities—the smaller the residual, the larger the posterior). By contrast, the other methods need to train a classifier (support vector machine) in a huge input space, incurring problems related to the curse of dimensionality because of the reduced number of training images per class.

These two aspects mainly explain why our method outperforms the reference methods.

We also compared our work to two other fusion strategies. Upon concatenation of the three feature vectors and normalization within $[0, 1]$, the first method is based on a support vector machine classifier, whereas the second technique makes use of the CS by taking the class of the smallest residual. The fivefold-based classification results are reported in Table 11.3. It emerges that the proposed strategies outperform the two fusion techniques, which appear to perform closely to each other.

Figure 11.5 depicts an example of image classification based on the SOS strategy. In this example, the test image belongs to the third class, Baseball Diamond. It can be observed that, for all three adopted features, the lowest residual points correctly to the third class, which is likewise observed in the final SOS-based residual. Another point to stress in this example is that the GLAC feature yields a residual vector of undesirable shape as compared with the other two features. Indeed, the ideal case would be a residual vector where the minimum pertains to the correct class, while the remaining part is much larger and uniformly distributed. The SOS fusion permits us to compensate for this issue by generating a global residual vector close to the desired shape, leading to a correct and more confident decision.

TABLE 11.3

Classification Accuracies Achieved by the Single-Feature CS Representation and by CS Fusion Strategies

	SVM	CS	SOS
Accuracy (%)	79.09	81.18	94.33

SVM, support vector machine.

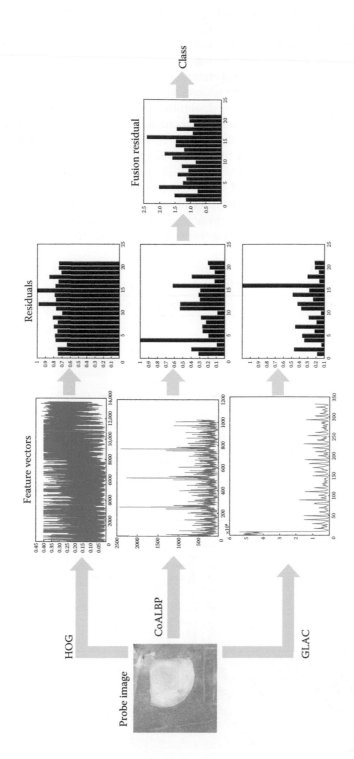

FIGURE 11.5
Image classification example showing the input feature vectors, the corresponding CS residual vectors, and the fusion-based residual vector.

11.4.2   Spectral fusion scenario

In Table 11.4, we report the results of our multispectral fusion scheme along with the two cases where (1) each color channel and (2) the gray-level version of the images are utilized individually. In the last two cases, only one residual vector is obtained. Hence, the decision follows the rule of the smallest residual, whose class is assigned to the test image.

It emerges clearly that fusing all the spectral channels significantly increases the accuracy over using only one single channel or a gray version of the image. That breaks down to the reason that when only one channel is employed, significant partial information (i.e., the reconstruction residuals generated by the remaining unemployed channels) used for the final decision is omitted. Therefore, making use of all available spectral information in the decision has emerged as the best choice to cope with the LU problem.

A further comparison with the works presented in [4,7] is also reported in Table 11.5. A significant increase has been demonstrated by our method over the one presented in [4], whereas a moderate but still interesting improvement has been achieved with respect to [7].

For a detailed analysis, we also provide average per-class classification accuracies. In particular, Figure 11.6 reports the classification rates pertaining to the SOS strategy. It

TABLE 11.4

Classification Accuracies of All the Strategies

Scenario	Accuracy (%)
R channel	69.09
G channel	71
B channel	68.57
Gray level	68.23
Spectral fusion	87.38

TABLE 11.5

Comparison versus Reference Works

Method	Accuracy (%)
Yang and Newsam [4]	81.19
Zhao, Tang, and Huo [7]	86.64
Spectral fusion	87.38

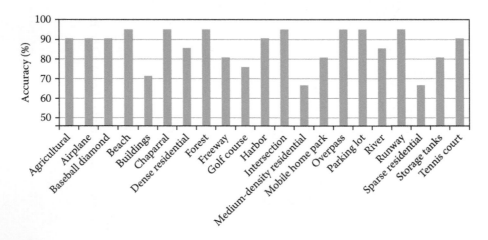

FIGURE 11.6

Per-class classification accuracies by means of the SOS fusion strategy.

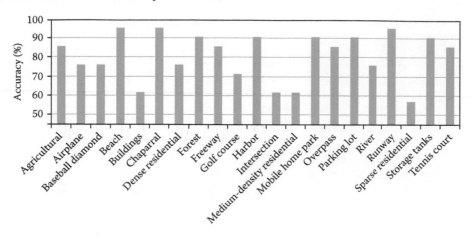

FIGURE 11.7
Per-class classification accuracies by means of the spectral fusion strategy.

can be observed that the lowest performances are scored for the classes Buildings, Medium Residential, and Sparse Residential. We believe that these three classes exhibit some spectral as well as spatial similarities with respect to each other, which entails a difficult interclass separability. The same behavior is manifested by means of the spectral fusion strategy (Figure 11.7.), in addition to the "intersection" class, whose spectral channels do not seem to offer advantageous properties to discriminate it from the other classes. In other words, we believe that its spectral layers might be close to each other in the feature space; fusing them is therefore unlikely to incur a noticeable plus.

An example demonstrating an input image belonging to Class 1 (i.e., agricultural) is depicted in Figure 11.8, along with its respective reconstruction coefficients in the form of a vector and the reconstruction residuals out of all 21 class dictionaries. In this example, the reconstruction coefficients are based on the dictionary associated with Class 1. The first peak refers to the closest image within the dictionary to the input image. Regarding the residuals, the first residual amounts for the smallest value making the probe image correctly assigned to the first class.

11.5   Conclusion

This chapter presents two sparse fusion schemes within the context of LU image classification. The underlying idea is to compactly represent LU images within a CS and two fusion frameworks. First, we assessed the feasibility of fusing opportune multiple features and showed that such a scenario indeed outperforms the case where single features are considered. Second, we further showed that taking advantage of all available spectral channels incurs tangible gains with respect to the standard gray-level representation.

Thus, the CS reconstruction residuals originating from different kinds of features and spectral channels are fused based on simple but efficient strategies. Indeed, as the results point out, promising outcomes have been obtained. Furthermore, the proposed methods exhibit substantial classification rate gains versus recent reference works.

Nevertheless, we believe that the current version of the schemes can undergo further improvements. For instance, we recommend the investigation of more elaborated fusion methods such as induced ordered weighted averaging operators [26], which has been

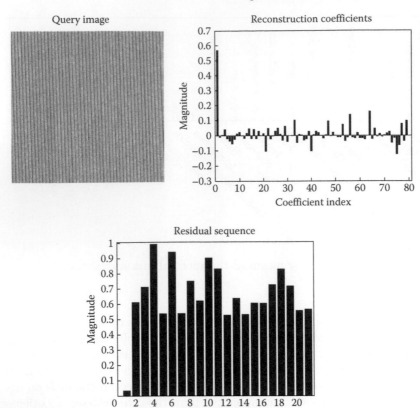

FIGURE 11.8
An instance of CS-based image reconstruction.

validated as a powerful fusion tool. The second element is related to the size of the input feature vectors, which can be reduced beforehand while maintaining or even improving classification performance. Fisher discriminant analysis could represent a good candidate for such a purpose [27]. Third, increasing the number of feature types in the ensemble (in this chapter we considered just three kinds) is another possible improvement. Finally, reformulating the proposed method under an active learning perspective would be another interesting avenue to explore [28].

Acknowledgments

The authors would like to thank D. Donoho and Y. Tsaig for providing the SparseLab toolbox (http://sparselab.stanford.edu/).

References

[1] C. H. Lin, P. H. Tsai, K. H. Lai, and J. Y. Chen, Cloud Removal from Multitemporal Satellite Images Using Information Cloning, *IEEE Transactions on Geoscience and Remote Sensing*, vol. 51, no. 1, pp. 232–241, 2013.

[2] F. P. S. Luus, F. Van den Bergh, and B. T. J. Maharaj, The Effects of Segmentation-Based Shadow Removal on Across-Date Settlement Type Classification of Panchromatic Quickbird Images, *IEEE Journal of Selected Topics in Applied Earth Observations and Remote Sensing*, vol. 6, no. 3, pp. 1274–1285, 2013.

[3] Y. Bazi, F. Melgani, and H. D. Al-Sharari, Unsupervised Change Detection in Multispectral Remotely Sensed Imagery with Level Set Methods, *IEEE Transactions on Geoscience and Remote Sensing*, vol. 48, no. 8, pp. 3178–3187, 2010.

[4] Y. Yang and S. Newsam, Bag-of-Visual-Words and Spatial Extensions for Land-Use Classification, in *Proceedings of the 18th ACM SIGSPATIAL International Conference on Advanced in Geographic Information Systems*, San Jose, CA, pp. 270–279, 2010.

[5] Y. Yang and S. Newsam, Spatial Pyramid Co-occurrence for Image Classification, in *Proceedings of the IEEE International Conference on Computer Vision*, pp. 1465–1472, 2011.

[6] C. Shizhi and Y. Tian, Pyramid of Spatial Relatons for Scene-Level Land Use Classification, *IEEE Transactions on Geoscience and Remote Sensing*, vol. 53, no. 4, pp. 1947–1957, 2015.

[7] L. Zhao, P. Tang, and L. Huo, Land-Use Scene Classification Using a Concentric Circle-Structured Multiscale Bag-of-Visual-Words Model, *IEEE Journal of Selected Topics in Applied Earth Observations and Remote Sensing*, vol. 8, pp. 4620–4631, 2015.

[8] T. Ahonen, A. Hadid, and M. Pietikainen, Face Description with Local Binary Patterns: Application to Face Recognition, *IEEE Transactions on Pattern Analysis and Machine Intelligence*, vol. 28, no. 12, pp. 2037–2041, 2006.

[9] D. L. Donoho, Compressed Sensing, *IEEE Transactions on Information Theory*, vol. 52, no. 4, pp. 1289–1306, 2006.

[10] E. J. Candès, J. Romberg, and T. Tao, Robust Uncertainty Principle-Exact Signal Reconstruction from Highly Incomplete Frequency Information, *IEEE Transactions on Information Theory*, vol. 52, no. 2, pp. 489–509, 2006.

[11] M. Aharon, M. Elad, and A. Bruckstein, K-SVD: An Algorithm for Designing Overcomplete Dictionaries for Sparse Representation, *IEEE Transactions on Signal Processing*, vol. 54, no. 11, pp. 4311–4322, 2006.

[12] J. Wright, A. Yang, A. Ganesh, S. Sastry, and Y. Ma, Robust Face Recognition via Sparse Representation, *IEEE Transactions on Pattern Analysis and Machine Intelligence*, vol. 30, no. 2, pp. 210–227, 2009.

[13] J. Yang, J. Wright, T. Huang, and Y. Ma, Image Super-Resolution via Sparse Representations, *IEEE Transactions on Image Processing*, vol. 19, no. 11, pp. 2861–2873, 2010.

[14] S. Rao, R. Tron, R. Vidal, and Y. Ma, Motion Segmentation via Robust Subspace Separation in the Presence of Outlying, Incomplete, and Corrupted Trajectories, *IEEE Transactions on Pattern Analysis and Machine Intelligence*, vol. 32, no. 10, pp. 1832–1845, 2008.

[15] J. Mairal, M. Elad, and G. Sapiro, Sparse Representation for Color Image Restoration, *IEEE Transactions on Image Processing*, vol. 17, no. 1, pp. 53–69, 2008.

[16] B. Shen, W. Hu, Y. Zhang, and Y.-J. Zhang, Image Inpainting via Sparse Representation, in *IEEE International Conference on Acoustics, Speech and Signal Processing*, pp. 697–700, 2009.

[17] L. Lorenzi, F. Melgani, and G. Mercier, Missing Area Reconstruction in Multispectral Images Under a Compressive Sensing Perspective, *IEEE Transactions on Geoscience and Remote Sensing*, vol. 51, no. 7, pp. 3998–4008, 2013.

[18] A. Quattoni, M. Collins, and T. Darrell, Transfer Learning for Image Classification with Sparse Prototype Representation, in *IEEE International Conference on Computer Vision and Pattern Recognition*, pp. 1–8, 2008.

[19] J. Wright, Y. Ma, J. Mairal, G. Sapiro, T. S. Huang, and S. Yan, Sparse Representation for Computer Vision and Pattern Recognition, *Proceedings of the IEEE*, vol. 98, no. 6, pp. 1031–1044, 2010.

[20] E. J. Candès and T. Tao, Decoding by Linear Programming, *IEEE Transactions on Information Theory*, vol. 51, no. 12, pp. 4203–4215, 2005.

[21] D. L. Donoho, Y. Tsaig, I. Drori, and J. L. Starck, Sparse Solution of Underdetermined Systems of Linear Equations by Stagewise Orthogonal Matching Pursuit, *IEEE Transactions on Information Theory*, vol. 58, no. 2, pp. 1094–1121, 2012.

[22] N. Dalal and B. Triggs, Histograms of Oriented Gradients for Human Detection, in *Proceedings of the IEEE Conference on Computer Vision and Pattern Recognition*, vol. 1, pp. 886–893, 2005.

[23] T. Moranduzzo and F. Melgani, Detecting Cars in UAV Images with a Catalog-Based Approach, *IEEE Transactions on Geoscience and Remote Sensing*, vol. 52, no. 10, pp. 6353–6367, 2013.

[24] R. Nosaka, Y. Ohkawa, and K. Fukui, Feature Extraction Based on Co-occurrence of Adjacent Local Binary Patterns, *Advances in Image & Video Technology*, vol. 7088, pp. 82–91, 2012.

[25] T. Kobayashi and N. Otsu, Image Feature Extraction using Gradient Local Auto-Correlations, in *Proceedings of the 10th European Conference on Computer Vision, Part I. LNCS*, vol. 5302, pp. 346–358, 2008.

[26] Y. Bazi, N. Alajlan, F. Melgani, H. AlHichri, and R. Yager, Robust Estimation of Water Chlorophyll Concentrations with Gaussian Process Regression and IOWA Aggregation Operators, *IEEE Journal of Selected Topics in Applied Earth Observations and Remote Sensing*, vol. 7, no. 7, pp. 3019–3028, 2014.

[27] H. Huang, H. Feng, and C. Peng, Complete Local Fisher Discriminant Analysis with Laplacian Score Ranking for Face Recognition, *Neurocomputing*, vol. 89, pp. 64–77, 2012.

[28] B. Demir and L. Bruzzone, A Novel Active Learning Method in Relevance Feedback for Content-Based Remote Sensing Image Retrieval, *IEEE Transactions on Geoscience and Remote Sensing*, vol. 53, no. 5, pp. 2323–2334, 2015.

12

Compressive Sensing for Reconstruction, Classification, and Detection of Hyperspectral Images

Bing Zhang, Wei Li, Lianru Gao, and Xu Sun

CONTENTS

12.1 Introduction

Hyperspectral imagery (HSI) acquired by remote-sensing systems (e.g., a spaceborne or an airborne sensor) typically records reflectance values over a wide region of the electromagnetic spectrum [1,2]. Each pixel in an HSI represents many contiguous and narrow spectral bands (e.g., a spectral range of 0.4 to 2.4 μm). This enables HSI to potentially provide rich information about the materials in the image scene. HSI has become a widely available modality used in a wide variety of applications, including urban-growth analysis, biological and chemical detection, environmental monitoring, mineral exploration, etc.

In conventional HSI sensing systems, transformations such as discrete wavelet transforms (DWT) [3,4] or principal component analysis (PCA) [5–7] combined with the JPEG2000 [8,9] are employed for spectral decorrelation and compression at the sender side of a remote-sensing system. This paradigm has a disadvantage in that practical implementations of these transformations (e.g., PCA) are typically computationally expensive, particularly insofar as the sender or encoder is concerned. Due to the emergence of compressive sensing [10–12], reconstruction from random projections has been widely studied. This onboard dimensionality reduction would dramatically reduce the communication burden for the sender (remote sensor) since the signal-sensing process is accomplished with a simple measurement (several sensing architectures have been proposed in [13,14]). The data recovery procedure is transferred to receiver which assumes that signal can be sparsely represented with some transform and recovered from the projections via a nonlinear reconstruction.

Traditional classification algorithms are always introduced from pattern recognition and based on the spectral information [15]. In recent years, the limitation of using only spectral features has been recognized, and the need for incorporating contextual information has also been addressed. The spectral–spatial classifiers could improve the performance and reduce the salt-and-pepper appearance that exits when only spectral information is used for classification. Zhang and Li proposed an adaptive Markov Random Field (MRF) approach and introduced an appropriate weighting coefficient for determining the spatial contribution in the classification [16,17]. The neighborhood-constrained k-means (NC-k-means) [18] algorithm incorporated a pure neighborhood index into the traditional k-means algorithm to achieve better classification accuracy with significant spatial autocorrelations among neighboring pixels. Yu and Gao integrated the subspace-based support vector machines (SVM) and MRF for hypspectral classification [19]. Target detection in hyperpectral image is based on the invariance of material spectral and the difference between target and background spectral. Several classical supervised target detection algorithms were widely used [2] and effectively improved [20–22]. Another kind of target detection is known as anomaly detection, in which RX detector is the most famous algorithm [23]. Different versions of RX, such as kernel-RX [24], weight-RX [25], and linear-filter-RX [26], were proposed to improve the performance of RX detector.

In recent years, sparse representation, related to compressive sensing theory, has also been successfully applied in other scientific fields, such as classification and target detection. Sparse representation–based classification (SRC) [27], which does not assume any data density distribution, was originally developed for face recognition. The essence of SRC is built on the concept that a testing pixel can be sparsely represented as a linear combination of labeled data via ℓ_0- or ℓ_1-norm minimization. It does not require a training–process, which is obviously dissimilar to the training-testing fashion in conventional classifiers (e.g., SVM [28]). In SRC, the class label of a testing pixel is determined to be that of the class whose labeled samples provide the smallest representation error.

In this chapter, compressive sensing techniques for reconstruction, classification, and detection in HSI are discussed. The remainder of this work is organized as follows. Section 12.2 introduces several methods on reconstruction from random projections. Section 12.3 primarily describes the sparse representation classifier and its extensions. Section 12.4 mainly presents the sparse representation detector and some improved versions. Section 12.5 makes some concluding remarks.

12.2 Reconstruction from random projections

Figure 12.1a illustrates a conventional HSI sensing system. Consider a zero-mean dataset with samples $\mathbf{X} = \{\mathbf{x}_i\}_{i=1}^{M}$ in $\mathbb{R}^N$ (N is the number of bands and M is the total number of samples)—the covariance matrix is $\mathbf{\Sigma} = \mathbf{X}\mathbf{X}^T/M$, which needs MN^2 multiplications. At the sender, the eigen-decomposition of $\mathbf{\Sigma}$ is significantly time consuming, especially when the spectral dimensionality and the spatial size of the HSI increase (the computational complexity is $\mathcal{O}(N^3)$). This signal-acquisition paradigm hence imposes a heavy burden on severely resource-constrained environments such as an airborne or a satellite platform.

Figure 12.1b shows an alternate approach to such traditional transform-based sensing systems. The signal-acquisition device of such a lightweight sender applies an $N \times S$ orthonormal projection matrix $\mathbf{P}$ (for simplicity, the matrix is generated with independent Gaussian random variables followed by Gram–Schmidt orthogonalization) to obtain random projections $\mathbf{Y} = \mathbf{P}^T\mathbf{X}$ in $\mathbb{R}^S$ ($S \ll N$, S/N is referred as the *subrate*). Note that the computational complexity of the sender is much less than that of traditional signal-acquisition platforms. Compressive sensing (CS) [10,29] and compressive-projection PCA (CPPCA) [30] are recently developed methods that can provide accurate reconstruction from such data-independent random projection. Both CS and CPPCA* effectively shift the computational burden from a resource-constrained sender to a presumably more capable receiver (base station).

12.2.1 MT-BCS

Under the traditional CS paradigm [10,29], we attempt to recover $\mathbf{X}$ from the compressive random projection $\mathbf{Y}$. CS theory establishes that if the dataset $\mathbf{X}$ can be sparsely represented with some $N \times N$ transform $\mathbf{\Psi}$, then we can recover $\mathbf{X}$ from the projections via a nonlinear reconstruction, that is,

$$\widetilde{\mathbf{X}}^* = \arg\min_{\widetilde{\mathbf{X}}} \left\|\widetilde{\mathbf{X}}\right\|_1, \quad \text{such that } \mathbf{Y} = \mathbf{P}^T\mathbf{\Psi}\widetilde{\mathbf{X}}. \tag{12.1}$$

There are many algorithms that exploit this central idea of CS. One popular algorithm is multi-task Bayesian compressive sensing (MT-BCS) [31], which has shown promising performance in reconstructing signals with noisy measurements and provided a tighter

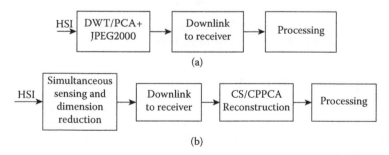

FIGURE 12.1
(a) Traditional dataflow pipeline for HSI and (b) pipeline with onboard dimensionality reduction via random projections.

---

*http://www.ece.msstate.edu/~fowler/CPPCA/

approximation to the Equation 12.1. MT-BCS* can be used to reconstruct the hyperspectral data which possess high inter-feature correlation.

12.2.2   CPPCA

In [30,32], CPPCA has been shown to be superior to MT-BCS insofar as the reconstruction performance and execution speeds are concerned. Here, we briefly review the CPPCA algorithm, which forms the basis for the reconstruction strategy fully exploiting the spectral correlation. Consider M zero-mean vectors $\mathbf{X}$. The covariance matrix is $\boldsymbol{\Sigma} = \mathbf{X}\mathbf{X}^T/M$, and traditional PCA seeks a linear transformation $\mathbf{W}$ of eigenvectors resulting from the eigen-decomposition of $\boldsymbol{\Sigma}$, that is,

$$\boldsymbol{\Sigma} = \mathbf{W}\boldsymbol{\Lambda}\mathbf{W}^T, \tag{12.2}$$

where $\mathbf{W}$ contains the N unit eigenvectors of $\boldsymbol{\Sigma}$ columnwise. The PCA transform of $\mathbf{X}$ is then $\widetilde{\mathbf{X}} = \mathbf{W}^T\mathbf{X}$. In CPPCA setup, the receiver knows only projection $\mathbf{Y}$ and the projection matrix $\mathbf{P}$, which means conventional methods to produce eigenvectors via Equation 12.2 will not work.

The projected vectors have a covariance matrix

$$\widetilde{\boldsymbol{\Sigma}} = \mathbf{Y}\mathbf{Y}^T/M = \mathbf{P}^T\mathbf{X}\mathbf{X}^T\mathbf{P}/M = \mathbf{P}^T\boldsymbol{\Sigma}\mathbf{P}. \tag{12.3}$$

CPPCA [30] is a reconstruction algorithm which employs a projection-onto-convex-sets (POCS) [33] procedure to recover the principal eigenvectors based on an eigen-decomposition of $\widetilde{\boldsymbol{\Sigma}}$. Specifically, the sender splits M zero-mean vectors $\mathbf{X}$ into J partitions $\mathbf{X}^{(j)}(j = 1, 2, \cdots, J)$, each associated with its own randomly chosen projection matrix $\mathbf{P}^{(j)}$, $1 \leq j \leq J$, such that we have $\mathbf{Y}^{(j)} = \mathbf{P}^{(j)T}\mathbf{X}^{(j)}$. In [30], a modulo partition is used for this partitioning, such that $\mathbf{X}^{(j)} = \{\mathbf{x}_i \in \mathbf{X} | (i - 1) \bmod J = j - 1\}$, with each $\mathbf{X}^{(j)}$ closely resembling the entire dataset $\mathbf{X}$ statistically and hence possessing approximately the same eigen-decomposition of its covariance matrix. At the CPPCA receiver, covariance matrices $\widetilde{\boldsymbol{\Sigma}}^{(j)}$ are calculated from $\mathbf{Y}^{(j)}$. The POCS procedure is repeated to obtain approximations of the first L principal eigenvectors in $\mathbf{W}$ by using the first L Ritz vectors $\mathbf{u}_s$ [34] calculated from $\widetilde{\boldsymbol{\Sigma}}^{(j)}$. The recovered L principal eigenvectors are assembled into an $N \times L$ matrix $\boldsymbol{\Psi} = \begin{bmatrix} \widehat{\mathbf{w}}_1 & \cdots & \widehat{\mathbf{w}}_L \end{bmatrix}$. The CPPCA receiver then proceeds to recover the PCA coefficients by using the pseudoinverse

$$\widetilde{\mathbf{X}}^{(j)} = (\mathbf{P}^{(j)T}\boldsymbol{\Psi})^{\dagger}\mathbf{Y}^{(j)}, \tag{12.4}$$

where L is determined by round $(\alpha S/\log N)$, α being a constant ($\alpha = 1.0$ being a simple case) in [35]. As in [30], we use $J = 20$ for the work here.

12.2.3   Class-dependent CPPCA

Class-dependent CPPCA (CD-CPPCA) [36] is an improved version of CPPCA reconstruction algorithm. For this approach, the sender-side sensing procedure is exactly the same as that employed in original CPPCA. At the receiver, the random projections $\mathbf{Y}$ are first classified into K groups using a pixel-wise classification algorithm, such as an SVM classifier [37,38], in the projected subspace. After the grouping procedure, the CPPCA reconstruction algorithm is employed for each group independently, which is shown in Figure 12.2. Specifically, a small set of "exemplars" is first chosen randomly from the projections $\mathbf{Y}^{(j)}(j = 1, \ldots, J)$ to be recovered using MT-BCS (which is efficient at recovering a small number of samples [39], unlike CPPCA which requires more samples to

---

*http://www.ee.duke.edu/~lcarin/BCS.html

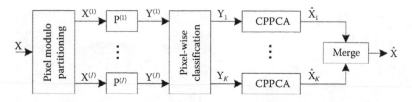

FIGURE 12.2
Brief flowchart of the class-dependent CPPCA algorithm in Li et al. (2013).

be effective). These reconstructed samples are then clustered into different groups, producing a "pseudo" *a priori* label information (training data). A trained pixel-wise classifier is then employed to classify each pixel in each $\mathbf{Y}^{(j)}$ into one of K classes. Thus, each $\mathbf{Y}^{(j)}$ is further partitioned into K groups, $\mathbf{Y}_k^{(j)}$, $k = 1, \ldots, K$, based on this classifier. Next, CPPCA reconstruction is employed independently on each of these class partitions. That is, for each k, we calculate $\widetilde{\boldsymbol{\Sigma}}_k^{(j)} = \mathbf{Y}_k^{(j)}\mathbf{Y}_k^{(j)^T}/M_{k,j}$ as well as Ritz vectors $\mathbf{u}_{k,s}$, where $M_{k,j}$ are the number of vectors in $\mathbf{Y}_k^{(j)}$. Finally, PCA coefficients are calculated using the following pseudoinverse

$$\widetilde{\mathbf{X}}_k^{(j)} = (\mathbf{P}^{(j)^T}\boldsymbol{\Psi}_k)^{\dagger}\mathbf{Y}_k^{(j)}, \tag{12.5}$$

where the L recovered eigenvectors for class k form $\boldsymbol{\Psi}_k = \begin{bmatrix} \widehat{\mathbf{w}}_{k,1} & \cdots & \widehat{\mathbf{w}}_{k,L} \end{bmatrix}$.

In CD-CPPCA (as shown in Figure 12.2), after obtaining random projections of the HSI from the sender, each pixel is first labeled at the receiver as representing a distinct class in the scene. After the grouping procedure, the CPPCA [30] recovery algorithm is employed for reconstruction of sets of pixels belonging to each class in this partition independently. The key motivation behind this strategy is that the original CPPCA algorithm recovers the primary eigenvectors representing the entire dataset; however, when the data distribution in the original input space is distinctly multimodal, this approach may be suboptimal. On the contrary, CD-CPPCA recovers the primary eigenvectors for each distinct cluster separately, using them to recover pixels in that cluster independently of the others. In doing so, the overall receiver-side signal reconstruction from random projections is more accurate, since it is based on statistics which are representative of each of the dominant mixtures in the scene.

12.2.4 Experimental results and analysis

The hyperspectral imagery which is employed to quantify the performance of reconstruction in the randomly projected subspace is a Hyperspectral Digital Image Collection Experiment (HYDICE) image. This image scene covers the Washington, DC, Mall area [1] which represents an urban scenario with 1280×307 pixels with 191 spectral bands. Training data for supervised class-dependent CPPCA are extracted from five classes (Roofs, Streets/Paths, Grass, Trees, and Water). Approximately 2000 samples are employed for training (383, 402, 405, 405, and 404 for each class, respectively).

Another hyperspectral dataset [40] we used covers an agricultural (experimental corn crop) field in Brooksville, Mississippi. It was collected using the Pro-SpecTIR-VNIR sensor on June 6, 2008. The images have 128 spectral bands with a spectral coverage of 0.4 to 0.994 μm region and a spatial resolution of 1.0 m. This dataset represents the corn crop under varying degrees of chemical-induced stress—each degree of stress in the crop is a unique class in this dataset. The scene has a spatial size of 283×213 pixels and a total of three classes are employed in our experiment. There are approximately 1300 samples for training (356, 538, and 402 for each class, respectively).

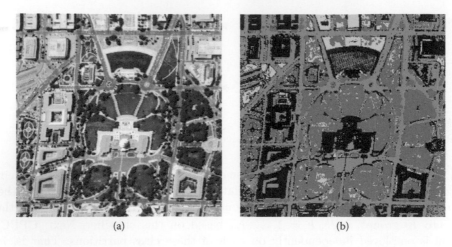

FIGURE 12.3
(a) Pseudo-color image and (b) classification map for the Washington, DC, Mall datatset.

FIGURE 12.4
(a) Pseudo-color image and (b) classification map for the corn field datatset.

Figures 12.3 and 12.4 provide the pseudo-color image and corresponding classification map obtained by using an SVM classifier for subrate 0.3. In these maps, each color represents a unique class/cluster. In CD-CPPCA, CPPCA performs reconstruction for each class/cluster independently.

Efficacy of all the reconstruction strategies is measured by studying the signal-to-noise ratio (SNR) over a wide range of subrates [32]. In this work, a vector-based SNR (measured in dB) is employed,

$$\text{SNR}(\mathbf{x}_i, \mathbf{x}_j) = 10 \log_{10} \frac{\text{var}(\mathbf{x}_i)}{\text{MSE}(\mathbf{x}_i, \mathbf{x}_j)}, \tag{12.6}$$

where $\text{var}(\mathbf{x}_i)$ is the variance of the data vector $\mathbf{x}_i$, the mean square error (MSE) is,

$$\text{MSE}(\mathbf{x}_i, \mathbf{x}_j) = \frac{1}{N}\|\mathbf{x}_i - \mathbf{x}_j\|^2, \tag{12.7}$$

and N is the dimensionality of feature space. We then average the vector-based SNR over all the vectors. Another popular evaluation method to measure the quality of reconstructed HSI is the average spectral angle [32], which is also commonly employed for detecting anomalous pixels in HSI.

Figures 12.5 and 12.6 illustrate the reconstruction performance for the aforementioned methods using the two experimental datasets. It is apparent that CD-CPPCA and CPPCA can significantly outperform MT-BCS over a range of practical subrates, and

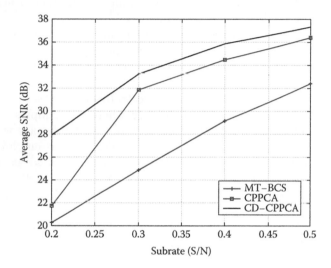

FIGURE 12.5
Reconstruction performance for the Washington, DC, Mall dataset.

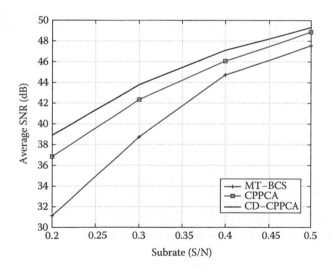

FIGURE 12.6
Reconstruction performance for the corn field dataset.

CD-CPPCA always has the best performance. For the urban Washington, DC, Mall dataset, CD-CPPCA even can achieve a 2 to 6 dB improvement in SNR when compared with CPPCA. For the agriculture corn filed scene, CD-CPPCA approximately provides a 1 to 2 dB improvement when compared with CPPCA.

12.3 Sparse representation–based classification

SRC has been proved to be effective in classification [27]. However, it has been argued that it is the "collaborative" nature of the approximation instead of the "competitive" nature imposed by sparseness constraint that actually improves the classification accuracy. In collaborative representation–based classification (CRC) [41], all the atoms collaborate on the representation of a single pixel, and each atom has an equal chance to participate in representation. It is solved with an ℓ_2-norm regularized least-squares formulation.

Besides, extensions of SRC and CRC have already been investigated. In [42,43], kernel version of SRC (KSRC), representing the data in a high-dimensional kernel-induced feature space, was presented. In [44], KSRC using spatial–spectral features was further discussed. In [45,46], the original CRC was extended to the case where the samples were weighted based on the locality information. In [47,48], nearest regularized subspace with Tikhonov regularization (CRT) classifier was presented. CRT was designed to employ an ℓ_2-norm penalty in the style of a distance-weighted Tikhonov regularization so as to adjust the regularization penalty based on the similarity between the testing pixel and the within-class labeled samples, which is denoted as within-class CRC. In [49], joint within-class CRC was introduced, which approximated the testing samples using the class-specific labeled samples separately instead of using all the labeled data. Kernel-based CRC (KCRC) was further investigated for HSI classification in [50].

12.3.1 Sparse representation–based classification

Let a dataset have labeled samples $\mathbf{X} = \{\mathbf{x}_i\}_{i=1}^M \in \mathbb{R}^N$ (N is the number of spectral bands) with class labels $\omega_i \in \{1, 2, \ldots, C\}$, where C represents the number of classes, and M is the total number of samples. Let M_l be the number of labeled samples in the lth class, and $\sum_{l=1}^C M_l = M$. An approximation of a testing pixel $\mathbf{y}$ is represented via a linear combination of all labeled samples as a dictionary, $\mathbf{X}$. In SRC [27], the objective is to find a sparse representation vector $\boldsymbol{\alpha}^{(SRC)}$ such that $\left\|\mathbf{y} - \mathbf{X}\boldsymbol{\alpha}^{(SRC)}\right\|_2^2$ is minimized with a sparse constraint term $\left\|\boldsymbol{\alpha}^{(SRC)}\right\|_1$. The objective function can be expressed as

$$\arg \min_{\boldsymbol{\alpha}^{(SRC)}} \left\|\mathbf{y} - \mathbf{X}\boldsymbol{\alpha}^{(SRC)}\right\|_2^2 + \lambda\left\|\boldsymbol{\alpha}^{(SRC)}\right\|_1, \tag{12.8}$$

where λ is a regularization parameter. After obtaining $\boldsymbol{\alpha}^{(SRC)}$, $\mathbf{X}$ and $\boldsymbol{\alpha}^{(SRC)}$ are separated into l class-specific subsets according to the given class labels of the labeled samples, that is, $\{\mathbf{X}_l\}_{l=1}^C \in \mathbb{R}^{N \times M_l}$ and $\left\{\boldsymbol{\alpha}_l^{(SRC)}\right\}_{l=1}^C \in \mathbb{R}^{M_l \times 1}$. Class label of the testing sample is then determined according to the class that minimizes the residual, that is,

$$r_l^{SRC}(\mathbf{y}) = \left\|\mathbf{X}_l\boldsymbol{\alpha}_l^{(SRC)} - \mathbf{y}\right\|_2, \tag{12.9}$$

and class label $SRC(\mathbf{y}) = \arg \min_{l=1,\ldots,C} r_l^{SRC}(\mathbf{y})$.

12.3.2 Collaborative representation–based classification

In CRC [41], the objective is to find a representation vector $\boldsymbol{\alpha}^{(CRC)}$ such that $\left\|\mathbf{y} - \mathbf{X}\boldsymbol{\alpha}^{(CRC)}\right\|_2^2$ is minimized under the constraint $\left\|\boldsymbol{\alpha}^{(CRC)}\right\|_2^2$ is also minimized. The objective function can be expressed as

$$\arg\min_{\boldsymbol{\alpha}^{(CRC)}} \left\|\mathbf{y} - \mathbf{X}\boldsymbol{\alpha}^{(CRC)}\right\|_2^2 + \lambda\left\|\boldsymbol{\alpha}^{(CRC)}\right\|_2^2. \tag{12.10}$$

Taking derivative with regard to $\boldsymbol{\alpha}^{(CRC)}$ and setting the resultant equation to zero yields

$$\boldsymbol{\alpha}^{(CRC)} = \left(\mathbf{X}^T\mathbf{X} + \lambda\mathbf{I}\right)^{-1}\mathbf{X}^T\mathbf{y}. \tag{12.11}$$

After obtaining $\boldsymbol{\alpha}^{(CRC)}$, class label of the testing pixel is then determined according to the minimum residue $r_l^{CRC}(\mathbf{y})$. Note that compared with SRC, the solution of CRC is a closed form as indicated in Equation 12.10, resulting in high computational efficiency.

Different from SRC, CRC has closed-form solutions. Figure 12.7a and b illustrates an example about the weight coefficients of SRC and CRC using a testing pixel from the experimental data. It is obvious that in Figure 12.7a, the weight coefficients are sparse, and

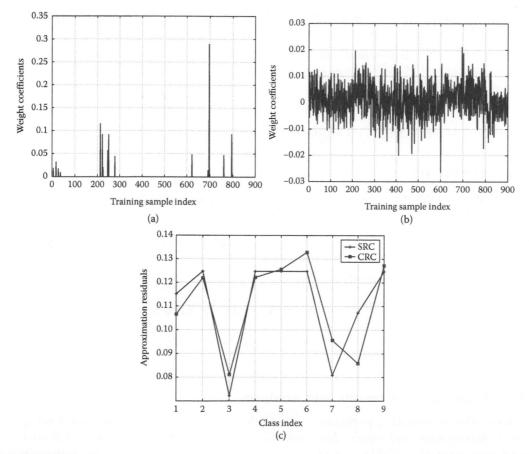

FIGURE 12.7

One example (testing pixel from class 3) of SRC and CRC using 100 training samples per class using the University of Pavia data (9 classes in total): (a) weight coefficients for SRC, (b) weight coefficients for CRC, and (c) residuals for SRC and CRC.

in Figure 12.7b, the weight coefficients are dense. Figure 12.7c further depicts the residuals of the testing pixel (from class 3, i.e., *Bitumen*) with respect to the training data of each class. In this example, the minimum residuals of SRC and CRC are both from class 3, which indicates the effectiveness of SRC and CRC.

12.3.3   MNF-SRC

Here, minimum noise fraction (MNF) [51], as a preprocessing, is combined with SRC and CRC, respectively. MNF-SRC and MNF-CRC are implemented by the following two main steps: (1) dimensionality reduction, where training data $\mathbf{X}$ and a testing pixel $\mathbf{y}$ are transformed using MNF and (2) sparse and collaborative representation with which the approximation of the pixel are obtained. The final output is determined by the class that minimizes the residual.

In MNF-SRC, training data $\mathbf{X}$ and the testing pixel $\mathbf{y}$ are first projected via an MNF transformation. The transformed samples can be expressed as $\phi(\mathbf{X})$ and $\phi(\mathbf{y})$. Then, an objective function similar to SRC is used to find a sparse weight vector $\boldsymbol{\alpha}^{(MNF-SRC)}$ such that $\phi(\mathbf{y})$ can be represented by a linear combination of $\phi(\mathbf{X})$ with a sparse constraint term $\left\|\boldsymbol{\alpha}^{(MNF-SRC)}\right\|_1$. The objective function can be expressed as

$$\arg\min_{\boldsymbol{\alpha}^{(MNF-SRC)}}\left\|\phi(\mathbf{y}) - \phi(\mathbf{X})\boldsymbol{\alpha}^{(MNF-SRC)}\right\|_2^2 + \lambda\left\|\boldsymbol{\alpha}^{(MNF-SRC)}\right\|_1. \tag{12.12}$$

Once $\boldsymbol{\alpha}^{(MNF-SRC)}$ is obtained, the class label of $\mathbf{y}$ is determined according to the minimum residual $r_l^{MNF-SRC}(\mathbf{y})$ computed by the class-specific $\phi(\mathbf{X}_l)\boldsymbol{\alpha}_l^{(MNF-SRC)}$ and $\phi(\mathbf{y})$, that is,

$$r_l^{MNF-SRC}(\mathbf{y}) = \left\|\phi(\mathbf{X}_l)\boldsymbol{\alpha}_l^{(MNF-SRC)} - \phi(\mathbf{y})\right\|_2, \tag{12.13}$$

and class label $MNF - SRC(\mathbf{y}) = \arg\min_{l=1,...,C} r_l^{MNF-SRC}(\mathbf{y})$.

12.3.4   MNF-CRC

In MNF-CRC, the objective function is minimized by an ℓ_2-norm regularization, which can be expressed as

$$\arg\min_{\boldsymbol{\alpha}^{(MNF-CRC)}}\left\|\phi(\mathbf{y}) - \phi(\mathbf{X})\boldsymbol{\alpha}^{(MNF-CRC)}\right\|_2^2 + \lambda\left\|\boldsymbol{\alpha}^{(MNF-CRC)}\right\|_2^2. \tag{12.14}$$

The solution can be analytically derived as

$$\boldsymbol{\alpha}^{(MNF-CRC)} = \left(\phi(\mathbf{X})^T\phi(\mathbf{X}) + \lambda\mathbf{I}\right)^{-1}\phi(\mathbf{X})^T\phi(\mathbf{y}). \tag{12.15}$$

After obtaining $\boldsymbol{\alpha}^{(MNF-CRC)}$, class label of the testing pixel $\mathbf{y}$ is then determined according to the minimum residue $r_l^{MNF-CRC}(\mathbf{y})$.

12.3.5   Experimental results and analysis

To validate the classification performance, the first data[*] employed were acquired using National Aeronautics and Space Administration's (NASA) Airborne Visible/Infrared Imaging Spectrometer (AVIRIS) sensor [52] and were collected over northwest Indiana's Indian Pine test site in June 1992. The image represents a classification scenario with 145×145 pixels and 220 bands in 0.4 to 2.45 μm region of visible and infrared spectrum

---

[*]http://www.ehu.eus/ccwintco/index.php?title=Hyperspectral_Remote_Sensing_Scenes

with a spatial resolution of 20 m. The scenario contains two-thirds agriculture, and one-third forest. In this work, a total of 200 bands are used after removal of water-absorption bands. There are 16 different land-cover classes but not all mutually exclusive in the designated ground truth map, and 8 of them are used for following experiments. The total number of training and testing samples are summarized in Table 12.1.

The second experimental dataset was collected by the reflective optics system imaging spectrometer (ROSIS) sensor [53]. The image scene, with a spatial coverage of 610 × 340 pixels covering the city of Pavia, Italy, was collected under the HySens project managed by Deutsches zentrum für Luft- und raumfahrt (DLR) (the German Aerospace Agency). The dataset has 103 spectral bands prior to water-band removal with a spectral coverage from 0.43 to 0.86 μm and a spatial resolution of 1.3 m. Approximately 42,776 labeled pixels with nine classes are from the ground truth map. More detailed information of the number of training and testing samples are summarized in Table 12.2.

Here, we firstly report the performance of representation-based classifiers toward λ to demonstrate the sensitivity effect to these representation-based classifiers. In general, leave-one-out cross-validation (LOOCV) strategy based on available training samples is considered for parameter tuning. Figures 12.8 and 12.9 show the varying λ of SRC and CRC for the

TABLE 12.1

Overall Accuracy (%) and Class-Specific Accuracy for AVIRIS Indian Pines Dataset

Class	Classification Algorithms					
	SVM	MLRsub	CRC	SRC	MNF-CRC	MNF-SRC
Corn-no till	47.73	55.39	80.27	75.62	**88.70**	84.79
Corn-min till	50.64	78.95	56.12	80.10	47.12	**87.29**
Grass/pasture	83.67	88.14	95.37	97.38	95.57	**98.39**
Hay-windrowed	99.09	99.54	99.80	**100.00**	**100.00**	**100.00**
Soybean-no till	60.57	53.92	67.46	69.52	70.14	**81.71**
Soybean-min till	55.46	45.37	48.58	62.03	56.81	**62.48**
Soybean-clean till	60.64	83.33	81.11	81.76	84.20	**90.88**
Woods	85.61	99.76	98.53	**99.77**	96.14	99.46
OA	63.02	67.29	72.21	78.18	75.30	**82.58**

Note: The bold values represent the best results (50 samples per class).

TABLE 12.2

Overall Accuracy (%) and Class-Specific Accuracy for ROSIS University of Pavia Dataset

Class	Classification Algorithms					
	SVM	MLRsub	CRC	SRC	MNF-CRC	MNF-SRC
Asphalt	70.91	48.51	18.47	72.61	36.16	**72.78**
Bare Soil	79.17	65.35	**81.77**	76.67	70.20	81.32
Bitumen	81.10	68.24	**95.95**	80.85	90.04	84.85
Bricks	85.59	84.10	**97.06**	94.97	86.03	89.78
Gravel	99.02	99.76	99.93	99.93	**100.00**	**100.00**
Meadows	70.60	63.70	20.06	80.65	81.59	**86.22**
Metal sheets	88.43	83.39	89.25	**91.20**	88.35	82.93
Shadows	**81.35**	64.53	19.15	58.09	33.89	56.98
Tress	**100.00**	99.30	99.47	96.73	99.68	99.37
OA	78.97	66.18	62.30	78.06	67.40	**80.29**

Note: The bold values represent the best results (70 samples per class).

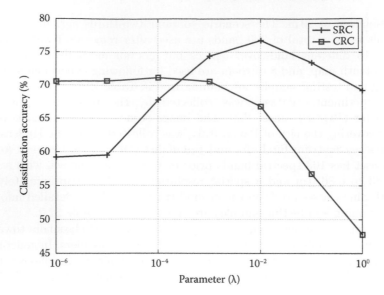

FIGURE 12.8
Parameter tuning (i.e., λ) of representation-based classifiers for the Indian Pines data.

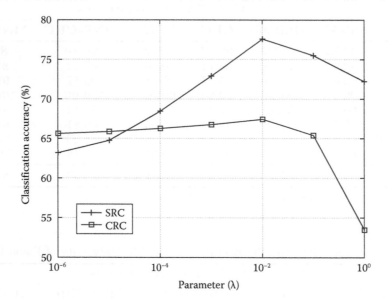

FIGURE 12.9
Parameter tuning (i.e., λ) of representation-based classifiers for the University of Pavia data.

Indian Pines data and the University of Pavia data, respectively. In the Indian Pines data, the optimal λ is 0.01 for SRC and 0.001 for CRC, respectively. In the University of Pavia data, the optimal λ is 0.01 for SRC and 0.01 for CRC, respectively.

To further evaluate the performance of SRC, CRC, MNF-SRC, and MNF-CRC, traditional SVM and subspace-based multinomial logistic regression (MLRsub) are added as comparisons using the Indian Pines dataset. In this test, the dimensionality kept after MNF transformation is adopted as 10. Table 12.1 reports the overall accuracy (OA) and class-specific accuracy of 50 samples per class randomly selected with a total of 400 training samples. In this comparison, representation-based classifiers generally perform better than SVM and MLRsub. As a whole, MNF-SRC yields the best OA of 82.58% (approximately 4.4% higher than SRC). Most importantly, MNF-SRC and MNF-CRC provide higher accuracies than SRC and CRC, which proves the effectiveness of the combination of dimensionality reduction technique like MNF and representation-based classifiers. The corresponding classification maps are given in Figure 12.10.

As for the University of Pavia data, the dimensionality kept after MNF transformation is adopted as 10. Table 12.2 reports the OA and class-specific accuracy of 70 samples per class randomly selected with a total of 630 training samples. As reported, SVM and MLRsub obtain considerable results compared with representation-based classifiers. CRC provides poor performance with relatively low class-specific accuracy for class 1 (*Asphalt*) and class 6 (*Meadows*), which can also be seen in classification maps in Figure 12.11. Nevertheless, MNF-SRC and MNF-CRC perform better than SRC and CRC, respectively. MNF-SRC

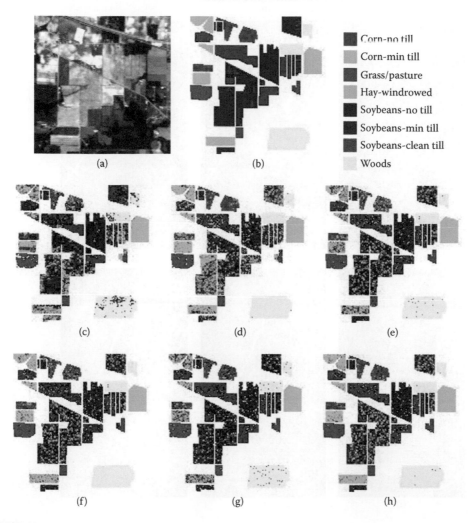

Corn-no till
Corn-min till
Grass/pasture
Hay-windrowed
Soybeans-no till
Soybeans-min till
Soybeans-clean till
Woods

FIGURE 12.10
Classification maps for AVIRIS Indian Pines dataset: (a) true-color composition, (b) ground truth map, (c) SVM (63.02%), (d) MLRsub (67.29%), (e) CRC (72.21%), (f) SRC (78.18%), (g) MNF-CRC (75.30%), and (h) MNF-SRC (82.58%).

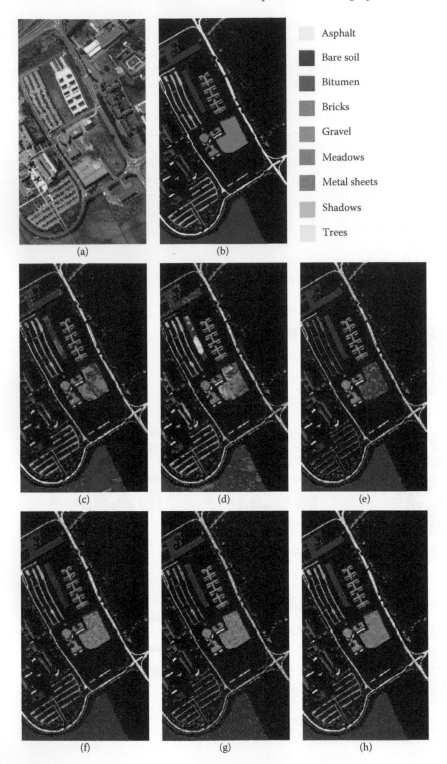

FIGURE 12.11
Classification maps for ROSIS University of Pavia dataset: (a) true-color composition,
(b) ground truth map, (c) SVM (78.97%), (d) MLRsub (66.18%), (e) CRC (62.30%),
(f) SRC (78.06%), (g) MNF-CRC (67.40%), and (h) MNF-SRC (80.29%).

obtains the best OA of 82.29% (approximately 2.2% higher than SRC), which affirms the robustness of the classifier. The corresponding classification maps are illustrated in Figure 12.11.

12.4 Sparse representation–based target detection

Recently, a target detection algorithm based on sparse representation [54,55] has been proposed for HSI analysis. The target detection algorithm is based on the concept that a pixel can be represented as a sparse linear combination of available "labeled" samples. The algorithm fully exploits the discriminative characteristics of sparse representation to determine its label (i.e., background or target pixel).

With or without *a priori* knowledge, target detection algorithms can be generally grouped into two categories, that is, unsupervised anomaly detection and supervised target detection. In [56], collaborative representation was proposed for anomaly detection in HSI. The algorithm fully utilized the fact that each pixel in background can be approximately represented by its spatial neighboring pixels, while anomalies cannot. The collaborative representation was reinforced by ℓ_2-norm minimization, resulting in high computational efficiency. In [55], a joint sparsity model (JSM) was used to consider spatial correlation to increase the accuracy of sparse reconstruction so as to improve detection performance. The detection output is simply generated according to the difference between residuals corresponding to background and target dictionaries. In [57], a combined sparse and collaborative representation–based algorithm was proposed. In the method, for each testing pixel, sparse representation of known target signatures was estimated via the ℓ_1-regularization, while collaborative representation of background atoms in a local window was obtained using the ℓ_2-regularization. The final target detection was achieved by evaluating the difference of these two representation residuals.

In this section, we mainly investigate the detectors which are designed to detect one (target) class and reject other background classes. Thus, training data consist only of the target class. This is useful in practical applications to detect targets from unknown background without labeled samples. The idea can be used to build one-class representation-based target detector, including one-class sparse representation–based detector (OCSRD) and one-class collaborative representation–based detector (OCCRD).

12.4.1 One-class SRD

Consider hyperspectral data with labeled (target) samples $\mathbf{X}_t = \{\mathbf{x}_i\}_{i=1}^{M_t} \in \mathbb{R}^N$ (N is the number of spectral bands), n_t is the total number of target samples (usually one representative signature for each class). The key in the presented OCSRD and OCCRD is to use $\mathbf{X}_t$ to obtain approximation of each pixel with sparse representation and collaborative representation, respectively. In OCSRD, to find weight vector $\boldsymbol{\alpha}^{(SRD)}$ of a testing pixel $\mathbf{y}$, the objective function can be expressed as

$$\arg \min_{\boldsymbol{\alpha}^{(SRD)}} \left\| \mathbf{y} - \mathbf{X}_t \boldsymbol{\alpha}^{(SRD)} \right\|_2^2 + \lambda \left\| \boldsymbol{\alpha}^{(SRD)} \right\|_1, \tag{12.16}$$

and the residual of the pixel is described

$$r^{SRD}(\mathbf{y}) = \left\| \mathbf{X}_t \boldsymbol{\alpha}^{(SRD)} - \mathbf{y} \right\|_2. \tag{12.17}$$

The output of OCSRD is presented as $D^{SRD}(\mathbf{y}) = -r^{SRD}(\mathbf{y})$. Thus, if $D^{SRD}(\mathbf{y})$ is larger than a prescribed threshold, then the testing pixel $\mathbf{y}$ is determined as a target pixel; otherwise, $\mathbf{y}$ belongs to the background.

12.4.2   One-class CRD

In OCCRD, the objective function is solved by an ℓ_2-norm regularization

$$\arg \min_{\boldsymbol{\alpha}^{(CRD)}} \left\| \mathbf{y} - \mathbf{X}_t \boldsymbol{\alpha}^{(CRD)} \right\|_2^2 + \lambda \left\| \boldsymbol{\alpha}^{(CRD)} \right\|_2^2, \qquad (12.18)$$

and the residual of the pixel is directly calculated

$$r^{CRD}(\mathbf{y}) = \left\| \mathbf{X}_t \boldsymbol{\alpha}^{(CRD)} - \mathbf{y} \right\|_2$$
$$= \left\| \mathbf{X}_t \left[\left(\mathbf{X}_t^T \mathbf{X}_t + \lambda \mathbf{I} \right)^{-1} \mathbf{X}_t^T \mathbf{y} \right] - \mathbf{y} \right\|_2. \qquad (12.19)$$

The identity matrix $\mathbf{I}$ can also be replaced by a biasing Tikhonov matrix designed as

$$\boldsymbol{\Gamma}_\mathbf{y} = \begin{bmatrix} \left\| \mathbf{y} - \mathbf{x}_1 \right\|_2 & & 0 \\ & \ddots & \\ 0 & & \left\| \mathbf{y} - \mathbf{x}_{M_t} \right\|_2 \end{bmatrix}. \qquad (12.20)$$

12.4.3   MNF-OCSRD

MNF-OCSRD and MNF-OCCRD are implemented by the following two main steps: (1) dimensionality reduction, where $\mathbf{X}_t$ and a testing pixel $\mathbf{y}$ are transformed using MNF and (2) sparse and collaborative representation, with which the approximation and the residual of the pixel are obtained. The final output is based on a comparison with a prescribed threshold.

In MNF-OCSRD, training data $\mathbf{X}_t$ and a testing pixel $\mathbf{y}$ are firstly projected via an MNF transformation. The transformed samples can be expressed as $\phi(\mathbf{X}_t)$ and $\phi(\mathbf{y})$. Then, an objective function similar to OCSRD is used to find weight vector $\boldsymbol{\alpha}^{(MNF-OCSRD)}$

$$\arg \min_{\boldsymbol{\alpha}^{(MNF-OCSRD)}} \left\| \phi(\mathbf{y}) - \phi(\mathbf{X}_t) \boldsymbol{\alpha}^{(MNF-OCSRD)} \right\|_2^2 + \lambda \left\| \boldsymbol{\alpha}^{(MNF-OCSRD)} \right\|_1, \qquad (12.21)$$

and the residual of the pixel is described as,

$$r^{MNF-OCSRD}(\mathbf{y}) = \left\| \phi(\mathbf{X}_t) \boldsymbol{\alpha}^{(MNF-OCSRD)} - \phi(\mathbf{y}) \right\|_2. \qquad (12.22)$$

The output is presented as $D^{MNF-OCSRD}(\mathbf{y}) = -r^{MNF-OCSRD}(\mathbf{y})$. Similar to OCSRD, the comparison result of $D^{MNF-OCSRD}(\mathbf{y})$ and a predefined threshold determine whether the testing pixel belongs to a target or the background.

12.4.4   MNF-OCCRD

In MNF-OCCRD, the objective function is solved by an ℓ_2-norm regularization

$$\arg \min_{\boldsymbol{\alpha}^{(MNF-OCCRD)}} \left\| \phi(\mathbf{y}) - \phi(\mathbf{X}_t) \boldsymbol{\alpha}^{(MNF-OCCRD)} \right\|_2^2 + \lambda \left\| \boldsymbol{\alpha}^{(MNF-OCCRD)} \right\|_2^2, \qquad (12.23)$$

and the residual of the pixel is calculated as,

$$r^{MNF-OCCRD}(\mathbf{y}) = \left\| \phi(\mathbf{X}_t) \boldsymbol{\alpha}^{(MNF-OCCRD)} - \phi(\mathbf{y}) \right\|_2$$
$$= \left\| \phi(\mathbf{X}_t) \left[(\phi(\mathbf{X})_t)^T \phi(\mathbf{X}_t) + \lambda \mathbf{I} \right]^{-1} \phi(\mathbf{X}_t)^T \phi(\mathbf{y}) \right] - \phi(\mathbf{y}) \right\|_2. \qquad (12.24)$$

12.4.5   Experimental results and analysis

The first experimental data employed were also collected by the HYDICE sensor. This scene consists of 80×100 pixels for an urban area [58]. The spatial resolution is approximately 1 m. A total of 175 bands with a spectral coverage of 0.4–2.5 μm are remained after removal of water vapor absorption bands. There are approximately 21 anomalous pixels, representing cars and roof. In our experiments, five chosen target signatures are used for training. The scene and the ground truth map of anomalies are shown in Figure 12.12.

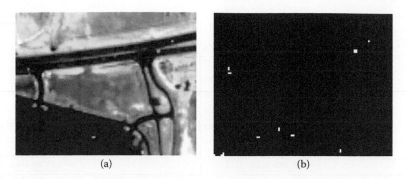

(a)　　　　　　　　　　　　　(b)

FIGURE 12.12

(a) False-color image and (b) ground truth map for the HYDICE Urban scene with 21 target pixels.

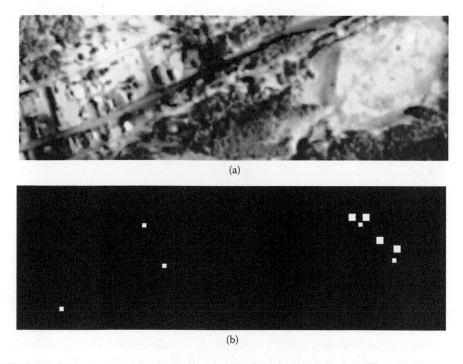

(a)

(b)

FIGURE 12.13

(a) False-color image and (b) ground truth map for the HyMap scene with 145 target pixels.

The second dataset[*] was acquired by the HyMap [59] airborne hyperspectral imaging sensor, which provides 126 spectral bands spanning the wavelength interval 0.4 to 2.5 μm. The image dataset, covering one area of Cooke City, Montana, was collected on July 4, 2006, with the spatial size 200×800 pixels. Each pixel has approximately 3 m of ground resolution. Seven types of targets, including four fabric panel targets (i.e., Red Cotton, Yellow Nylon, Blue Cotton, and Red Nylon) and 3 two-vehicle targets (i.e., 1993 Chevy Blazer, 1997 Toyota T100, and 1985 Subaru GL Wagon), were deployed in the region of interest. In our experiment, we crop a sub-image of size 100×300 pixels, including all 145 anomalies as depicted in Figure 12.13. Seven target signatures are used for training.

---

[*]http://dirsapps.cis.rit.edu/blindtest/

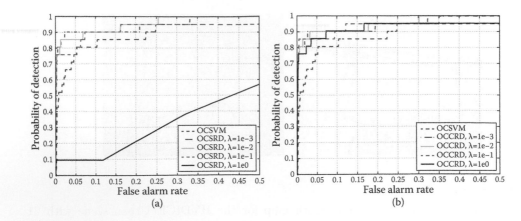

FIGURE 12.14
ROC performance of OCSVM, OCSRD, and OCCRD for the HYDICE Urban scene: (a) OCSRD and (b) OCCRD.

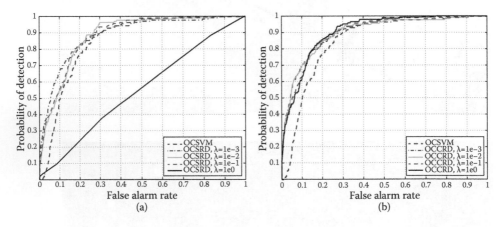

FIGURE 12.15
ROC performance of OCSVM, OCSRD, and OCCRD for the HyMap scene: (a) OCSRD and (b) OCCRD.

Receiver operating characteristic (ROC) curve [60] is usually employed to quantitatively evaluate the detection ability. Figures 12.14 and 12.15 illustrate the ROC curves as a function of regularized parameter λ, reflecting the sensitivity of OCCRD and OCSRD over a wide range of the parameter space, that is, $\{1e - 3, 1e - 2, 1e - 1, 1e0\}$ for the HYDICE Urban scene and the HyMap scene, respectively. Table 12.3 further lists the values of area under ROC (AUC). Apparently, the performance may degenerate rapidly if inappropriate λ is chosen, such as OCSRD when $\lambda = 1$. We also compare the performance of OCCRD and OCSRD with the typical one-class SVM (OCSVM) [61,62]. It is interesting to observe that OCCRD and OCSRD are significantly superior to the traditional OCSVM. Take the HyMap scene for example, AUC (%) values of OCCRD are 90.04, 89.97, 89.73, and 90.31 for various λ values; however, the one of OCSVM is 85.93, worse than all of them. Figure 12.16 further illustrates the detection maps of P_d (i.e., probability of detection) when P_f (i.e., false alarm rate) is fixed to 0.002. It shows that OCSRD and OCCRD outperform the traditional OCSVM even when P_f is fixed to a small value.

Figures 12.17 and 12.18 illustrate the ROC curves as a function of regularized parameter λ, reflecting the sensitivity of MNF-OCSRD and MNF-OCSRD toward λ for

TABLE 12.3
AUC (%) for Several Target Detectors
Using These Two Experimental Data

	HYDICE Urban	**HyMap**
OCSRD		
$\lambda = $ 1e-3	97.16	88.83
$\lambda = $ 1e-2	97.49	88.71
$\lambda = $ 1e-1	95.02	88.87
$\lambda = $ 1e0	55.24	54.64
OCCRD		
$\lambda = $ 1e-3	97.23	90.04
$\lambda = $ 1e-2	97.40	89.97
$\lambda = $ 1e-1	97.19	89.73
$\lambda = $ 1e0	95.74	90.31
OCSVM	94.79	85.93

(a) (b) (c)

FIGURE 12.16
Detection maps for HYDICE Urban data when P_f is fixed to 0.002: (a) OCSVM:
Pd=33.33%, (b) OCSRD: Pd=71.43%, and (c) OCCRD: Pd=76.19%.

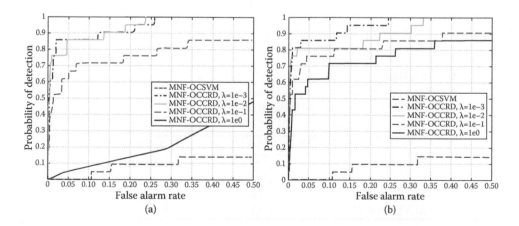

FIGURE 12.17
ROC performance of OCSVM, MNF-OCSRD, and MNF-OCCRD for the HYDICE Urban
scene: (a) MNF-OCSRD and (b) MNF-OCCRD.

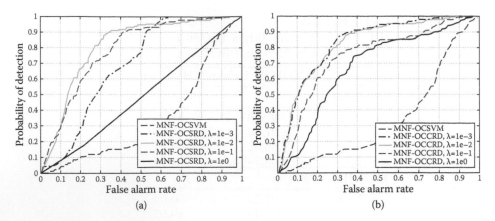

FIGURE 12.18
ROC performance of OCSVM, MNF-OCSRD, and MNF-OCCRD for the HyMap scene:
(a) MNF-OCSRD and (b) MNF-OCCRD.

TABLE 12.4
AUC (%) for MNF-OCSRD, MNF-OCCRD, and
MNF-OCSVM Using Two Experimental Datasets

	HYDICE Urban	HyMap
MNF-OCSRD		
$\lambda = $ 1e-3	96.66	68.91
$\lambda = $ 1e-2	96.66	81.07
$\lambda = $ 1e-1	84.07	78.45
$\lambda = $ 1e0	50.22	48.47
MNF-OCCRD		
$\lambda = $ 1e-3	97.21	82.89
$\lambda = $ 1e-2	94.82	82.18
$\lambda = $ 1e-1	90.40	74.27
$\lambda = $ 1e0	83.55	67.03
MNF-OCSVM	18.59	33.90

TABLE 12.5
Computational Cost (in *seconds*) for OCSRD,
OCCRD, MNF-OCSRD, and MNF-OCCRD
Using Two Experimental Datasets

	HYDICE Urban	HyMap
OCSRD	6.12	20.43
OCCRD	6.34	22.17
MNF-OCSRD	4.43	16.96
MNF-OCCRD	4.85	18.64

the HYDICE Urban and the HyMap datasets. Table 12.4 lists the values of area under
ROC (AUC). The dimensionality reduced in MNF is adopted as 25 and 5 to the HYDICE
Urban scene and the HyMap scene, respectively. As reported in Figure 12.17a, it proves
again that the result may be degenerated if inappropriate λ is chosen, such as MNF-
OCSRD when λ=1. The poor performance obtained by SVM using MNF transformed

pixels proves that it may not be suitable to combine dimensionality reduction like MNF with OCSVM. As reported in Table 12.5, it is interesting to observe that MNF-OCSRD and MNF-OCCRD obtain considerable results with less computation complexity compared with the results of OCSRD and OCCRD reported before. Take the HYDICE scene for example, AUC (%) values of MNF-OCCRD are 97.21, 94.82, 90.40, and 83.55 for various λ values; however, the one of OCSVM combined with MNF is 18.59, significantly worse than all of them. Therefore, it proves that the integration of MNF with sparse representation and collaborative representation, that is, MNF-OCSRD and MNF-OCCRD, is effective and robust with higher computational speed.

12.5 Conclusion

In this chapter, compressive sensing techniques for reconstruction, classification, and detection in hyperspectral imagery were individually investigated. Specially, we mainly introduced three reconstruction methods from random projections, that is, MT-BCS, CPPCA, and CD-CPPCA; three representation-based classifiers and their extensions, that is, SRC, CRC, MNF-SRC, and MNF-CRC; and two one-class representation-based target detectors and their improved versions, that is, OCSRD, OCCRD, MNF-OCSRD, and MNF-OCCRD. To validate the effectiveness of these state-of-the-art methods, several real hyperspectral data are employed. Experimental results demonstrated that compressive sensing theory (especially for sparse representation) can be successfully applied in these HSI applications. More researches on this topic are very worth being further investigated in the future.

Acknowledgment

This work was supported by the National Natural Science Foundation of China under Grant No. 41325004.

References

[1] D. A. Landgrebe, Hyperspectral image data analysis, *IEEE Signal Processing Magazine*, vol. 19, no. 1, pp. 17–28, 2002.

[2] D. Manolakis and G. Shaw, Detection algorithms for hyperspectral imaging applications, *IEEE Signal Processing Magazine*, vol. 19, no. 1, pp. 29–43, 2002.

[3] J. B. Boettcher, Q. Du, and J. E. Fowler, Hyperspectral image compression with the 3D dual-tree wavelet transform, in *Proceedings of the International Geoscience and Remote Sensing Symposium*, Barcelona, Spain, July 2007, pp. 1033–1036.

[4] E. Christophe, C. Mailhes, and P. Duhamel, Hyperspectral image compression: Adapting SPIHT and EZW to anisotropic 3-D wavelet coding, *IEEE Transactions on Image Processing*, vol. 17, no. 12, pp. 2334–2346, 2008.

[5] S. Subramanian, N. Gat, A. Ratcliff, and M. Eismann, Real-time hyperspectral data compression using principal components transformation, in *Proceedings of the AVIRIS Earth Science & Applications Workshop*, Pasadena, CA, February 2000.

[6] Q. Du, J. E. Fowler, and W. Zhu, On the impact of atmospheric correction on lossy compression of multispectral and hyperspectral imagery, *IEEE Transactions on Geoscience and Remote Sensing*, vol. 47, no. 1, pp. 130–132, 2009.

[7] B. Penna, T. Tillo, E. Magli, and G. Olmo, Transform coding techniques for lossy hyperspectral data compression, *IEEE Transactions on Geoscience and Remote Sensing*, vol. 45, no. 5, pp. 1408–1421, 2007.

[8] Q. Du and J. E. Fowler, Hyperspectral image compression using JPEG2000 and principal component analysis, *IEEE Geoscience and Remote Sensing Letters*, vol. 4, no. 2, pp. 201–205, 2007.

[9] Q. Du and J. E. Fowler, Low-complexity principal component analysis for hyperspectral image compression, *International Journal of High Performance Computing Applications*, vol. 22, no. 4, pp. 438–448, 2008.

[10] E. J. Candès and M. B. Wakin, An introduction to compressive sampling, *IEEE Signal Processing Magazine*, vol. 25, no. 2, pp. 21–30, 2008.

[11] M. A. Davenport, P. T. Boufounos, M. B. Wakin, and R. G. Baraniuk, Signal processing with compressive measurements, *IEEE Journal of Selected Topics in Signal Processing*, vol. 4, no. 2, pp. 445–460, 2010.

[12] T. T. Do, L. Gan, N. H. Nguyen, and T. D. Tran, Fast and efficient compressive sensing using structurally random matrices, *IEEE Transactions on Signal Processing*, vol. 60, no. 1, pp. 139–154, 2012.

[13] M. E. Gehm, R. John, D. J. Brady, R. M. Willett, and T. J. Schulz, Single-shot compressive spectral imaging with a dual-disperser architecture, *Optics Express*, vol. 15, no. 21, pp. 14013–14027, 2007.

[14] A. Wagadarikar, R. John, R. Willett, and D. Brady, Single disperser design for coded aperture snapshot spectral imaging, *Applied Optics*, vol. 47, no. 10, pp. B44–B51, 2008.

[15] J. A. Richards and X. Jia, *Remote Sensing Digital Image Analysis*, 4th ed., Berlin, Germany: Springer-Verlag, 2006.

[16] S. Li, B. Zhang, D. Chen, L. Gao, and M. Peng, Adaptive support vector machine and markov random field model for classifying hyperspectral imagery, *Journal of Applied Remote Sensing*, vol. 5, no. 1, pp. 053538–053538–11, 2011.

[17] B. Zhang, S. Li, X. Jia, L. Gao, and M. Peng, Adaptive markov random field approach for classification of hyperspectral imagery, *Geoscience and Remote Sensing Letters, IEEE*, vol. 8, no. 5, pp. 973–977, 2011.

[18] B. Zhang, S. Li, C. Wu, L. Gao, W. Zhang, and M. Peng, A neighbourhood-constrained k-means approach to classify very high spatial resolution hyperspectral imagery, *Remote Sensing Letters*, vol. 4, no. 2, pp. 161–170, 2013.

[19] H. Yu, L. Gao, J. Li, S. S. Li, B. Zhang, and J. O. N. A. Benediktsson, Spectral-spatial hyperspectral image classification using subspace-based support vector machines and adaptive markov random fields, *Remote Sensing*, vol. 8, no. 4, p. 355, 2016.

[20] S. Li, B. Zhang, L. Gao, and M. Peng, Research of hyperspectral target detection algorithms based on variance minimum, *Guangxue Xuebao/Acta Optica Sinica*, vol. 30, no. 7, pp. 2116–2122, 2010.

[21] L. Gao, B. Yang, Q. Du, and B. Zhang, Adjusted spectral matched filter for target detection in hyperspectral imagery, *Remote Sensing*, vol. 7, no. 6, pp. 6611–6634, 2015.

[22] B. Zhang and Y. Wei, Real-time target detection in hyperspectral images based on spatial-spectral information extraction, *EURASIP Journal on Advances in Signal Processing*, vol. 2012, no. 1, pp. 1–15, 2012.

[23] I. S. Reed and X. Yu, Adaptive multiple-band CFAR detection of an optical pattern with unknown spectral distribution, *IEEE Transactions on Acoustics, Speech and Signal Processing*, vol. 38, no. 10, pp. 1760–1770, 1990.

[24] H. Kwon and N. M. Nasrabadi, Kernel rx-algorithm: A nonlinear anomaly detector for hyperspectral imagery, *IEEE Transactions on Geoscience and Remote Sensing*, vol. 43, no. 2, pp. 388–397, 2005.

[25] Q. Guo, B. Zhang, Q. Ran, L. Gao, J. Li, and A. Plaza, Weighted-rxd and linear filter-based rxd: Improving background statistics estimation for anomaly detection in hyperspectral imagery, *IEEE Journal of Selected Topics in Applied Earth Observations and Remote Sensing*, vol. 7, no. 6, pp. 2351–2366, 2014.

[26] L. Gao, Q. Guo, A. Plaza, J. Li, and B. Zhang, Probabilistic anomaly detector for remotely sensed hyperspectral data, *Journal of Applied Remote Sensing*, vol. 8, no. 1, p. 083538, 2014.

[27] J. Wright, A. Y. Yang, A. Ganesh, S. S. Sastry, and Y. Ma, Robust face recognition via sparse representation, *IEEE Transactions on Pattern Analysis and Machine Intelligence*, vol. 31, no. 2, pp. 210 227, 2009.

[28] R. Archibald and G. Fann, Feature selection and classification of hyperspectral images with support vector machines, *IEEE Geoscience and Remote Sensing Letters*, vol. 4, no. 4, pp. 674–677, 2007.

[29] D. L. Donoho, Compressed sensing, *IEEE Transactions on Information Theory*, vol. 52, no. 4, pp. 1289–1306, 2006.

[30] J. E. Fowler, Compressive-projection principal component analysis, *IEEE Transactions on Image Processing*, vol. 18, no. 10, pp. 2230–2242, 2009.

[31] S. Ji, D. Dunson, and L. Carin, Multitask compressive sensing, *IEEE Transactions on Signal Processing*, vol. 57, no. 1, pp. 92–106, 2009.

[32] J. E. Fowler and Q. Du, Reconstructions from compressive random projections of hyperspectral imagery, in *Optical Remote Sensing: Advances in Signal Processing and Exploitation Techniques*, S. Prasad, L. M. Bruce, and J. Chanussot, Eds., Springer, New York, 2011, pp. 31–48.

[33] P. L. Combettes, The foundations of set theoretic estimation, *Proceedings of the IEEE*, vol. 81, no. 2, pp. 182–208, 1993.

[34] B. N. Parlett, *The Symmetric Eigenvalue Problem*, Philadelphia, PA: Society for Industrial and Applied Mathematics, 1998.

[35] W. Li and J. E. Fowler, Decoder-side dimensionality determination for compressive-projection principal component analysis of hyperspectral data, in *Proceedings of the International Conference on Image Processing*, Brussels, Belgium, September 2011, pp. 329–332.

[36] W. Li, S. Prasad, and J. E. Fowler, Classification and reconstruction from random projections for hyperspectral imagery, *IEEE Transactions on Geoscience and Remote Sensing*, vol. 51, no. 2, pp. 833–843, 2013.

[37] R. Calderbank, S. Jafarpour, and R. Schapire, *Compressive Learning: Universal Sparse Dimensionality in the Measurement Domain*, Technical Report, Princeton University, New Jersey, 2009.

[38] C. Cortes and V. N. Vapnik, Support vector networks, *Machine Learning*, vol. 20, no. 3, pp. 273–297, 1995.

[39] J. E. Fowler and Q. Du, Anomaly detection and reconstruction from random projections, *IEEE Transactions on Image Processing*, vol. 21, no. 1, pp. 184–195, 2012.

[40] S. Prasad and L. M. Bruce, Decision fusion with confidence-based weight assignment for hyperspectral target recognition, *IEEE Transactions on Geoscience and Remote Sensing*, vol. 46, no. 5, pp. 1448–1456, 2008.

[41] L. Zhang, M. Yang, and X. Feng, Sparse representation or collaborative representation: Which helps face recognition? in *Proceedings of the International Conference on Computer Vision*, Barcelona, Spain, November 2011, pp. 471–478.

[42] J. Yin, Z. Liu, Z. Jin, and W. Yang, Kernel sparse representation based classification, *Neurocomputing*, vol. 77, no. 1, pp. 120–128, 2012.

[43] L. Zhang, W. Zhou, P. Chang, J. Liu, Z. Yan, T. Wang, and F. Li, Kernel sparse representation-based classifier, *IEEE Transactions on Signal Processing*, vol. 60, no. 4, pp. 1684–1695, 2012.

[44] J. Liu, Z. Wu, Z. Wei, L. Xiao, and L. Sun, Spatial-spectral kernel sparse representation for hyperspectral image classification, *IEEE Journal of Selected Topics in Applied Earth Observations and Remote Sensing*, vol. 6, no. 6, pp. 2462–2471, 2013.

[45] J. Chen and L. Jiao, Hyperspectral imagery classification using local collaborative representation, *International Journal of Remote Sensing*, vol. 36, no. 3, pp. 734–748, 2015.

[46] R. Timofte and L. V. Gool, Adaptive and weighted collaborative representations for image classification, *Pattern Recognition Letters*, vol. 43, pp. 127–135, 2014.

[47] W. Li, E. W. Tramel, S. Prasad, and J. E. Fowler, Nearest regularized subspace for hyperspectral classification, *IEEE Transactions on Geoscience and Remote Sensing*, vol. 52, no. 1, pp. 477–489, 2014.

[48] W. Li, K. Liu, and H. Su, Wavelet-based nearest-regularized subspace for noise-robust hyperspectral image classification, *Journal of Applied Remote Sensing*, vol. 8, p. 083665, 2014.

[49] W. Li and Q. Du, Joint within-class collaborative representation for hyperspectral image classification, *IEEE Journal of Selected Topics in Applied Earth Observations and Remote Sensing*, vol. 7, no. 6, pp. 2200–2208, 2014.

[50] W. Li, Q. Du, and M. Xiong, Kernel collaborative representation with tikhonov regularization for hyperspectral image classification, *IEEE Geoscience and Remote Sensing Letters*, vol. 12, no. 1, pp. 48–52, 2015.

[51] A. A. Nielsen, Kernel maximum autocorrelation factor and minimum noise fraction transformations, *IEEE Transactions on Image Processing*, vol. 20, no. 3, pp. 612–624, 2011.

[52] F. A. Kruse, Comparison of ATREM, ACORN, and FLAASH atmospheric corrections using low-altitude AVIRIS data of Boulder, CO, in *Proceedings of the 13th JPL Airborne Geoscience Workshop*, Pasadena, CA, March 2004.

[53] P. Gamba, A collection of data for urban area characterization, in *Proceedings of the International Geoscience and Remote Sensing Symposium*, vol. 1, Anchorage, AK, September 2004, pp. 69–72.

[54] Y. Chen, N. M. Nasrabadi, and T. D. Tran, Sparse representation for target detection in hyperspectral imagery, *IEEE Journal of Selected Topics in Signal Processing*, vol. 5, no. 3, pp. 629–640, 2011.

[55] Y. Chen, N. M. Nasrabadi, and T. D. Tran, Simulanetous joint sparsity model for target detection in hyperspectral imagery, *IEEE Geoscience and Remote Sensing Letters*, vol. 8, no. 4, pp. 676–680, 2011.

[56] W. Li and Q. Du, Collaborative representation for hyperspectral anomaly detection, *IEEE Transactions on Geoscience and Remote Sensing*, vol. 53, no. 3, pp. 1463–1474, 2015.

[57] W. Li, Q. Du, and B. Zhang, Combined sparse and collaborative representation for hyperspectral target detection, *Pattern Recognition*, vol. 48, pp. 3904–3916, 2015.

[58] L. Ma, M. M. Crawford, and J. Tian, Local manifold learning-based k-nearest-neighbor for hyperspectral image classification, *IEEE Transactions on Geoscience and Remote Sensing*, vol. 48, no. 11, pp. 4099–4109, 2010.

[59] D. Snyder, J. Kerekes, I. Fairweather, R. Crabtree, J. Shive, and S. Hager, Devlopment of a web-based application to evaluate target finding algorithms, in *Proceedings of the International Geoscience and Remote Sensing Symposium*, Boston, MA, July 2008, pp. 915–918.

[60] J. A. Hanley and B. J. McNeil, A method of comparing the areas under receiver operating characteristic curves derived from the same cases, *Radiology*, vol. 148, no. 3, pp. 839–843, 1983.

[61] J. Muñoz-Marí, F. Bovolo, L. Gomez-Chova, L. Bruzzone, and G. Camps-Valls, Semisupervised one-class support vector machines for classification of remote sensing data, *IEEE Transactions on Geoscience and Remote Sensing*, vol. 48, no. 8, pp. 3188–3197, 2010.

[62] G. Bilgin, S. Ertürk, and T. Yıldırım, Segmentation of hyperspectral immages via subtractive clustering and cluster validation using one-class support vector machines, *IEEE Transactions on Geoscience and Remote Sensing*, vol. 49, no. 8, pp. 2936–2944, 2011.

13

Structured Abundance Matrix Estimation for Land Cover Hyperspectral Image Unmixing

Paris V. Giampouras, Konstantinos E. Themelis, Athanasios A. Rontogiannis, and Konstantinos D. Koutroumbas

CONTENTS

13.1 Introduction

Spectral unmixing (SU) of hyperspectral images (HSIs) has been in the spotlight of both research and applications during the recent years [1]. In a nutshell, SU refers to the process of (a) detecting the pure material spectra (called endmembers) that are present in the scene and (b) estimating their corresponding fractional proportions in each pixel (called abundances). This unmixing process is commonly addressed in literature as two distinct tasks. That said, several endmembers' extraction algorithms have been proposed in literature, (e.g. [2,3]) focusing on the identification of the materials' spectra. On the contrary, abundance estimation algorithms [4,5], presuppose that the dictionary of the endmembers is already known, concentrating on the accurate estimation of the abundance values.

Abundance estimation algorithms hinge on the assumption that the underlying mixing process is properly modeled. Along those lines, the linear mixing model (LMM) has been widely adopted by numerous unmixing algorithms. Its main premise is that the spectral signatures of the pixels result from the linear combination of different endmembers existing in the scene. That said, unmixing can be viewed as a linear regression problem. Since abundances correspond to fractional proportions, nonnegativity and sum-to-one constraints naturally arise, rendering unmixing a *constrained linear regression problem*.

Besides the nonnegativity and the sum-to-one constraints, additional assumptions have been espoused in the unmixing literature. In this spirit, sparsity of the vector of abundances has been lately adopted by disparate unmixing methods [4–6]. The rationale behind this assumption is that when the dictionary consists of a large number of endmembers, then only a subset of those materials is active in the formation of a given single pixel. As a consequence, abundance vectors shall have only few nonzero entries. Spatial correlation is also prevalent in land cover HSIs and has been subject to exploitation for enhancing abundance estimation accuracy. In light of this, joint-sparse unmixing algorithms were proposed in [7,8]. Therein, it is assumed that in homogeneous regions of HSIs, it is likely that neighboring pixels will be composed by the interaction of the same endmembers. Put it differently, the abundance vectors of spatially correlated pixels will share the same support set. As a result, the abundance matrix that contains those vectors presents a joint-sparse structure. In a similar vein, in [9], a window-based approach was put forth. Therein, for the unmixing of each single pixel, a window containing the spectral information of its adjacent pixels was employed. The abundance vectors corresponding to the pixels contained in the window were supposed to be linearly dependent, thus paving the way for the pursuit of low-rank abundance matrices. In [10], another window-based approach was proposed whereby sparsity and spatial correlation were simultaneously taken into account, via exploiting at the same time sparsity and low-rankness. In this regime, the unmixing task can be viewed as a multiple constrained optimization problem.

Inspired by the work of Giampouras et al. [10], in this chapter we present a new unmixing algorithm that imposes concurrently the two aforementioned constraints, that is, sparsity and low-rankness. Departing from the localized way of incorporating spatial correlation via the utilization of a sliding window for the unmixing of a single pixel, we account for the correlation among all pixels of the image[*]. The unmixing problem is thus formulated as a matrix regression problem with the corresponding abundance matrix constrained to be simultaneously sparse and low-rank. These two constraints are efficiently imposed by regularizing an initial least-squares cost function with a sparsity inducing weighted ℓ_1 norm and an upper bound of the low-rank promoting nuclear norm [12]. The formed minimization problem is attacked by a low-complexity alternating coordinate descent-type algorithm which splits the original nonconvex cost function into mutually dependent convex subproblems. The proposed algorithm is robust and converges after a few iterations. We demonstrate the usefulness of the proposed algorithm by conducting experiments on both synthetic and real land cover hyperspectral data. Experimental results reveal that the proposed method has superior performance compared with competing state-of-the-art unmixing schemes.

The remainder of this chapter is structured as follows: First, we formulate unmixing as a simultaneously sparse and low-rank matrix estimation problem in Section 13.2. Section 13.3 presents the proposed coordinate descent style method to tackle this problem. The experimental results of the proposed scheme are presented in Section 13.4.

13.2 System model and problem formulation

Let $\mathbf{Y} = [\mathbf{y}_1, \mathbf{y}_2, \ldots, \mathbf{y}_K]$ denote the $L \times K$ observation matrix, consisting of the L-band spectral signatures of the total K pixels of the examined HSI. Herein, we assume that the observed spectral signatures are produced following a linear regression process; that is, the linear mixing model (LMM) is espoused for $\mathbf{Y}$, thus

$$\mathbf{Y} = \mathbf{\Phi W} + \mathbf{E}. \tag{13.1}$$

---

[*]The algorithm presented in this chapter is a modified version of the window-based scheme proposed in [11].

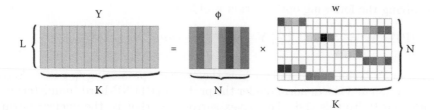

FIGURE 13.1
Graphical illustration of the proposed unmixing scheme. Abundance matrix is considered sparse and low-rank (in this example, rank = 4). White cells in matrix $\mathbf{W}$ represent zero values.

Matrix $\mathbf{\Phi} = [\phi_1, \phi_2, \ldots, \phi_N] \in \mathcal{R}_+^{L \times N}$ is the dictionary containing the spectra of N endmembers as its columns, and the $N \times K$ nonnegative matrix $\mathbf{W}$ is composed of the abundance column vectors of the K pixels of the HSI. Finally, it is assumed that additive zero-mean Gaussian independent and identically distributed (i.i.d.) noise corrupts the data with its entries composing matrix $\mathbf{E}$.

With regard to the abundance matrix $\mathbf{W}$, two widely adopted constraints in the unmixing literature are the *nonnegativity* and *sum-to-one* constraints expressed as

$$\mathbf{W} \geq 0, \text{ and } \mathbf{1}^T\mathbf{W} = \mathbf{1}^T, \tag{13.2}$$

where $\mathbf{1}$ is the all-ones vector. Herein, we neglect the sum-to-one constraint for reasons detailed also in [13]. In this work, our objective is to estimate the nonnegative abundance matrix $\mathbf{W}$, given the pixels' spectra matrix $\mathbf{Y}$ and the endmembers' dictionary $\mathbf{\Phi}$. Therefore, SU can be deemed as a constrained linear matrix regression problem.

The sparsity assumption has been broadly adopted in numerous unmixing algorithms that have come to the scene in the recent years [5,13]. The main premise of this is that when a large number of endmembers is contained in the endmembers' library, only a subset of these is present in the mixing process of a given pixel. That said, the abundance values corresponding to the remaining absent endmembers shall be zero. In an effort to bring into play and exploit the spatial correlation that exists in homogeneous regions of HSIs, unmixing algorithms that assume a joint-sparse structure on the abundance matrix have been recently put forth [7–9]. Actually those algorithms are based on the fact that it is likely that the same endmembers are present in an homogeneous region of an HSI. On the contrary, in [9], spatial correlation was alternatively taken into account by assuming that it gives rise to low-rank abundance matrices. More specifically, in the algorithm proposed in [9], a sliding window is employed for the unmixing of each single pixel. Due to the presupposed spatial correlation, the abundance vectors corresponding to the pixels in the window can be considered to be linearly dependent, thus leading to low-rank abundance matrices. Interestingly, this way the unmixing process boils down to a *reduced rank regression* problem. In what follows, we aim at efficiently exploiting *simultaneously* spatial correlation and sparsity. Toward this, it is assumed that the abundance matrix of HSIs that contain many homogeneous regions entail low-rank abundance matrices. At the same time, when the endmembers' dictionary consists of a large number of spectra, sparsity still characterizes the formed abundance matrix (see Figure 13.1). To this end, the unmixing problem is now formulated as a *sparse reduced-rank regression problem*. This formulation gives rise to the pursuit of an abundance matrix $\mathbf{W}$ that (a) fits well to the data matrix $\mathbf{Y}$ with respect to a given dictionary $\mathbf{\Phi}$, (b) has low rank, and (c) has only a few nonzero, positive elements. Abundance matrix $\mathbf{W}$ is henceforth estimated by solving the following optimization problem,

$$(\text{P1}) : \hat{\mathbf{W}} = \underset{\mathbf{W} \in \mathcal{R}_+^{N \times K}}{\text{argmin}} \left\{ \frac{1}{2} \| \mathbf{Y} - \mathbf{\Phi} \mathbf{W} \|_F^2 + rank(\mathbf{W}) + \| \mathbf{W} \|_0 \right\},$$

where $\| \cdot \|_F$ is the Frobenius norm and $\| \cdot \|_0$ is the sparsity promoting ℓ_0 norm. Actually, both the rank function and the ℓ_0 norm render the problem (P1) NP-hard (nondeterministic polynomial-time hard). To this end, the convex surrogates; that is, the nuclear norm and the ℓ_1 norm are put in place of the rank function and the ℓ_0 norm, respectively. This leads to the following convex minimization problem,

$$(\text{P2}) : \hat{\mathbf{W}} = \underset{\mathbf{W} \in \mathcal{R}_+^{N \times K}}{\text{argmin}} \left\{ \frac{1}{2} \| \mathbf{Y} - \mathbf{\Phi} \mathbf{W} \|_F^2 + \lambda_* \| \mathbf{W} \|_* + \lambda_1 \| \mathbf{W} \|_1 \right\}.$$

Note that λ_* and λ_1 are regularization parameters that control the importance of low-rankness and sparsity of the related terms.

Abundance matrix estimation through solving a minimization problem similar to (P2) was studied in [10]. It should be emphasized though that the minimization of the nuclear norm requires a singular value decomposition (SVD) step, whose complexity when applied on a matrix of size $N \times K$ is $\mathcal{O}(K^3 + N^2 K + NK^2)$. It is easily observed that this can be a serious impediment when either the number N of spectral in the dictionary or the number K of the pixels in the image are large. For this reason, in [10], a sliding window approach similar to that in [9] was followed. This way, the unmixing of a single pixel involved a small number K of the neighboring pixels, making the unmixing process a computationally feasible task.

Following an alternative path, herein we depart from the sliding window approach of [10] and attempt to exploit the spatial correlation of all pixels of the HSI. Consequently, as K now corresponds to the total number of pixels, it takes huge values, thus rendering the SVD operation computationally cumbersome. Capitalizing on this, we propose a different way for minimizing the rank of $\mathbf{W}$. Specifically, we adopt an explicit low-rank parameterization for $\mathbf{W}$ [12], which assumes that an upper bound r for the rank of $\mathbf{W}$ ($r \geq rank(\mathbf{W})$ and $r << K, N$) is *a priori* available. According to this parameterization, $\mathbf{W}$ is now written as the product of two matrices $\mathbf{P} \in \mathcal{R}^{N \times r}$ and $\mathbf{Q} \in \mathcal{R}^{K \times r}$, that is, $\mathbf{W} = \mathbf{P}\mathbf{Q}^T$. The gain from this parameterization is that now it holds,

$$\| \mathbf{W} \|_* \leq \frac{1}{2} \left(\| \mathbf{P} \|_F^2 + \| \mathbf{Q} \|_F^2 \right), \tag{13.3}$$

that is, $\| \mathbf{W} \|_*$ is a tight lower bound of the average of the squared Frobenious norms of $\mathbf{P}$ and $\mathbf{Q}$. Substituting this upper bound of the nuclear norm in (P2), we end up with the following optimization problem:

$$(\text{P3}) : \left\{ \hat{\mathbf{W}}, \hat{\mathbf{P}}, \hat{\mathbf{Q}} \right\} = \underset{\mathbf{P}, \mathbf{Q}, \mathbf{W}}{\text{argmin}} \, \mathcal{L}(\mathbf{P}, \mathbf{Q}, \mathbf{W}), \tag{13.4}$$

where

$$\mathcal{L}(\mathbf{P}, \mathbf{Q}, \mathbf{W}) = \left\{ \frac{1}{2} \| \mathbf{Y} - \mathbf{\Phi}\mathbf{P}\mathbf{Q}^T \|_F^2 + \frac{\lambda_*}{2} \left(\| \mathbf{P} \|_F^2 + \| \mathbf{Q} \|_F^2 \right) + \lambda_1 \| \mathbf{W} \|_1 + \frac{\mu}{2} \| \mathbf{W} - \mathbf{P}\mathbf{Q}^T \|_F^2 \right\}. \tag{13.5}$$

It should be noted that the term $\| \mathbf{W} - \mathbf{P}\mathbf{Q}^T \|_F^2$ that appears in (13.5) imposes the fitting between $\mathbf{W}$ and its low-rank representation $\mathbf{P}\mathbf{Q}^T$, with μ being the Lagrange multiplier parameter. Contrary to the convex minimization problem (P2), (P3) is nonconvex with respect to matrices $\mathbf{W}$, $\mathbf{P}$, and $\mathbf{Q}$. Moreover, the presence of the ℓ_1 norm induces a nonsmooth behavior of the objective function that must be suitably handled. In the following, a novel block coordinate descend (BCD) algorithm [14] is presented that solves (P3) efficiently and provides an estimate of the abundance matrix $\mathbf{W}$.

13.3   The proposed algorithm

The separability of the above-defined nonconvex and nonsmooth minimization problem (P3) permits us to split it into three distinct subproblems. First, we define these subproblems

$$(\text{P3a}) : \hat{\mathbf{P}} = \underset{\mathbf{P}}{\arg\min} \mathcal{L}(\mathbf{P}, \mathbf{Q}, \mathbf{W}),$$

$$(\text{P3b}) : \hat{\mathbf{Q}} = \underset{\mathbf{Q}}{\arg\min} \mathcal{L}(\mathbf{P}, \mathbf{Q}, \mathbf{W}),$$

$$(\text{P3c}) : \hat{\mathbf{W}} = \underset{\mathbf{W}}{\arg\min} \mathcal{L}(\mathbf{P}, \mathbf{Q}, \mathbf{W}),$$

which are convex and can be solved independently following a block coordinate descent strategy. As is shown below, the solutions of the above-described problems are interrelated, thus paving the way for an iterative scheme, which converges to a local minimum of the cost function defined in Equation 13.5. In the following sections, we examine these minimization problems one by one.

13.3.1   Solution of (P3a)

Since $\mathcal{L}(\mathbf{P}, \mathbf{Q}, \mathbf{W})$ is differentiable with respect to $\mathbf{P}$, $\mathbf{P}$ can be obtained as the solution of the following equation:

$$\hat{\mathbf{P}} : \frac{\partial \mathcal{L}(\mathbf{P}, \mathbf{Q}, \mathbf{W})}{\partial \mathbf{P}} = \mathbf{0}. \tag{13.6}$$

Calculating the derivative in (13.6) yields

$$\left(\mathbf{\Phi}^T \mathbf{\Phi} + \mu \mathbf{I}_N\right) \mathbf{P} \mathbf{Q}^T \mathbf{Q} + \lambda_* \mathbf{P} = \left(\mathbf{\Phi}^T \mathbf{Y} + \mu \mathbf{W}\right) \mathbf{Q}, \tag{13.7}$$

where $\mathbf{I}_N$ denotes the identity matrix of size N. Next, we set $\mathbf{A} = \mathbf{\Phi}^T \mathbf{\Phi} + \mu \mathbf{I}_N$, $\mathbf{B} = \mathbf{Q}^T \mathbf{Q}$ and $\mathbf{C} = \left(\mathbf{\Phi}^T \mathbf{Y} + \mu \mathbf{W}\right) \mathbf{Q}$. Equation 13.7 can be compactly written as

$$\mathbf{A} \mathbf{P} \mathbf{B} + \lambda_* \mathbf{P} = \mathbf{C}. \tag{13.8}$$

Equation 13.8 belongs to the class of the so-called Stein matrix equations that, among others, have been widely used in the field of control theory [15]. To solve this Stein Equation 13.8, we adopt the robust algorithm proposed in [16]. In this algorithm, matrix $\mathbf{A}$ is reduced to its Hessenberg form $\mathbf{H} = \mathbf{U} \mathbf{A} \mathbf{U}^T$ and matrix $\mathbf{B}$ is suitably replaced by its Schur representation $\mathbf{S} = \mathbf{V} \mathbf{B} \mathbf{V}^T$, where $\mathbf{U}$ and $\mathbf{V}$ are orthogonal matrices. Favorably, the symmetry of both $\mathbf{A}$ and $\mathbf{B}$ leads to matrices $\mathbf{H}$ and $\mathbf{S}$ that are tri-diagonal and diagonal, respectively. By multiplying both sides of Equation 13.8 from the left and the right by $\mathbf{U}^T$ and $\mathbf{V}$, respectively, and defining $\mathbf{X} = \mathbf{U} \mathbf{P} \mathbf{V}^T$, Equation 13.8 is rewritten as

$$\mathbf{H} \mathbf{X} \mathbf{S} + \lambda_* \mathbf{X} = \mathbf{F}, \tag{13.9}$$

where $\mathbf{F} = \mathbf{U}^T \mathbf{C} \mathbf{V}$. Let us denote by $\mathbf{x}_i$, $\mathbf{f}_i$ the ith columns of $\mathbf{X}, \mathbf{F}$, respectively and by s_{ii} the ith diagonal element of $\mathbf{S}$. Then, we get from Equation 13.9 the following system of equations

$$\left(s_{ii} \mathbf{H} + \lambda_* \mathbf{I}_N\right) \mathbf{x}_i = \mathbf{f}_i, \tag{13.10}$$

which can be solved for $\mathbf{x}_i$ with only $\mathcal{O}(N)$ operations, due to the tri-diagonal form of $\mathbf{H}$ [17]. After estimating $\mathbf{X}$, column by column, matrix $\mathbf{P}$ is obtained by the inverse transform

$$\hat{\mathbf{P}} = \mathbf{U}^T \mathbf{X} \mathbf{V}. \tag{13.11}$$

13.3.2   Solution of (P3b)

Similarly to (P3a), the minimization problem (P3b) can be solved as

$$\hat{\mathbf{Q}} : \frac{\partial \mathcal{L}(\mathbf{P}, \mathbf{Q}, \mathbf{W})}{\partial \mathbf{Q}} = \mathbf{0}. \tag{13.12}$$

Utilizing $\mathbf{A}, \mathbf{C}$ defined above, Equation 13.12 results in the closed form expression

$$\hat{\mathbf{Q}} = \mathbf{C}^T \mathbf{P} \left(\mathbf{P}^T \mathbf{A} \mathbf{P} + \lambda_* \mathbf{I}_r \right)^{-1} \tag{13.13}$$

that requires the computation of a small sized $(r \times r)$ matrix inversion.

13.3.3   Solution of (P3c)

The optimization problem (P3c) is employed in order to estimate matrix $\mathbf{W}$. Considering $\mathbf{P}$ and $\mathbf{Q}$ as constants, minimization of $\mathcal{L}(\mathbf{P}, \mathbf{Q}, \mathbf{W})$ with respect to $\mathbf{W}$ leads to

$$\hat{\mathbf{W}} = \underset{\mathbf{W}}{\operatorname{argmin}} \left\{ \|\mathbf{W}\|_1 + \frac{\mu}{2\lambda_1} \|\mathbf{W} - \mathbf{P}\mathbf{Q}^T\|_F^2 \right\}, \tag{13.14}$$

which is the proximal operator of the ℓ_1 norm on $\mathbf{P}\mathbf{Q}^T$. Writing Equation 13.14 as

$$\hat{\mathbf{W}} = \underset{\mathbf{W}}{\operatorname{argmin}} \sum_{n=1}^{N} \sum_{k=1}^{K} \left(|w_{nk}| + \frac{\mu}{2\lambda_1} \left(w_{nk} - \mathbf{p}_n^T \mathbf{q}_k \right) \right), \tag{13.15}$$

where $\mathbf{p}_n^T$ denotes the nth row of matrix $\mathbf{P}$ and $\mathbf{q}_k^T$ the kth row of $\mathbf{Q}$, $\hat{\mathbf{W}}$ can be determined via elementwise soft-threshoding [18]. Thus, we have that

$$\hat{w}_{nk} = \operatorname{SHR}_{\lambda_1/\mu}(\mathbf{p}_n^T \mathbf{q}_k), \tag{13.16}$$

where $\operatorname{SHR}_\lambda(x) = \operatorname{sign}(x)\max(0, |x| - \lambda)$. In this work, in an attempt to enhance the imposition of sparsity, we use the weighted version of the ℓ_1 norm, thus matrix $\mathbf{D}$ is employed, whose elements are given by

$$d_{nk} = \frac{1}{|\hat{w}_{nk}| + \epsilon}. \tag{13.17}$$

$\hat{w}_{nk}$ in Equation 13.17 is the estimate of w_{nk} at the previous iteration of the algorithm and ϵ is a very small constant. The solution of Equation 13.14 can be written in a more compact form as

$$\hat{\mathbf{W}} = \operatorname{SHR}_{\mathbf{D}(\lambda_1/\mu)}(\mathbf{P}\mathbf{Q}^T). \tag{13.18}$$

The concluding scheme dubbed ALternating Minimization Sparse Low-Rank Unmixing (ALMSpLRU) algorithm is summarized in Algorithm 1. It should be noted that the aforementioned nonnegativity constraint is imposed by projecting $\mathbf{W}$ onto the nonnegative orthant of $\mathcal{R}^{N \times K}$, denoted as $\mathcal{R}_+^{N \times K}$ (step 9). In Algorithm 1, the computational complexity of each step is also included. We observe that step 3 is the most computationally demanding step, requiring $\mathcal{O}(NKr)$ operations per iteration. Interestingly, this computational cost is much lower compared with that of the SVD-based schemes of [10], which depend on the number of pixels that is taken into account each time in the unmixing process. Finally, as verified by extensive simulations, the proposed algorithm is robust, and converges after a few of iterations.

Algorithm 13.1: ALMSpLRU

Inputs $\mathbf{Y}, \boldsymbol{\Phi}$
Initialize parameters $\lambda_1, \lambda_*, \mu$
Initialize $\mathbf{P}, \mathbf{Q}, \mathbf{W} = \mathbf{PQ}^T$
Set $\mathbf{A} = \boldsymbol{\Phi}^T\boldsymbol{\Phi} + \mu\mathbf{I}_N$
Set $[\mathbf{U}, \mathbf{H}] = \text{hess}(\mathbf{A})$ ▷ Hessenberg form of $\mathbf{A}$
Set $\mathbf{T} = \boldsymbol{\Phi}^T\mathbf{Y}$
repeat
  1: $\mathbf{B} = \mathbf{Q}^T\mathbf{Q}$, $\mathcal{O}(Nr^2)$
  2: $[\mathbf{V}, \mathbf{S}] = \text{schur}(\mathbf{B})$,   ▷ Schur form of $\mathbf{B}$ $\mathcal{O}(r^3)$
  3: $\mathbf{C} = (\mathbf{T} + \mu\mathbf{W})\mathbf{Q}$, $\mathcal{O}(NKr)$
  4: $\mathbf{F} = \mathbf{U}^T\mathbf{CV}$, $\mathcal{O}(N^2r)$
  5: $(s_{ii}\mathbf{H} + \lambda_*\mathbf{I}_N)\mathbf{x}_i = \mathbf{f}_i$,   ▷ $i = 1, 2, \ldots, r$ $\mathcal{O}(rN)$
  6: $\mathbf{P} = \mathbf{UXV}^T$, $\mathcal{O}(N^2r)$
  7: $\mathbf{Q} = \mathbf{C}^T\mathbf{P}(\mathbf{P}^T\mathbf{AP} + \lambda_*\mathbf{I}_r)^{-1}$, $\mathcal{O}(N^2r)$
  8: $\mathbf{W} = \text{SHR}_{\mathbf{D}(\lambda_1/\mu)}(\mathbf{PQ}^T)$, $\mathcal{O}(NK)$
  9: Project $\mathbf{W}$ onto $\mathcal{R}_+^{N \times K}$ $\mathcal{O}(NK)$
until convergence
Output $\hat{\mathbf{W}}$

13.4   Experimental results

In this section, we aspire to demonstrate the effectiveness of the proposed algorithm on both simulated and real data experiments. To better illustrate the advantages of the proposed algorithm in terms of estimation accuracy, we compare it with three state-of-the-art unmixing algorithms, namely the constrained sparse unmixing by variable splitting and augmented Lagrangian algorithm CSUnSAl+ [4], the nonnegatively constrained joint-sparse method MMV-ADMM [9], and the fast Bayesian inference iterative conditional expectations BiICE [5]. Notably, as shown in Table 13.1, the proposed ALMSpLRU algorithm presents lower computational complexity per iteration as compared with its rivals, albeit it exploits both spatial correlation and sparsity. Actually, this property favorably results from the aforementioned nonconvex parameterization of the abundance matrix.

TABLE 13.1
Computational Complexity per Iteration for K Pixels

Algorithm	ALMSpLRU	CSUnSAl [4]	MMV-ADMM [9]	BiICE [19]
Computational complexity	$\mathcal{O}(KNr)$	$\mathcal{O}(KN^2)$	$\mathcal{O}(K^2N^2 + K^2LN)$	$\mathcal{O}(KN^2)$

13.4.1   Simulated data experiments

For synthetic data experiments, we compare the performance of all the tested algorithms in terms of the root mean square error (RMSE), defined as

$$\text{RMSE} = \sqrt{\frac{1}{Nn} \sum_{i=1}^{n} \|\hat{\mathbf{w}}_i - \mathbf{w}_{true}\|_2}, \tag{13.19}$$

where $\hat{\mathbf{w}}_i$ and $\mathbf{w}_{true}$ represent the estimated and actual abundance vectors, respectively, n is the number of the pixels, and N stands for the number of endmembers, as has already been mentioned in previous sections. The signal-to-reconstruction error (SRE) [13] is also utilized in our experiments, which is defined as the ratio between the power of signal and the error, that is,

$$\text{SRE} = 10\log_{10} \left(\frac{\frac{1}{n}\sum_{i=1}^{n} \|\hat{\mathbf{w}}_i\|_2^2}{\frac{1}{n}\sum_{i=1}^{n} \|\hat{\mathbf{w}}_i - \mathbf{w}_{true}\|_2^2} \right). \tag{13.20}$$

The sparsity and joint-sparsity imposing parameters of ALMSpLRU, CSUnSAl+, and MMV-ADMM are fine tuned with 12 different values, that is, $\lambda = \{0, 10^{-10}, 10^{-9}, \ldots, 1\}$. The same values are used also for the low-rank regularization parameter of ALMSpLRU. Finally, the Lagrange multiplier parameter μ is set to 0.1 for all deterministic algorithms.

13.4.1.1   Sparse and low-rank abundance matrix estimation

In the sequel, we study the performance of the proposed ALMSpLRU algorithm in estimating simultaneously sparse and low-rank abundance matrices. To this end, we randomly select $N = 60$ endmembers contained in the USGS spectral library $\mathbf{Z} \in \mathcal{R}_+^{224 \times 498}$ [20]. Then, we generate an $N \times K$ abundance matrix of rank 12 and with $K = 1000$. The sparsity level (defined as the ratio number of nonzero entries/NK) is 0.1. Next the spectral signatures are linearly produced via the LMM, and noise of SNR = 35dB contaminates the data. The superior performance of ALMSpLRU is illustrated in Figure 13.2. As is clearly shown both in terms of RMSE and visually from the recovered abundance matrices and the residual errors (i.e., $|\mathbf{W} - \hat{\mathbf{W}}|$), ALMSpLRU is proven more effective than its competing schemes in estimating the sought abundance matrix. This is so, since ALMSpLRU accounts for both sparsity and low-rankness contrary to the rest of the sparsity-aware unmixing algorithms.

13.4.1.2   Sensitivity to noise

In this experiment, we test the robustness of the proposed ALMSpLRU algorithm and the competing schemes to noise corruption. Toward this, the same process detailed above is followed for generating $K = 60$ linearly mixed pixels, out of $N = 60$ randomly selected from the USGS library endmembers, utilizing $N \times K$ simultaneously sparse and low-rank abundance matrices of rank 4 and sparsity level 0.1. The experiment is executed 10 times for SNR values 15, 20, 25, 30, and 35. In Table 13.2, the average RMSE and SRE values corresponding to each SNR are given. As it is easily observed, ALMSpLRU is again proven competent in estimating more accurately the simultaneously sparse and low-rank abundance matrix than the other state-of-the-art unmixing algorithms, in all tested cases corresponding to corruption of data by disparate noise levels. This is verified for both performance metrics utilized, that is, RMSE and SRE.

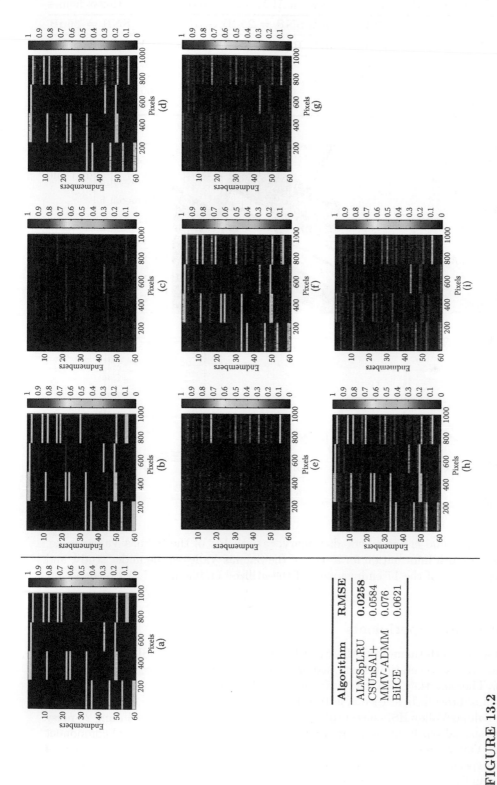

Algorithm	RMSE
ALMSpLRU	**0.0258**
CSUnSAl+	0.0584
MMV-ADMM	0.076
BiICE	0.0621

FIGURE 13.2

Performance comparison of ALMSpLRU and CSUnSAl+, MMV-ADMM, and BiICE in estimating simultaneously sparse and low-rank abundance matrices: (a) W, Ground truth; (b) $\hat{W}$, ALMSpLRU; (c) residual, ALMSpLRU; (d) $\hat{W}$, BiICE; (e) residual, BiICE; (f) $\hat{W}$, CSUnSAl+; (g) residual, CSUnSAl+; (h) $\hat{W}$, MMV-ADMM; and (i) residual, MMV-ADMM.

TABLE 13.2
RMSE and SRE versus SNR Comparison between ALMSpLRU and the Competing Schemes

Algorithm	SNR = 15dB		SNR = 20dB		SNR = 25dB		SNR = 30dB		SNR = 35dB	
	RMSE	SRE	RMSE	SRE	RMSE	SRE	RMSE	SRE	RMSE	SRE
ALMSpLRU	0.141	2.30	0.116	3.80	0.096	5.45	0.064	8.90	0.058	9.84
CSunSAL	0.174	0.467	0.148	1.73	0.134	2.61	0.093	5.63	0.068	8.37
MMV-ADMM	0.146	1.95	0.1246	3.25	0.122	3.40	0.093	5.60	0.074	7.71
BiICE	0.221	−1.78	0.165	0.26	0.134	2.50	0.089	5.78	0.087	6.36

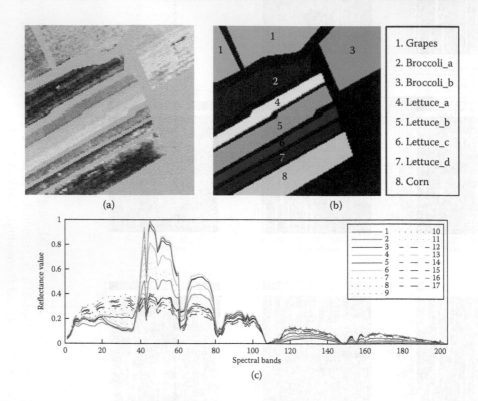

FIGURE 13.3
Salinas Valley image and endmembers' dictionary: (a) 5th PC of the Salinas Valley scene, (b) rough ground truth information for a part of the Salinas Valley scene under study, and (c) spectral signatures of the 17 endmembers of the utilized endmembers' dictionary.

13.4.2 Real data experiments

Herein, we test the performance of ALMSpLRU on a real land cover HSI, namely a part of the Salinas Valley vegetation scene acquired by AVIRIS sensor over Salinas Valley in California. Therein, they can be found in eight different vegetation species, namely grapes, brocolli_a, brocolli_b, lettuce_a, lettuce_b, lettuce_c, lettuce_d, and corn, as shown in Figure 13.3b. Salinas Valley HSI consists of $L = 204$ spectral bands (after excluding 20 noisy bands) and its spatial resolution is 3.7 m. Figure 13.3a shows the fifth principal component of the Salinas Valley image which provides a rough information about the formation of the vegetation species over the scene. The endmembers dictionary consists of 17 spectral signatures manually selected from the scene as in [21].

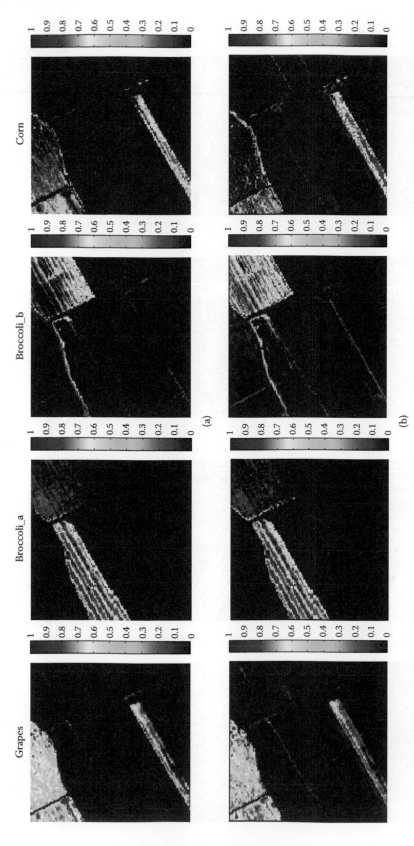

FIGURE 13.4

Abundance maps of Salinas Valley hyperspectral image: (a) ALMSpLRU and (b) CSUnSAL+.

(Continued)

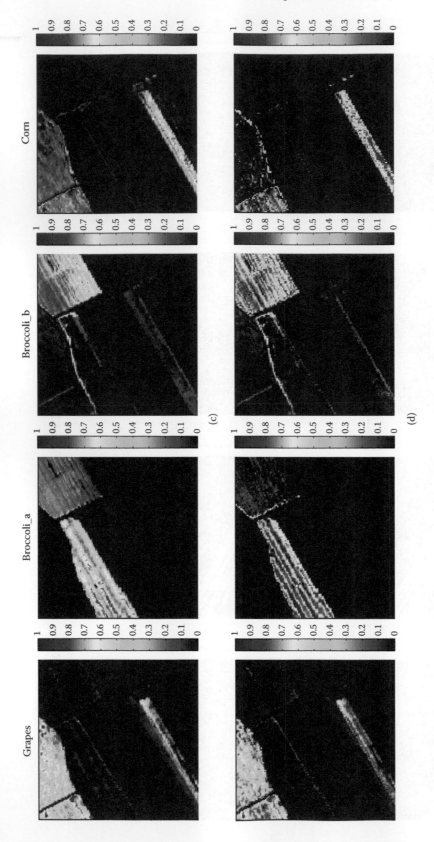

FIGURE 13.4 (Continued)
Abundance maps of Salinas Valley hyperspectral image: (c) MMV-ADMM and (d) BiICE.

Figure 13.4 illustrates the abundance maps obtained by ALMSpLRU, CSUnSAl+, MMV-ADMM, and BiICE for four different vegetation species, that is, grapes, broccoli_a, broccoli_b, and corn. It is easily observed that all the algorithms offer comparable results, with ALMSpLRU presenting maps closer to those corresponding to the first five principal components of the HSI provided in [21], especially when it comes to the abundance maps of grapes and broccoli_b. More concretely, ALMSpLRU detects in a more detailed way the distribution of broccoli_b in the upper corner of the image. Moreover, as it can be observed by visually inspecting Figures 13.3a and 13.4, it effectively eliminates the erroneous detection of those species. Hence, we may safely conclude that the proposed method provides enhanced abundance estimation results that are suitable for use in land cover mapping applications.

References

[1] W. Ma, J. Bioucas-Dias, T.-H. Chan, N. Gillis, P. Gader, A. Plaza, A. Ambikapathi, and C.-Y. Chi, A signal processing perspective on hyperspectral unmixing: Insights from remote sensing, *IEEE Signal Processing Magazine*, vol. 31, no. 1, pp. 67–81, 2014.

[2] J. Nascimento and J. Bioucas-Dias, Vertex component analysis: A fast algorithm to unmix hyperspectral data, *IEEE Transactions on Geoscience and Remote Sensing*, vol. 43, no. 4, pp. 898–910, 2005.

[3] J. Li and J. Bioucas-Dias, Minimum volume simplex analysis: A fast algorithm to unmix hyperspectral data, in *IEEE International Geoscience and Remote Sensing Symposium (IGARSS)*, vol. 3, pp. III – 250–III – 253, July 2008.

[4] J. Bioucas-Dias and M. Figueiredo, Alternating direction algorithms for constrained sparse regression: Application to hyperspectral unmixing, in *2nd Workshop on Hyperspectral Image and Signal Processing: Evolution in Remote Sensing (WHISPERS)*, June 2010, pp. 1–4.

[5] K. E. Themelis, A. A. Rontogiannis, and K. D. Koutroumbas, A novel hierarchical Bayesian approach for sparse semi-supervised hyperspectral unmixing, *IEEE Transactions on Signal Processing*, vol. 60, no. 2, pp. 585–599, 2012.

[6] K. E. Themelis, A. A. Rontogiannis, and K. D. Koutroumbas, Semi-supervised hyperspectral unmixing via the weighted lasso, in *IEEE International Conference on Acoustics, Speech and Signal Processing (ICASSP)*, March 2010, pp. 1194–1197.

[7] M.-D. Iordache, J. Bioucas-Dias, and A. Plaza, Collaborative sparse regression for hyperspectral unmixing, *IEEE Transactions on Geoscience and Remote Sensing*, vol. 52, no. 1, pp. 341–354, 2014.

[8] P. V. Giampouras, K. E. Themelis, A. A. Rontogiannis, and K. D. Koutroumbas, A variational Bayes algorithm for joint-sparse abundance estimation, in *6th Workshop on Hyperspectral Image and Signal Processing: Evolution in Remote Sensing (WHISPERS)*, June 2014.

[9] Q. Qu, N. Nasrabadi, and T. Tran, Abundance estimation for bilinear mixture models via joint sparse and low-rank representation, *IEEE Transactions on Geoscience and Remote Sensing*, vol. 52, no. 7, pp. 4404–4423, 2014.

[10] P. V. Giampouras, K. E. Themelis, A. A. Rontogiannis, and K. D. Koutroumbas, Simultaneously sparse and low-rank abundance matrix estimation for hyperspectral image unmixing, *IEEE Transactions on Geoscience and Remote Sensing*, vol. 54, no. 8, pp. 4775–4789, 2016.

[11] P. V. Giampouras, A. A. Rontogiannis, K. D. Koutroumbas, and K. E. Themelis, A sparse reduced-rank regression approach for hyperspectral image unmixing, in *3rd International Workshop on Compressed Sensing Theory and Its Applications to Radar, Sonar and Remote Sensing (CoSeRa), 2015*, June 2015, pp. 139–143.

[12] N. Srebro and A. Shraibman, *Rank, Trace-Norm and Max-Norm*. Berlin, Heidelberg: Springer Berlin Heidelberg, 2005, p. 545–560. [Online]. Available: http://dx.doi.org/ 10.1007/11503415_37

[13] M.-D. Iordache, J. Bioucas-Dias, and A. Plaza, Sparse unmixing of hyperspectral data, *IEEE Transactions on Geoscience and Remote Sensing*, vol. 49, no. 6, pp. 2014–2039, 2011.

[14] D. P. Bertsekas, *Nonlinear Programming*, Athena Scientific, Belmont, MA, 1999.

[15] H. Abou-Kandil, G. Freiling, G. Jank, and V. Ionescu. *Matrix Riccati Equation in Control and Systems Theory*, Basel, Boston, Berlin: Birkhäser Verlag. 2003.

[16] G. Golub, S. Nash, and C. Van Loan, A Hessenberg-Schur method for the problem $ax + xb = c$, *IEEE Transactions on Automatic Control*, vol. 24, no. 6, pp. 909–913, 1979.

[17] S. D. Conte, and C. deBoor, *Elementary Numerical Analysis: An Algorithmic Approach*, 2nd. edition, McGraw-Hill, New York, 1972.

[18] N. Parikh and S. Boyd, Proximal algorithms, *Foundations and Trends in Optimization*, vol. 1, no. 3, pp. 123–231, 2013.

[19] A. A. Rontogiannis, K. E. Themelis, and K. D. Koutroumbas, A fast variational Bayes algorithm for sparse semi-supervised unmixing of Omega/Mars express data, in *5th Workshop on Hyperspectral Image and Signal Processing: Evolution in Remote Sensing (WHISPERS)*, June 2013, pp. 974–978.

[20] R. N. Clark, G. A. Swayze, R. Wise, K. E. Livo, T. M. Hoefen, R. F. Kokaly, and S. J. Sutley, *USGS Digital Spectral Library*, 2007, http://speclab.cr.usgs.gov/spectral. lib06/ds231/datatable.html.

[21] E. Mylona, O. Sykioti, K. Koutroumbas, and A. A. Rontogiannis, Joint spectral unmixing and clustering for identifying homogeneous regions in hyperspectral images, in *IEEE International Conference Geoscience and Remote Sensing Symposium (IGARSS)*, July 2015, pp. 2409–1412.

14

Parallel Coded Aperture Method for Hyperspectral Compressive Sensing on GPU

Gabriel Mart´n, José Nascimento, and José Bioucas-Dias

CONTENTS

14.1 Introduction

Hyperspectral imaging instruments allow data collection in hundreds or even thousands of spectral bands (at different wavelength channels) for the same area on the surface of the Earth [1]. For instance, NASA is continuously gathering imagery data with instruments such as the Jet Propulsion Laboratory's Airborne Visible Infra-Red Imaging Spectrometer (AVIRIS), which is able to record the visible and near-infrared spectrum (wavelength region from 0.4 to 2.5 micrometers) of reflected light in an area of 2 to 12 kilometers wide and several kilometers long, using 224 spectral bands [2]. The resulting multidimensional data cube typically comprises several GBs per flight. As a result, the computational requirements needed to store, manage, and process these images are enormous [3].

Due to the extremely large volumes of data collected by imaging spectrometers, hyperspectral data compression has received considerable interest in recent years [4,5], mainly due to their capability to simplify the hardware and software requirements of the hyperspectral acquisition systems [6–9]. These data are usually acquired by a satellite or an airborne instrument and sent to a ground station on Earth for subsequent processing. Usually, the bandwidth connection between the satellite/airborne platform and the ground station is reduced, which limits the amount of data that can be transmitted. As a result, there is a clear need for (either lossless or lossy) hyperspectral data compression techniques that can be applied onboard the imaging instrument [10–12]. Furthermore, usually the computational resources onboard the imaging instrument are reduced, which limits the applicability of complex compression techniques onboard.

Compressive sensing (CS) aims to reduce the number of measurements needed to recover a signal of interest [13,14]. This is possible since CS measurements are obtained by performing random projections that destroy the typical correlation of the measurements while preserving the signal information coded in the random projected measurements. Exploiting the sparsity nature of the signal makes it possible to recover it with great accuracy from a much smaller number of CS measurements. In CS, the original data are inferred from the measurements by solving a convex optimization problem. A necessary condition to obtain good inferences is that original data admit a sparse or compressible representation in a given basis or frame. This means that most of the coefficients of the representation in that basis or frame are zero or small and, thus, the data can be well approximated with just a small number of large coefficients. It happens that hyperspectral images (HSI) are often highly compressible owing to a very high spatial and spectral correlation. Therefore, this imaging modality is a perfect candidate to apply the CS technology. Similarly to CS techniques, the Random Projections (RP) techniques aim at recovering the original signal from the random measurements. In CS and RP, the measurement process is based on performing random projections with known random vectors over the scene of interest. This process is very simple, and it may be implemented directly on the optics of the acquisition system or in the electronic hardware. This is a very light process from the computational point of view. However, the reconstruction process is more computationally demanding. In CS model, the reconstruction is performed in the ground station, where a variety of powerful computational resources may be available. This chapter is focused on the implementation of two reconstruction algorithms in graphics processing units (GPUs). This hardware is able to exploit data parallelism through a single instruction multiple data (SIMD) computing architecture, and provide unprecedented computational power.

This chapter describes the computationally efficient implementations of two different approaches for CS and RP on GPU: hyperspectral coded aperture (HYCA) algorithm for CS and spectral compressive acquisition (SpeCA) as a form of RP technique. This chapter shows the implementation of these algorithms with several optimizations for accelerating their computational performance while maintaining their accuracy. These parallel implementations exploit the GPU architecture at low level, using shared memory and coalesced access to memory.

The considered implementations are intercompared in the context of real hyperspectral imaging applications. NVidia GeForce GTX 980 and GTX TITAN platforms have been used to test the proposed implementations with both synthetic and real hyperspectral scenes. Our study reveals that the implementations on GPU can provide real-time performance. The implementations on GPUs have been carried out using NVidia CUDA and the cuBLAS library[*]. The cuFFT library[†] is only used on the parallel HYCA (P-HYCA) implementation, and the cuSOLVER library[‡] has been used only in the parallel SpeCA (P-SpeCA) implementation.

The remainder of the chapter is organized as follows. Section 14.2 describes the algorithms considered in this chapter. Section 14.3 describes the proposed GPU implementations. Section 14.4 presents an experimental evaluation of the proposed implementations in terms of both accuracy and parallel performance using synthetic and

---

[*]http://developer.nvidia.com/cuBLAS
[†]http://developer.nvidia.com/cuFFT
[‡]http://developer.nvidia.com/cuBLAS

real hyperspectral data sets on two GPU platforms. Finally, Section 14.5 concludes with some remarks and hints at plausible future research lines.

14.2   Methods description

In this section, HYCA and SpeCA methods are introduced. Next subsections describe each method.

14.2.1   HYCA—Hyperspectral coded aperture method

The original HYCA method for CS was developed in [15]. This approach compresses the data on the acquisition process, then the compressed signal is sent to Earth and stored in compressed form. Later, the original signal can be recovered by taking advantage of two key properties of HSIs: (i) the spectral vectors live systematically in low-dimensional subspaces [16] and (ii) the spectral bands present a high correlation in the spatial domain. The former property allows to represent the data vectors using a reduced set of spectral endmembers due to the mixing phenomenon and also exploits the high spatial correlation of the fractional abundances associated to the spectral endmembers.

Let $\mathbf{x}_i \in \mathbb{R}^{n_b}$, for $i = 1, \ldots, n_p$, denotes the $n_p := n_r \times n_c$ spectral vectors of an HSI, where n_r, n_c, and n_b denote, respectively, the number of rows, columns, and bands of the HSI, and $\mathbf{x} := [\mathbf{x}_1, \ldots, \mathbf{x}_{n_p}]^T$ (the operator $(\cdot)^T$ stands for transpose) denotes, in a vector format, the HSI. In order to perform the compression of the original signal $\mathbf{x}$, and as in [15], for each pixel $i \in \{1, \ldots, n_p\}$, a set of q inner products between $\mathbf{x}_i$ and samples of i.i.d. Gaussian random vectors is performed. The total number of measurements is therefore $q \times n_p$ yielding an undersampling factor of q/n_b. This measurement operation can be represented as a matrix multiplication

$$\mathbf{y} = \mathbf{A}\mathbf{x}, \tag{14.1}$$

where $\mathbf{A}$ is a block-diagonal matrix containing the matrices $\mathbf{A}_i \in \mathbb{R}^{q \times n_b}$ acting on the pixel $\mathbf{x}_i$, for $i \in \{1, \ldots, n_p\}$. For reasons linked with a) the computational management of the sampling process and b) the spatial correlation length of HSIs (see [15] for more details), matrices $\mathbf{A}_i$ are organized into spatial windows of size $ws \times ws = m$. Each window contains the same set of matrices. All windows have the same spatial configuration of $\mathbf{H}_j$, for $j = 1, \ldots, m$.

As referred before, HYCA method also takes advantage of the fact that the hyperspectral vectors $\mathbf{x}_i$ generally live in a low-dimensional subspace. This fact can be modeled by $\mathbf{x}_i = \mathbf{E}\mathbf{z}_i$, where $\mathbf{E} \in \mathbb{R}^{n_b \times p}$ is a matrix whose columns span the signal subspace and $\mathbf{z}_i \in \mathbb{R}^p$ denotes the vector of coordinates with respect to the columns of $\mathbf{E}$. Defining $\mathbf{z} := [\mathbf{z}_1^T, \ldots, \mathbf{z}_{n_p}^T]^T$, for $i = 1, \ldots, n_p$, we have

$$\mathbf{x} = (\mathbf{I} \otimes \mathbf{E})\mathbf{z}, \tag{14.2}$$

where $\mathbf{E} \in \mathbb{R}^{n_b \times p}$, with $p \ll n_b$, $\otimes$ stands for Kronecker product, and $\mathbf{I}$ is the identity matrix of suitable size. In this work, we assume that the linear mixing model is a good approximation to the spectral vectors $\mathbf{x}_i$ [17] and, therefore, matrix $\mathbf{E}$ contains in its columns the spectral signatures of the p endmembers. In this work, VCA algorithm [18] is used to infer $\mathbf{E}$. Since $\mathbf{E}$ is the mixing matrix, $\mathbf{z}$ contains the fractional abundances associated to each pixel.

Let $\mathbf{K} = \mathbf{A}(\mathbf{I} \otimes \mathbf{E})$. If matrices $\mathbf{E}$ and $\mathbf{A}$ are available, one can formulate the estimation of $\mathbf{z}$ from $(q \times n_x)$-dimensional vector of measurements. Since the fractional abundances in HSIs exhibit a high spatial correlation, we exploit this feature for estimating $\mathbf{z}$ using the following optimization problem:

$$\min_{\mathbf{z}} \ (1/2)\|\mathbf{y} - \mathbf{Kz}\|^2 + \lambda_{TV}\mathrm{TV}(\mathbf{z}) \tag{14.3}$$

$$\text{subject to: } \mathbf{z} \geq \mathbf{0},$$

where $\mathrm{TV}(\mathbf{z})$ stands for the sum of nonisotropic total variations (TVs) [19,20] associated with $\mathbf{z}$, one per image of abundance. Defined as

$$\mathrm{TV}(\mathbf{z}) := \boldsymbol{\phi}(\mathbf{Dz}),$$

where $\mathbf{D} := [\mathbf{D}_h^T \mathbf{D}_v^T]^T$, $\mathbf{D}_h, \mathbf{D}_v$ computes the horizontal and vertical backward differences, assuming a cyclic boundary, and

$$\boldsymbol{\phi}(\boldsymbol{\vartheta}) := \sum_{i=1}^{p} \sum_{j=1}^{n_p} \|\boldsymbol{\vartheta}[i,j]\|,$$

with $\boldsymbol{\vartheta} := [\boldsymbol{\vartheta}_h^T, \boldsymbol{\vartheta}_v^T]$, $\boldsymbol{\vartheta}_h$ standing for horizontal differences and $\boldsymbol{\vartheta}_v$ standing for vertical differences. The TV regularizer promotes piecewise abundance images $\mathbf{z}$. Therefore, the minimization of Equation 14.3 aims at finding a solution which is a compromise between the fidelity to the measured data, enforced by the quadratic term $(1/2)\|\mathbf{y} - \mathbf{Kz}\|^2$, and the properties enforced by the TV regularizer, that is, piecewise smooth image of abundances. The relative weight between the two characteristics of the solution is set the regularization paremeter $\lambda_{TV} > 0$.

Algorithm 14.1: Pseudocode of HYCA algorithm.

    1. Initialization: set $k = 0$, choose $\mu > 0$, $\mathbf{E}$, $\mathbf{z}^{(0)}$,
$$\mathbf{v}_1^{(0)}, \mathbf{v}_2^{(0)}, \mathbf{v}_3^{(0)}, \mathbf{v}_4^{(0)}, \mathbf{d}_1^{(0)}, \mathbf{d}_2^{(0)}, \mathbf{d}_3^{(0)}, \mathbf{d}_4^{(0)}$$

    2. repeat:

        3. $\mathrm{aux} = \mathbf{v}_1^{(k)} + \mathbf{d}_1^{(k)} + \mathbf{v}_2^{(k)} + \mathbf{d}_2^{(k)}$
$$\mathbf{z}^{(k+1)} \leftarrow (\mathbf{D}^T\mathbf{D} + 2\mathbf{I})^{-1} \times \left(\mathrm{aux} + \mathbf{D}_h^T(\mathbf{v}_3^{(k)} + \mathbf{d}_3^{(k)}) + \mathbf{D}_v^T(\mathbf{v}_4^{(k)} + \mathbf{d}_4^{(k)})\right)$$

        4. $\mathbf{v}_1^{(k+1)} \leftarrow (\mathbf{K}^T\mathbf{K} + \mu\mathbf{I})^{-1} \times \left(\mathbf{K}^T\mathbf{y} + \mu(\mathbf{z}^{(k+1)} - \mathbf{d}_1^{(k)})\right)$

        5. $\mathbf{v}_2^{(k+1)} \leftarrow \max\left(0, \mathbf{z}^{(k+1)} - \mathbf{d}_2^{(k)}\right)$

        6. $\mathbf{v}_3^{(k+1)} \leftarrow \mathrm{soft}\left(\mathbf{D}_h(\mathbf{z}^{(k+1)}) - \mathbf{d}_3^{(k)}, \lambda_{TV}/\mu\right)$

        7. $\mathbf{v}_4^{(k+1)} \leftarrow \mathrm{soft}\left(\mathbf{D}_v(\mathbf{z}^{(k+1)}) - \mathbf{d}_4^{(k)}, \lambda_{TV}/\mu\right)$

        8. Update Lagrange multipliers:
$$\mathbf{d}_1^{(k+1)} \leftarrow \mathbf{d}_1^{(k)} - \mathbf{z}^{(k+1)} + \mathbf{v}_1^{(k+1)}$$
$$\mathbf{d}_2^{(k+1)} \leftarrow \mathbf{d}_2^{(k)} - \mathbf{z}^{(k+1)} + \mathbf{v}_2^{(k+1)}$$
$$\mathbf{d}_3^{(k+1)} \leftarrow \mathbf{d}_3^{(k)} - \mathbf{D}_h\mathbf{z}^{(k+1)} + \mathbf{v}_3^{(k+1)}$$
$$\mathbf{d}_4^{(k+1)} \leftarrow \mathbf{d}_4^{(k)} - \mathbf{D}_v\mathbf{z}^{(k+1)} + \mathbf{v}_4^{(k+1)}$$

        9. Update iteration: $k \leftarrow k + 1$

    10. until $k = \mathrm{MAX\_ITERATIONS}$

    11. Reconstruction $\hat{\mathbf{x}} = (\mathbf{I} \otimes \mathbf{E})\mathbf{z}^k$

To solve the convex optimization problem in Equation 14.3, a methodology closely related with the one presented in [21] is adopted. The solution of this problem is obtained by an instance of the alternating direction method of multipliers (ADMM) [22], which decomposes very hard problems into a cyclic sequence of simpler problems. With this in mind, an equivalent way of writing the optimization problem in Equation 14.3 is

$$\min_{\mathbf{z}} \frac{1}{2}\|\mathbf{y} - \mathbf{Kz}\|^2 + \lambda_{TV}\, \boldsymbol{\phi}(\mathbf{Dz}) + \iota_{R+}(\mathbf{z}), \tag{14.4}$$

where $\iota_{R+}(\mathbf{z}) = \sum_{i=1}^{pn_p} \iota_{R+}(\mathbf{z}_i)$ is the indicator function ($\mathbf{z}_i$ represents the ith element of $\mathbf{z}$ and $\iota_{R+}(\mathbf{z}_i)$ is zero if $\mathbf{z}_i$ belongs to the nonnegative orthant and $+\infty$ otherwise). Given the objective function in Equation 14.4, we can write the following equivalent formulation:

$$\min_{\mathbf{z},\mathbf{v}_1,\mathbf{v}_2,\mathbf{v}_3,\mathbf{v}_4} \quad \frac{1}{2}\|\mathbf{y} - \mathbf{Kv}_1\|^2 + \iota_{R+}(\mathbf{v}_2) + \lambda_{TV}\, \boldsymbol{\phi}(\mathbf{Dz})$$

$$\text{subject to} \quad \mathbf{v}_1 = \mathbf{z}$$

$$\mathbf{v}_2 = \mathbf{z} \tag{14.5}$$

$$(\mathbf{v}_3, \mathbf{v}_4) = \mathbf{Dz}.$$

Algorithm 14.1 shows the pseudocode of the HYCA algorithm to solve the problem in Equation 14.5 and how to reconstruct the data using Equation 14.2.

14.2.2 SpeCA—Spectral compressive acquisition method

Let's define the data matrix $\mathbf{X} \in \mathbb{R}^{n \times n_b}$ that holds the spectral vectors $\mathbf{x}_i$ by columns, with $n := n_r n_c$. Due to the fact that spectral vectors generally live in a subspace, we may write $\mathbf{X} = \mathbf{EZ}$, where $\mathbf{E}$ is a matrix whose columns span the signal subspace and $\mathbf{Z} \in \mathbb{R}^{n \times p}$ holds the coefficients $\mathbf{z}_i$ by columns. The measurement process in the case of SpeCA is different than in the case of HYCA. Let's define the measurement matrices $\mathbf{C} \in \mathbb{R}^{m_c \times n_b}$ and $\mathbf{B}_k \in \mathbb{R}^{m_b \times n_b}$ whose elements are independently drawn at random from a Gaussian $\mathcal{N}(0,1)$ distribution. Matrix $\mathbf{C}$ acts on the HSI spectral domain generating m_c measurements per pixel. The measurements obtained with matrix $\mathbf{C}$ are $\mathbf{Y}_c = \mathbf{CX}$. The measurements obtained with the matrices $\mathbf{B}_k$ are $\mathbf{Y}_b = [\mathbf{y}_{b,1}, \ldots, \mathbf{y}_{b,n_v}] \in \mathbb{R}^{m_b \times n_v}$, where $\mathbf{y}_{b,k} = \mathbf{B}_k \mathbf{x}_{i_k} \in \mathbb{R}^{m_b}$, for $k = 1, \ldots, n_v$. Matrices $\mathbf{B}_k$ act on pixels with indexes $i_1, i_2, \ldots, i_{n_v}$ randomly chosen from the set $1, \ldots, n$, producing m_b measurements per sample pixel. Therefore, the total number of measurements per block of n pixels is $m_c n + m_b n_v$ yielding a measurement rate per pixel of $m_c + m_b n_v/n$.

The SpeCA recovery algorithm [23] starts by finding a factorization $\mathbf{Y}_c = \mathbf{FZ}$, where $\mathbf{F} = \mathbf{CE} \in \mathbb{R}^{m_c \times p}$, $\mathbf{C} \in \mathbb{R}^{m_c \times n_b}$ is the measurement matrix defined above, and $\mathbf{E} \in \mathbb{R}^{n_b \times p}$ holds a basis for the subspace $\mathcal{S}_p$. In this work, we have considered VCA endmember extraction algorithm to find $\mathbf{F}$ matrix [18], parameterized with p endmembers.

Assuming that $\mathbf{F}$ is full column rank, then the solution of $\mathbf{Y}_c = \mathbf{FZ}$ with respect to $\mathbf{Z}$ is unique and given by $\mathbf{Z} = (\mathbf{F}^T\mathbf{F})^{-1}\mathbf{F}^T\mathbf{Y}_c$. Unfortunately, we cannot recover $\mathbf{E}$ from the equation $\mathbf{F} = \mathbf{CE}$ because the underlying systems of equations are undetermined. To obtain $\mathbf{E}$, we use the measurements $\mathbf{Y}_b$ jointly with $\mathbf{Z}$.

Using the properties of the *vec* and the *Kronecker* operators, we have

$$\mathbf{y}_{b,k} = \mathbf{B}_k \mathbf{x}_{i_k} \tag{14.6}$$

$$= \mathbf{B}_k \mathbf{E} \mathbf{z}_{i_k} \tag{14.7}$$

$$= (\mathbf{z}_{i_k}^T \otimes \mathbf{B}_k)\text{vec}(\mathbf{E}), \quad k = 1 \ldots n_v. \tag{14.8}$$

By stacking all the above equations, we obtain the linear system

$$\mathbf{D}\operatorname{vec}(\mathbf{E}) = \mathbf{y}_b, \tag{14.9}$$

with $\mathbf{y}_b = \operatorname{vec}(\mathbf{Y}_b)$ and $\mathbf{D} = [(\widehat{\mathbf{z}}_{i_1} \otimes \mathbf{B}_1^T), \dots, (\widehat{\mathbf{z}}_{i_{n_v}} \otimes \mathbf{B}_{n_v}^T)]^T$.

Algorithm 14.2 shows the SpeCA pseudocode. Lines 2 and 3 implement the measurements and lines 6 to 12 carry out the HSI reconstruction. Symbol $(\cdot)^\dagger$ in line 7 denotes the Moore—Penrose pseudoinverse.

Algorithm 14.2: Spectral Compressive Acquisition (SpeCA).

Input: $\mathbf{C} \in \mathbb{R}^{m_c \times n_b}$, $\mathbf{B}_k \in \mathbb{R}^{m_b \times n_b}$, $k = 1, \dots, n_v$
Input: p
Output: $\widehat{\mathbf{X}}$, $\widehat{\mathbf{E}}$, $\widehat{\mathbf{Z}}$

1 **begin** *Measurements*
2 $\quad$ $\mathbf{Y}_c := \mathbf{C}\mathbf{X}$;
3 $\quad$ $\mathbf{Y}_b := [\mathbf{y}_{b,1}, \dots, \mathbf{y}_{b,n_v}]$, $\quad \mathbf{y}_{b,k} := \mathbf{B}_k \mathbf{x}_{i_k}$, $\quad k = 1, \dots, n_v$;
4 **end**
5 **begin** *HSI Reconstruction*
6 $\quad$ $\mathbf{F} := VCA(\mathbf{Y}_c, p)$
7 $\quad$ $\widehat{\mathbf{Z}} := \mathbf{F}^\dagger \mathbf{Y}_c$
8 $\quad$ $\mathbf{D} := [(\widehat{\mathbf{z}}_{i_1} \otimes \mathbf{B}_1^T), \dots, (\widehat{\mathbf{z}}_{i_{n_v}} \otimes \mathbf{B}_{n_v}^T)]^T$
9 $\quad$ $\operatorname{vec}(\widehat{\mathbf{M}}) := \operatorname{solution}\{\mathbf{D}\operatorname{vec}(\mathbf{E}) = \operatorname{vec}(\mathbf{Y}_b)\}$
10 $\quad$ $\widehat{\mathbf{E}} := \operatorname{vec}^{-1}(\operatorname{vec}(\widehat{\mathbf{E}}))$
11 $\quad$ $\widehat{\mathbf{X}} := \widehat{\mathbf{E}}\widehat{\mathbf{Z}}$
12 **end**

14.3 GPU implementations

GPUs can be abstracted as an array of highly threaded streaming multiprocessors (SMs), where each multiprocessor is characterized by an SIMD architecture; that is, in each clock cycle each processor executes the same instruction while operating on multiple data streams. Each SM has a number of streaming processors that share a control logic and instruction cache and have access to a local shared memory and to local cache memories in the multiprocessor, while the multiprocessors have access to the global GPU (device) memory. Figure 14.1 presents a typical architecture and the data flow communication between CPU and GPU.

The algorithms are constructed by chaining the so-called *kernels* which operate on entire streams and which are executed by a multiprocessor, taking one or more streams as inputs and producing one or more streams as outputs. Thereby, data-level parallelism is exposed to hardware, and kernels can be concurrently applied without any sort of synchronization. The kernels can perform a kind of batch processing arranged in the form of a grid of blocks where each block is composed by a group of threads that share data efficiently through the shared local memory and synchronize their execution for coordinating accesses to memory. As a result, there are different levels of memory in the GPU for the thread, block, and grid concepts. There is also a maximum number of threads that a block can contain (depending on the GPU model); however, the number of threads that can be concurrently executed is much larger due to the fact that several blocks executed by the same kernel can be managed concurrently. With the above ideas in mind, the proposed implementations for HYCA and SpeCA are detailed in the following sections.

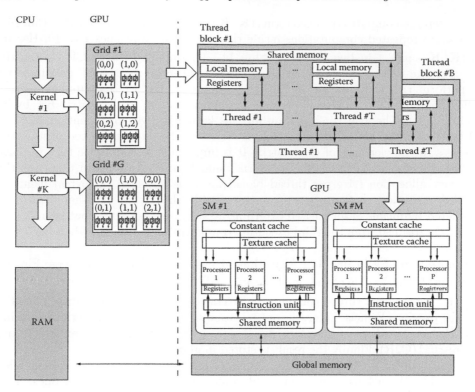

FIGURE 14.1
Typical NVidia GPU architecture, computation, and data transfer flow from/to CPU.

14.3.1 P-HYCA implementation

The implementation of P-HYCA algorithm starts with an initialization step. The `cuBLAS` and `cuFFT` libraries are first initialized. After that, the compressed HSI is loaded band by band from the hard disk to the main memory of the GPU. This arrangement intends to access consecutive pixels in the same wavelength in parallel by the processing kernels (coalesced accesses to memory). This means that the ith thread of a block will access the ith pixel for a given wavelength. This technique is used to maximize global memory and bandwidth and to minimize the number of bus transactions.

Once the original image is loaded in the global memory, the kernel `Compute_compression` performs the projections between the random vectors and the image pixels in order to compress the data. For this purpose, each thread will compute the multiplication of the matrix $\mathbf{H}_i$ with its corresponding pixel, so that the total number of threads will be equal to the number of pixels in the data set. The grid configuration of this kernel in the GPU will contain Num_b blocks with the maximum number of threads supported by the architecture (1024 for the considered GPU). Thus, the number of blocks Num_b will be given by the expression:

$$Num_b = \left\lceil \frac{n_p}{1024} \right\rceil. \tag{14.10}$$

Due to the fact that we have a small set of $\mathbf{H}_i$ matrices which are systematically multiplied by the image pixels located inside spatial windows, the compression matrices $\mathbf{H}_i$ are stored in shared memory in order to optimize the memory access to the $\mathbf{H}_i$ matrices. At the end of this process, the threads will store the compressed measurements $\mathbf{y}$ in the global memory.

The next step pre-computes the fixed terms $\left(\mathbf{K}^T\mathbf{K} + \mu\mathbf{I}\right)^{-1}$ and $\left(\mathbf{K}^T\mathbf{y}\right)$ of Algorithm 14.1 in order to avoid repeated computations inside the main loop from lines 2 to 10. Herein, the term $\left(\mathbf{K}^T\mathbf{K} + \mu\mathbf{I}\right)^{-1}$ in line 4 of Algorithm 14.1 is computed in the CPU using the LAPACK[*] package due to the fact that the size of these matrices is small and it is not worth to perform this computation in the GPU. However, $\left(\mathbf{K}^T\mathbf{y}\right)$ is computed using a kernel called Compute_KTY, which will perform the multiplication of the matrix $\mathbf{K}^T$ by its corresponding pixel. The grid configuration for this kernel is given by Equation 14.10 as explained before. In this kernel, the matrix $\mathbf{K}$ is stored in shared memory to optimize the memory access to the elements of this matrix. It is important to emphasize that we have declared shared memory dynamically in all kernel launch configurations. In this case, the shared memory allocation (size per thread block) must be specified (in bytes) using an optional third execution configuration parameter. Inside the kernel, the shared memory array is declared by means of an unsized extern array syntax, __extern__ double s[]. The size is simply determined from the third execution configuration parameter when the kernel is launched.

The optimization of $\mathbf{z}$ in line 3 of Algorithm 14.1 is carried out in two steps. First, a kernel computes the right side of the equation: $\mathbf{v}_1^{(k)} + \mathbf{d}_1^{(k)} + \mathbf{v}_2^{(k)} + \mathbf{d}_2^{(k)} + \mathbf{D}_h^T(\mathbf{v}_3^{(k)} + \mathbf{d}_3^{(k)}) + \mathbf{D}_v^T(\mathbf{v}_4^{(k)} + \mathbf{d}_4^{(k)})$; here each thread computes one element and the grid configuration is the same than the previous kernels. Later, the optimization with respect to $\mathbf{z}$ is performed using the cuFFT. Herein, two Fourier transform types were used: real-to-complex (R2C) and complex-to-real(C2R). Finally the result is stored in global memory.

The optimization of $\mathbf{v}_1$ in line 4 of Algorithm 14.1 is carried out with a kernel called Optimize_v1. This kernel uses the same grid configuration as the previous ones. This kernel also makes use of the shared memory to optimize the memory access to the matrix $\left(\mathbf{K}^T\mathbf{K} + \mu\mathbf{I}\right)^{-1}$, which was pre-computed before. The resulting $\mathbf{v}_1$ is stored in the global memory.

Line 5 in Algorithm 14.1 is carried out with a kernel called Optimize_v2, which computes the maximum between 0 and $\mathbf{z}^{(k+1)} - \mathbf{d}_2^{(k)}$. This kernel uses as many threads as the number of elements of the vector (n_p), with the same grid configuration as the previous kernels.

The optimization of $\mathbf{v}_3$ is carried out jointly by a single kernel called Optimize_v3_v4, which computes the lines 6 and 7 in Algorithm 14.1. This kernel uses the same configuration as the previous kernels with as many threads as the image pixels. In this kernel, each thread computes the horizontal and vertical differences of one element and performs the soft function for the corresponding element, where soft$(\cdot, \tau)$ denotes the application of the soft-threshold function $\mathbf{b} \mapsto \mathbf{b}\frac{\max\{\|\mathbf{b}\|_2 - \tau, 0\}}{\max\{\|\mathbf{b}\|_2 - \tau, 0\} + \tau}$.

Finally, Lagrange multipliers update is computed with two kernels called Compute_d12 and Compute_d34 which, respectively, compute the update of the variables $\mathbf{d}_1, \mathbf{d}_2$ and $\mathbf{d}_3, \mathbf{d}_4$.

The algorithm repeats this process until a number of iterations k is reached. Once the estimated fractional abundances $\mathbf{z}$ are computed, the algorithm reconstructs the original hyperspectral data set multiplying by the endmember matrix. This process is performed using the cuBLAS library. Specifically, the cublasSgemm function of cuBLAS was used to reconstruct the image $\hat{\mathbf{x}}$.

14.3.2   SpeCA implementation

In this subsection, we describe the parallel implementations of SpeCA on GPUs. We can divide SpeCA in two parts, in one hand we have the coder-side of the algorithm which computes the compressed measurements and on the other hand we have the decoder-side

---

[*]http://www.netlib.org/lapack/

which recovers the original data from the compressed measurements. In order to implement the coder-side, we have created one kernel with as many threads as the number of pixels, so that each thread computes the compressed measurement for each pixel. On the other hand, the decoder-side of the implementation starts applying the VCA algorithm to the $\mathbf{Y}_c$ measurements. In this work, we have used the parallel version of VCA for GPUs implemented in [24]. Then, the pseudoinverse of the matrix $\mathbf{F}$ is computed in the CPU, due to the fact that this matrix is very small and thus the GPU occupancy for this is very low, with the result that it is slower in GPU than in the CPU. After that, the computation of $\widehat{\mathbf{Z}}$ is performed in GPU using the *CublasDgemm* to perform the matrix multiplication between $\mathbf{F}^\dagger$ and $\mathbf{Y}_c$. The next step is to build the $\mathbf{D}$ matrix, we have implemented in two kernels. The first one is in charge of coping the abundances fractions associated with the random pixels $i_1, \dots, i_{n_v}$ from the matrix $\widehat{\mathbf{Z}}$ to another structure in memory in which these elements can be accessed in a coalesced way. The second one performs the Kronecker products between the abundances and the $\mathbf{B}_k$ random vectors. This second kernel uses $p \times n_b$ blocks and the maximum number of threads supported by the architecture. Each block computes one column of the matrix $\mathbf{D}$. Once the matrix $\mathbf{D}$ has been built, the solution of the linear system of equations is performed using the functions *cusolverDnSgeqrf* and *cusolverDnSormqr* from *cuSolver* library, which perform a QR factorization of the matrix $\mathbf{D}$ and solve the linear system of equations. Finally, the original data are recovered using again *CublasDgemm* to multiply the matrices $\widehat{\mathbf{E}}$ and $\widehat{\mathbf{Z}}$.

14.4 Experimental results

14.4.1 Hyperspectral image data

The experiments are carried out using two HSIs. The synthetic data set used in our experiments was generated from spectral signatures randomly selected from the United States Geological Survey (USGS)[*]. The image consists of a set of 30 signatures from the USGS library, and it is generated using the procedure described in [25] to simulate natural spatial patterns, composed by a total size of 600×512 pixels (275 MB). Figure 14.2 displays a grayscale composition and three examples of ground truth abundance maps from the simulated image.

The real HSI considered in experiments is the well-known AVIRIS Cuprite scene, available online in reflectance units after atmospheric correction[†]. This scene

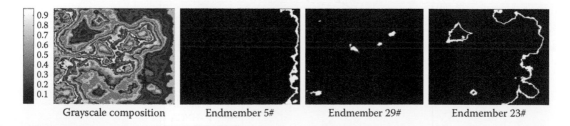

| Grayscale composition | Endmember 5# | Endmember 29# | Endmember 23# |

FIGURE 14.2
Grayscale composition and three examples of ground truth abundance maps of endmembers in the synthetic_2 hyperspectral data.

---

[*] http://speclab.cr.usgs.gov
[†] http://aviris.jpl.nasa.gov

has been widely used to validate the performance of endmember extraction algorithms. The portion used in experiments corresponds to a 250 × 190 pixels subset of the sector labeled as f970619t01p02_r02_sc03.a.rfl in the online data, which comprises 188 spectral bands in the range from 400 to 2500 nm and a total size of around 36 MB. Water absorption bands as well as bands with low signal-to-noise ratio (SNR) were removed prior to the analysis. The site is well understood mineralogically, and has several exposed minerals of interest, including *Alunite*, *Buddingtonite*, *Calcite*, *Kaolinite*, and *Muscovite*. For this subimage, the number of endmembers were estimated by Hysime method [16], and their signatures were extracted in a very fast way through VCA algorithm [18].

14.4.2   Analysis of accuracy

In order to evaluate the accuracy of the proposed implementations, the peak signal-to-noise ratio (PSNR) between the original image and the reconstructed image (after data compression and decompression) has been adopted as performance indicator, which is given by

$$\text{PSNR} = 10 \log_{10} \frac{\max(\mathbf{x})^2 \times n_p \times n_b}{||\widehat{\mathbf{x}} - \mathbf{x}||_F^2}, \tag{14.11}$$

where $\mathbf{x}$ and $\widehat{\mathbf{x}}$ denote the original and reconstructed HSIs, respectively, and $||.||_F$ denotes the Frobenius norm.

With the aim of evaluating the proposed implementations at different measurement rates, we have defined the number of measurements q and m_c in a range from 5 to 29 with a step size of 2 and the number of measurements $m_b = 1$ and $n_v = 20{,}000$ in the case of SpeCA. Thus, the measurements rates vary in a range from 0.02 to 0.16. The considered $ws = 2 \times 2$ in the case of HYCA, while the other parameters such as λ_{TV} were empirically set for optimal performance.

Figure 14.3 shows the PSNR achieved for the proposed implementations as a function of the measurement rates. This experiment shows that P-HYCA and P-SpeCA provide very

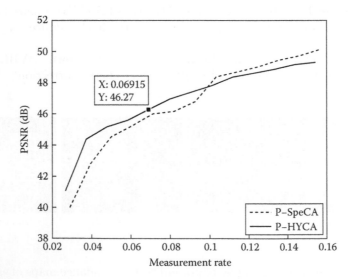

FIGURE 14.3
PSNR between the original and the reconstructed images for P-HYCA and for P-SpeCA as a function of different measurements rates for the Cuprite data set.

good results taking into account the very low measurement rates. P-HYCA provides better results when the measurement rate is lower, while SpeCA provides better results when the measurement rate is higher. This is expected as P-HYCA uses the ground truth endmembers which are specially helpful in a low measurement rate scenario, where P-SpeCA is not able to estimate them accurately, while the more the number of measurements the better P-SpeCA may estimate the endmembers, and therefore the results provided by P-SpeCA are similar or even better to those provided by P-HYCA.

14.4.3   Analysis of parallel performance

The proposed parallel versions have been tested on two different GPUs:

- The first GPU (denoted hereinafter as GPU1) is the *NVidia*^{TM} GeForce GTX 980 which features 2048 processor cores operating at 1.33 GHz, total dedicated memory of 4096 MB, 7.0 Gbps memory clock (with 256-bit GDDR5 interface), and memory bandwidth of 224 GB/s. This GPU is connected to a multi-core Intel i7-4790 CPU at 3.6 GHz with 4 physical cores and 32 GB of DDR3 RAM memory.

- The second GPU (denoted hereinafter as GPU2) used is the *NVidia*^{TM} GeForce GTX TITAN*, which features 2688 processor cores operating at 876 MHz, total dedicated memory of 6144 MB, 6.0 Gbps memory clock (with 384-bit GDDR5 interface), and memory bandwidth of 288.4 GB/s. This GPU is connected to a multi-core Intel i7-4770K CPU at 3.50 GHz with 4 physical cores and 32 GB of DDR3 RAM memory.

Before describing the parallel algorithm performance results, it is important to emphasize that the GPU versions provide exactly the same results as the serial versions of the implemented algorithms, using the gcc-4.8.2 (gnu compiler default) with optimization flags -O3 (for the single-core version) to exploit data locality and avoid redundant computations. Note that for the serial HYCA implementation, the 3.3.4 FFTW library† version has been used. Hence, the only difference between the serial and parallel versions is the time they need to complete their calculations. The serial algorithms were executed in one of the available CPU cores. For each experiment, 10 runs were performed and the mean values were reported (these times were always very similar, with differences on the order of a few milliseconds).

Tables 14.1 and 14.2 show the execution time of each part of the algorithms P-HYCA and P-SpeCA, respectivley, for the synthetic data set. The number of measurements was $q = p = 30$ in the case of P-HYCA and $m_c = p = 30$, $m_b = 1$, and $n_v = 20,000$ in the case of SpeCA. These number of measurements were high enough to provide a very good reconstruction result with similar accuracy in both cases.

These results report speedups higher than $68\times$, achieved on the GPU2 device in the case of P-HYCA and higher than $26\times$, achieved on the GPU2 device in the case of P-SpeCA. Notice that the speedup results for GPU2 are better than GPU1 because GPU2 has more processing cores with a higher memory bandwidth.

It should be also noted that the cross-track line scan time in AVIRIS, a push-broom instrument [2], is quite fast (8.3 milliseconds to collect 512 full pixel vectors). This introduces the need to process the considered scene in less than 4.980 seconds, in order to achieve real-time performance. As shown in Tables 14.1 and 14.2, the scene can be processed in real-time using method P-SpeCA on both GPU devices, while P-HYCA method could process the data set in near real-time.

---

*http://www.geforce.com/hardware/desktop-gpus/geforce-gtx-titan/specifications
†http://www.fftw.org/#documentation

TABLE 14.1
HYCA and P-HYCA Processing Times (in *seconds*) and Speedups for Two Different
Platforms with Synthetic Data Sets Experiment ($q = p = 30$)

HYCA	CPU	GPU1	GPU2
RAM $\rightarrow$ *GlobalMem.*	–	0.0448	0.0361
Compute compression	9.3532	0.0647	0.0577
RAM $\leftarrow$ *GlobalMem.*	–	0.0172	0.0140
Total time	9.3532	0.1267	0.1078
Speedup (CPU time/GPU time)	–	73.830×	86.788×
RAM $\rightarrow$ *GlobalMem.*	–	0.0123	0.0099
$(\mathbf{K^TK + mI})^{-1}$	0.0032	0.0034	0.0031
RAM $\rightarrow$ *GlobalMem.*	–	<0.0001	<0.0001
$(\mathbf{K}^T\mathbf{y})$	0.5933	0.0130	0.0084
Compute $\mathbf{z}$ (line 3)	140.694	1.723	1.4856
Compute $\mathbf{v}_1$ (line 4)	122.102	2.9557	1.9131
Compute $\mathbf{v}_2$ (line 5)	3.5172	0.2246	0.1494
Compute $\mathbf{v}_3$, $\mathbf{v}_4$ (lines 6–7)	29.245	0.4532	0.3323
Compute $\mathbf{d}$ (line 8)	82.1894	1.3446	0.8886
Reconstruction (line 11)	0.4229	0.0053	0.00043
RAM $\leftarrow$ *GlobalMem.*	–	0.0616	0.0508
Total time	378.767	7.6415	5.5429
Speedup (CPU/GPU time)	–	49.567×	68.334×

TABLE 14.2
SpeCA and P-SpeCA Processing Times (in *seconds*) and Speedups for Two Different
Platforms with Synthetic Data Sets Experiment ($m_c = 30$, $m_b = 1$, and $n_v = 20,000$)

SpeCA	CPU	GPU1	GPU2
RAM $\rightarrow$ *GlobalMem.*	–	0.1036	0.094
Random Projections	3.014	0.0564	0.051
RAM $\leftarrow$ *GlobalMem.*	–	0.0103	0.006
Total time	3.0144	0.1703	0.151
Speedup (CPU time/GPU time)	–	17.697×	19.963×
RAM $\rightarrow$ *GlobalMem.*	–	0.0612	0.065
VCA	3.7752	0.6384	0.5523
$(\mathbf{F}^T\mathbf{F})^{-1}\mathbf{F}^T$	<0.001	<0.001	<0.001
$\mathbf{F}^T\mathbf{Y_c}$	0.0881	0.001	0.002
Compute $\mathbf{D}$	1.040	0.0139	0.012
Solve QR	78.377	3.0538	2.5490
Reconstruction $\widehat{\mathbf{E}}\widehat{\mathbf{Z}}$	0.579	0.0046	0.0040
RAM $\leftarrow$ *GlobalMem.*	–	0.0615	0.048
Total time	86.875	3.877	3.263
Speedup	–	22.407×	26.624×

14.5   Conclusions and future research lines

In this work, we have developed computationally efficient implementations of HYCA and
SpeCA methods for hyperspectral CS and reconstruction from random projections on

GPU platforms. The significant speedups reported in the experiments are expected to bridge the gap toward real-time CS of hyperspectral data sets, which is a highly desirable requirement for many remote sensing applications. The performance of the proposed implementations has been evaluated (in terms of the quality of the solutions provided and their parallel performance) using a real data set collected by the AVIRIS instrument and synthetic scenarios. The experimental results reported in this chapter indicate that remotely sensed hyperspectral imaging can greatly benefit from the development of efficient implementations of CS algorithms in specialized hardware devices for better exploitation of high-dimensional data sets.

Although the results reported in this chapter are very encouraging, in future work we will continue exploring additional strategies for optimization, such as splitting the original HSI into subimages and applying a multi-GPU implementation to each of them. We are also investigating the use of OpenCL as a computing standard for multi-core architectures. Moreover, other high-performance computing architectures such as digital signal processors (DSPs) or field programmable gate arrays (FPGAs) will be also explored due to their capacity to be used as onboard processing modules in airborne and (particularly) spaceborne Earth observation missions.

Acknowledgment

The authors would like to thank Instituto de Telecomunicações and Portuguese Science and Technology Foundation for their support under Project UID/EEA/50008/2013 and grant SFRH/BPD/94160/2013.

References

[1] J. M. Bioucas-Dias, A. Plaza, G. Camps-Valls, P. Scheunders, N. Nasrabadi, and J. Chanussot, Hyperspectral remote sensing data analysis and future challenges, *IEEE Geoscience and Remote Sensing Magazine*, vol. 1, pp. 6–36, 2013.

[2] R. O. Green *et al.*, Imaging spectroscopy and the airborne visible/infrared imaging spectrometer (AVIRIS), *Remote Sensing of Environment*, vol. 65, no. 3, pp. 227–248, 1998.

[3] A. Plaza and C.-I. Chang, *High Performance Computing in Remote Sensing*, Boca Raton, FL: Taylor & Francis, 2007.

[4] G. Motta, F. Rizzo, and J. A. Storer, *Hyperspectral Data Compression*, Berlin: Springer, 2006.

[5] B. Huang, *Satellite Data Compression*, Berlin: Springer, 2011.

[6] C. Li, T. Sun, K. Kelly, and Y. Zhang, A compressive sensing and unmixing scheme for hyperspectral data processing, *IEEE Transactions on Image Processing*, vol. 21, no. 3, pp. 1200–1210, 2011.

[7] Q. Zhang, R. Plemmons, D. Kittle, D. Brady, and S. Prasad, Joint segmentation and reconstruction of hyperspectral data with compressed measurements, *Applied Optics*, vol. 50, no. 22, pp. 4417–4435, 2011.

[8] M. Golbabaee and P. Vandergheynst, Joint segmentation and reconstruction of hyperspectral data with compressed measurements, in *IEEE International Conference on Image Processing (ICIP'12)*, 2012.

[9] M. Golbabaee, S. Arberet, and P. Vandergheynst, "Compressive source separation: Theory and methods for hyperspectral imaging," *IEEE Transactions on Image Processing*, vol. 22, no. 12, pp. 5096–5110, 2013.

[10] Q. Du and J. E. Fowler, Low-complexity principal component analysis for hyperspectral image compression, *International Journal of High Performance Computing Applications*, vol. 22, pp. 273–286, 2009.

[11] C. Song, Y. Li, and B. Huang, A GPU-accelerated wavelet decompression system with SPIHT and Reed-Solomon decoding for satellite images, *IEEE Journal of Selected Topics in Applied Earth Observations and Remote Sensing*, vol. 4, no. 3, pp. 683–690, 2011.

[12] S.-C. Wei and B. Huang, GPU acceleration of predictive partitioned vector quantization for ultraspectral sounder data compression, *IEEE Journal of Selected Topics in Applied Earth Observations and Remote Sensing*, vol. 4, no. 3, pp. 677–682, 2011.

[13] D. Donoho, M. Elad, and V. Temlyakov, Stable recovery of sparse overcomplete representations in the presence of noise, *IEEE Transactions on Information Theory*, vol. 52, no. 1, pp. 6–18, 2006.

[14] J. E. Candès, T. Romberg, and Tao, Robust uncertainty principles: Exact signal reconstruction from highly incomplete frequency information, *Communications on Pure and Applied Mathematics*, vol. 59, no. 8, p. 1207, 2006.

[15] G. Martin, J. M. Bioucas-Dias, and A. Plaza, HYCA: A new technique for hyperspectral copressive sensing, *IEEE Transactions on Geoscience and Remote Sensing*, vol. 53, no. 5, pp. 2819–2831, 2014.

[16] J. M. Bioucas-Dias and J. M. P. Nascimento, Hyperspectral subspace identification, *IEEE Transactions on Geoscience and Remote Sensing*, vol. 46, no. 8, pp. 2435–2445, 2008.

[17] J. M. Bioucas-Dias, A. Plaza, N. Dobigeon, M. Parente, Q. Du, P. Gader, and J. Chanussot, Hyperspectral unmixing: Geometrical, statistical, and sparse regression-based approaches, *IEEE Journal of Selected Topics in Applied Earth Observations and Remote Sensing*, vol. 5, no. 2, pp. 354–379, 2012.

[18] J. M. P. Nascimento and J. M. Bioucas-Dias, Vertex component analysis: A fast algorithm to unmix hyperspectral data, *IEEE Transactions on Geoscience and Remote Sensing*, vol. 43, no. 4, pp. 898–910, 2005.

[19] L. Rudin, S. Osher, and E. Fatemi, Nonlinear total variation based noise removal algorithms, *Physica D: Nonlinear Phenomena*, vol. 60, no. 1–4, pp. 259–268, 1992.

[20] A. Chambolle, An algorithm for total variation minimization and applications, *Journal of Mathematical Imaging and Vision*, vol. 20, pp. 89–97, 2004.

[21] M. Afonso, J. Bioucas-Dias, and M. Figueiredo, An augmented Lagrangian approach to the constrained optimization formulation of imaging inverse problems, *IEEE Transactions on Image Processing*, vol. 20, no. 3, pp. 681–695, 2011.

[22] J. Eckstein and D. Bertsekas, On the Douglas–Rachford splitting method and the proximal point algorithm for maximal monotone operators, *Mathematical Programming*, vol. 5, pp. 293–318, 1992.

[23] G. Martin and J. Bioucas-Dias, Hyperspectral compressive sensing from spectral projections, in *IEEE International Geoscience and Remote Sensing Symposium (IGARSS), 2015*, IEEE, 2015, pp. 1000–1003.

[24] J. Nascimento, J. Bioucas-Dias, J. Rodriguez, V. Silva, and A. Plaza, Parallel hyperspectral unmixing on gpus, *IEEE Geoscience and Remote Sensing Letters*, vol. 11, no. 3, pp. 666–670, 2013.

[25] G. S. Miller, The definition and rendering of terrain maps, in *ACM SIGGRAPH Computer Graphics*, vol. 20, no. 4, pp. 39–48, ACM, 1986.

15

Algorithms and Prototyping of a Compressive Hyperspectral Imager

Alessandro Barducci, Giulio Coluccia, Donatella Guzzi, Cinzia Lastri, Enrico Magli, and Valentina Raimondi

CONTENTS

15.1 Introduction

Compressive sensing (CS) (Candes and Tao 2006) has recently emerged as an efficient technique for sampling a signal with fewer coefficients than dictated by classical Shannon/Nyquist theory. The assumption underlying this approach is that the signal to be sampled must have concise representation on a convenient basis, meaning that there exists a basis where the signal can be expressed with few large coefficients and many (close-to-)zero coefficients. In CS, sampling is performed by taking a number of linear projections of the signal onto pseudorandom sequences, whereas reconstruction exploits knowledge of a domain where the signal is "sparse."

    CS has also been used to develop innovative "compressive" imaging systems. A single-pixel camera (Compressive Sensing and Sparse Approximation s.d.) uses a single detector to sequentially acquire random linear measurements of a scene via light modulation. This kind of design is very interesting for imaging at wavelengths beyond the visible light spectrum, where manufacturing detectors is very expensive; CS could be used to design cheaper sensors or sensors providing better resolution for an equal number of detectors. This paradigm is even more appealing for spectral imaging (e.g., Barducci, et al. 2012, 2013, 2014; Magli et al. 2012), where the amount of data generated by the imaging sensor is very large and the system can benefit from a natively compressed imaging format; reconstruction of such large datasets may require *ad hoc* algorithms in order to fully exploit data redundancies (e.g., Coluccia et al. 2012; Kuiteing et al., 2014).

Although compressive hyperspectral imaging has been studied in simulation, there are very few practical implementations; the related background is described in detail in Section 15.1.2. In this chapter, we describe a prototype implementation of a compressive hyperspectral imager, highlighting design, and data quality issues.

15.1.1   Compressive sensing

In the standard CS framework, introduced by Candes and Tao (2006), a signal $x \in \mathbb{R}^{N \times 1}$ that has a sparse representation in some basis $\Psi \in \mathbb{R}^{N \times N}$, that is,

$$x = \Psi\vartheta, \|\vartheta\|_0 = K, K \ll N$$

can be recovered by a smaller vector $y \in \mathbb{R}^{M \times 1}, K < M < N$ of linear measurements $y = \Phi x$, where $\Phi \in \mathbb{R}^{M \times N}$ is the sensing matrix. The optimum solution, requiring at least $M = K + 1$ measurements, can be obtained as follows:

$$\vartheta = \arg\min_{\vartheta} \|\vartheta\|_0 \text{ s.t. } y = \Phi\Psi\vartheta.$$

Because the ℓ_0-norm minimization is a (NP-)hard problem, one can resort to a linear programming reconstruction by minimizing the ℓ_1 norm:

$$\vartheta = \arg\min_{\vartheta} \|\vartheta\|_1 \text{ s.t. } y = \Phi\Psi\vartheta$$

provided that M is large enough ($\sim K \log(N/K)$).

The same algorithm is used for signals that are not exactly sparse but rather compressible, meaning that they (or their representation θ in basis Ψ) can be expressed only by K significant coefficients, while the remaining ones are (close to) zero.

When the measurements are noisy, the ℓ_1 minimization with relaxed constraints is used for reconstruction:

$$\vartheta = \arg\min_{\vartheta} \|\vartheta\|_1 \text{ s.t. } \|y - \Phi\Psi\vartheta\|_2 < \varepsilon, \tag{15.1}$$

where ε bounds the amount of noise in the data. It has been shown (Baraniuk et al. 2008) that extracting the elements of Φ at random from a sub-Gaussian distribution allows correct reconstruction with overwhelming probability.

15.1.2   Compressive hyperspectral imaging instruments

CS theory has been applied to the development of several optical instruments, in particular optical cameras and imaging spectrometers. Optical and 3D cameras can benefit from the use of CS technologies in those spectral ranges where the production of 1D or 2D sensor arrays with a large number of elements and uniform optical properties is particularly expensive and challenging. Clear examples are cameras for infrared (IR) imaging, which are significantly more expensive than their counterparts in the visible spectrum. Over the last decade, CS has emerged as a potential means of producing inexpensive IR cameras. The approach is the CS single-pixel camera, where a single detector acquires coded measurements of a high-resolution image. In any case, the use of CS for visible or IR imaging, as in the single-pixel camera, implies trade-offs between cost, sensitivity, resolution, size, and speed. Several CS-based prototypes for imaging purposes have already been developed.

One example is the laboratory breadboard created by the Georgia Institute of Technology. The prototype is an interesting CS architecture that employs a standard 256×256 CMOS. The sensor and its interface circuitry have been combined with a complex computational circuitry that performs the domain transformation intrinsic in a CS system, instead of employing a light modulator in front of a single-pixel detector (Robucci et al. 2008). Another prototype involving CS technology was produced in the framework of the Compressive Optical MONTAGE Photography Initiative (COMP-I) funded under DARPA's MONTAGE program (COMP-I s.d.). The COMP-I project goals are to produce a miniaturized visible and longwave infrared (LWIR) camera (Brady et al. 2005). In order to obtain high spatial and temporal resolutions, an array of single-pixel cameras that sense in parallel was used to increase the measurement rate. A proof-of-concept prototype using a sensor with 64×64 pixels was developed on this idea. Each sensor pixel is the detector for a single-pixel camera, the light modulator being a larger Digital Micromirror Device (DMD), unique for the entire system (Chen et al. 2015). A different solution consisting of lensless compressive imaging was proposed for the visible (VIS) or IR ranges (Huang et al. 2013). The majority of the implemented prototypes are single-pixel, monochromatic cameras. The most well-known is the single-pixel camera developed at Rice University, Rice University, Houston, Texas (Compressive Sensing and Sparse Approximation s.d.). Research performed at Rice University has resulted in the creation of a commercial company, InView Technology Corporation (http://inviewcorp.com/), that produces and commercializes an SWIR camera based on the CS technique (Shortwave Infrared [SWIR] Cameras s.d.).

The prototype CS single-pixel hyperspectral imager with stereoscopic capabilities described in Lai et al. (2012) is an example of CS architecture applied to 3D imaging. CS reduces the number of pixels of the camera to one, so that a sensor array can be replaced by a one-pixel detector. The range of wavelengths to which such detectors are responsive is much wider than that of a conventional array sensor and the costs for them are far lower, especially in the spectral range outside visible. CS theory has been also utilized to reduce the acquisition time for 3D acquisitions, introducing an end-to-end measurement system for capturing spectral data on 3D objects (Kim et al. 2012) in the spectral range from the near-ultraviolet (359 nm) to near-infrared (NIR) (1003 nm) at 12 nm spectral resolution.

Different studies have faced the problem of the application of CS to spectroscopic measurements, in particular as far as CS application to hyperspectral imaging is concerned (Gehm et al. 2007; Sun and Kelly 2009; Wagadarikar et al. 2008). Results show some advantages but also several critical issues related to the use of CS in this field, in particular concerning the availability of large-size, high-speed modulators.

The interest in developing hyperspectral systems with CS technology is fed by the opportunity of reducing the data throughput of such sensors, which is often challenging especially for satellite missions. Simple spectrometers with no imaging requirements (Brady et al. 2006) could partly avoid the issues that arise with hyperspectral imaging instruments such as the need for large and fast optical modulators. CS-based imaging spectroscopy is appealing in a contest where measurements are expensive or subject to physical limitation. Spectroscopy of very weak sources or of sources in the challenging near- or mid-IR spectral ranges using exotic detector materials may provide situations in which CS measurement is attractive. As an example, group testing could be used to minimize the number of Indium Gallium Arsenide (InGaAs) detector elements needed to characterize narrow-band NIR sources (Potuluri et al. 2004) (Hamza and Brady 2006; Willett et al. 2011). Physical sciences is developing an LWIR CS hyperspectral imager based on a single-pixel architecture, with a substantial cost reduction (Dupuis et al. 2015). The imager employs a DMD converted for LWIR operation in place of the traditional cooled LWIR focal plane array.

Soldevila et al. (2013) proposed an optical system that performs spectral imaging in the VIS and partly in the IR. The system relies on single-pixel camera architecture where the first lens system produces an image on DMD. The light emerging from the bright pixels of the DMD is collected by a second lens system similar to the first one. This lens system feeds the light to a silica multimode fiber coupled to a commercial concave grating spectrometer operating in the VIS–NIR spectral range. Some developments have been also transposed into patents, like the Tel-Aviv digital snapshot spectral imager (Golub et al. 2009).

CS theory was applied to single-detector IR rosette scanning systems (Uzeler et al. 2015), where a single detector pseudo-imaging rosette scanning system scans the scene in a specific pattern and performs processing to estimate the target location without forming an image. The CS framework enables the reconstruction of the rosette scanning seeker detector outputs to form an image for target acquisition and tracking purposes, thus improving the seeker's performance.

The concept of the single-pixel camera has been extended to provide continuous real-time video (Edgar et al. 2015; Radwell et al. 2014) simultaneously in the VIS and SWIR.

IR radiation changes induced by human activity can provide important information about activity patterns. In Guan et al. (2016), six typical physical activities were classified using CS data obtained from a three-view IR motion-sensing system.

Application and instrumentation based on CS are definitely a hot topic for research, and many prototypes have been developed and studied. However, nowadays the only engineered and commercialized products are the ones produced by InView Technology Corporation, which also owns several patents on this technology (Inview Technology Portfolio s.d.).

15.2   Prototype of a compressive hyperspectral imaging instrument

15.2.1   Instrumentation

The prototype for investigating CS technology was developed at the IFAC premises in Florence, Italy in the frame of an ESA ITI-B project (ESA contract n.4000106941/12/NL/CO). The main instrument selected for this CS investigation is a push-broom imaging spectrometer manufactured by the firm Horiba Jobin Yvon, Kyoto, Japan (model iHR550) that reaches a spectral resolution around 0.02 nm Full Width Half Maximum (FWHM) when equipped with a 2400 gr/mm grating. This instrument gives rise in its output focal plane to a 2D spatial–spectral domain, that is, $x - \lambda$. The spectrometer has two input and two output ports that can be equipped with additional devices and detectors. The electro-optical elements necessary to perform the CS acquisition of the signal will be hosted by the secondary port of the output focal plane. Figure 15.1 shows a preliminary test measurement performed observing the down-welling irradiance from the laboratory. The instrument is equipped with a thermoelectrically cooled 2D detector having 1024×256 active pixels, and operating from 350 nm up to 1200 nm.

Figure 15.2 depicts the optical assembly being deployed on the secondary output focal plane of this instrument in order to perform the CS signal acquisition at the same time as standard signal measurement is carried out at the main exit port.

The selected spatial light modulator (SLM) is a liquid crystal panel whose characteristics are reported in Table 15.1.

The low-resolution detector matrix originally selected for the prototype was a silicon photomultiplier (SiPM) array having 12×12 pixels. The optical system in Figure 15.2 matches the 2 Mpix of the SLM with the 144 pixels of this detector, allowing the user to parallelize the CS acquisition of 144 subimages (image blocks), each of which is reconstructed independently of each other. The greatest advantage of SiPM devices is constituted by

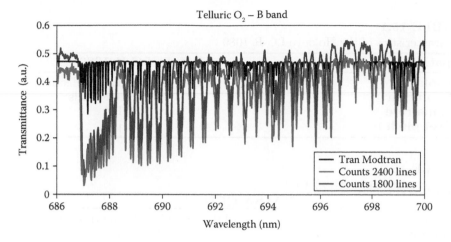

FIGURE 15.1

Transmission spectra in the B band of molecular oxygen. The blue line is the transmittance simulated with the Modtran 5.2 code at its maximum spectral resolution. The red and purple lines are the measurements collected with the imaging spectrometer using two gratings with 1800 gr/mm and 2400 gr/mm, respectively. Experimental results have been corrected for the effect of the non-unitary refraction index of air, which compresses the wavelength axis.

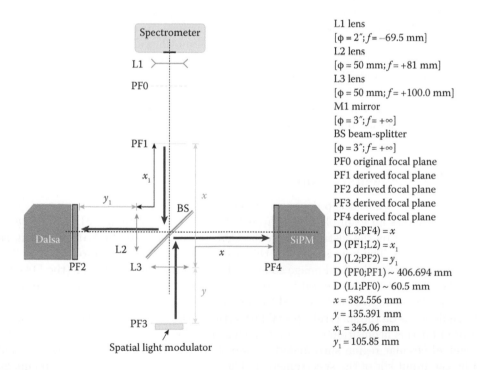

FIGURE 15.2

Layout of CS optical assembly being installed on the secondary exit port of the instrument.

TABLE 15.1

Characteristics of the Holoeye LC-R 1080

Spatial Light Modulator	Holoeye LC-R 1080
Display	LCoS microdisplay (reflective)
Resolution	1920×1200 pixels
Frame rate	60 HZ
Pixel pitch	8.1 μm

TABLE 15.2

Characteristics of DALSA CCD

DALSA CCD	
Detector	TH7888A CCD frame transfer
Number of pixel	1024×1024
Pixel size	14.0×14.0 μm
Binning	1, 2
Detector spectral range	430–1000 nm (full width @QE > 3%)
Maximum frame/rate	60 fps

CCD: Charge-Coupled Device; QE: Quantum Efficiency.

their high sampling frequency, which can easily accommodate the data sampling frequency requested to a remote sensing spectral imager. Unfortunately, this kind of detector is usually developed for scintillation measurements and suppresses the continuous (DC) signal component. In such a way, it is quite complex to retrieve the correct signal amplitude when adopting a CS architecture. To sum up, we decided to utilize a standard CCD detector for the prototype in lieu of the discussed SiPM chip. Table 15.2 shows the main technical specifications of the selected CCD.

Note that the spatial resolution allowed by this detector is higher than needed by the prototype (1 Mpix vs. 144 pix requested); indeed, the acquired image is undersampled after the acquisition and 2× binning is utilized in order to mitigate the redundancy of the acquired dataset.

15.2.2 CS experimental setup

Figure 15.3 shows the whole setup deployed at the IFAC laboratory.

Figure 15.4a shows an example of a spatial pattern displayed on the SLM. The same spatial pattern is seen by the instrument on the DALSA camera sees the same pattern, as shown in Figure 15.4b. The light source for this acquisition is a red laser diode.

It is worth noting that the final image collected by the prototype's detector (the DALSA camera) is the product of the spectral image output by the selected spectrometer and the modulating 2D signal impressed on the SLM. The limited visibility in Figure 15.4b of the SLM pattern depends on the spectral–spatial properties of the analyzed source. The red laser diode adopted for this acquisition exhibits its own spatial light distribution, constituted by a bright central circular region surrounded by two short segments in the vertical direction (selected by the input slit of the spectrometer). The brighter region in the prototype image of Figure 15.4b is the projection of the central circular spot of the laser source, which is expanded in the horizontal direction depending on the spectral density of power typical of this source. In this case, the laser has a bandwidth around 1–2 nm that is broadened due to the large aperture imposed to the spectrometer input slit (about 1 mm). This explains the illumination distribution of the image in Figure 15.4b.

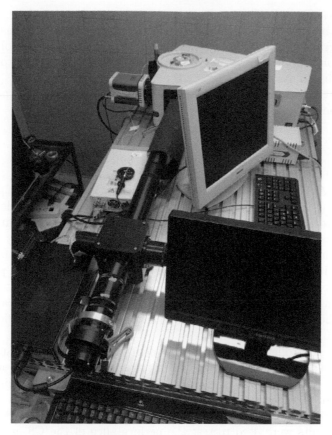

FIGURE 15.3
View of the CS prototype.

(a)   (b)

FIGURE 15.4
(a) Image impressed onto the liquid crystal spatial light modulator (SLM) adopted for the development of the prototype and (b) as "seen" by the DALSA camera of the prototype.

From the previous reasoning, it is easily understood that the developed CS arm produces the optical implementation of a 2D modulation operator, in which the 2D modulating signal is programmed by means of the SLM device. The modulated signal is the spectral image at the output port of an imaging spectrometer operating in the push-broom configuration.

Let $\rho_\lambda(x)$ indicate the ground (target) scan line at the input port of the push-broom spectrometer, x being the spatial coordinate along the spectrometer's input slit and λ

the electromagnetic wavelength. The input signal $\rho_\lambda(x)$ has an implicit dependence on the spectral wavelength, meaning that for a given point source x the electromagnetic (EM) power contributions related to different wavelengths λ coincide in the same physical point. The spectral dispersion introduced by the spectrometer expands the spectral domain of the input signal, originating on its output port a 2D copy $\rho(x, \lambda)$ of $\rho_\lambda(x)$.

The 2D image $m(\xi, \eta)$ displayed by the SLM directly modulates the signal $\rho(x, \lambda)$, as shown by the equation below.

$$
\begin{aligned}
s(\xi, \eta) &= m(\xi, \eta)\rho(x, \lambda) \\
(\xi, \eta) &= (x, \lambda)\mathbf{A} + (\xi_0, \eta_0).
\end{aligned}
\tag{15.2}
$$

In this equation, $\mathbf{A}$ is a generic coordinate transformation operator (e.g., a matrix) relating the domains (ξ, η) and (x, λ), whereas the term (ξ_0, η_0) is a simple origin shift. From a theoretical standpoint, the instrument can be tuned so that this operator is reduced to an affine transformation.

Following these considerations, an experimental procedure was developed for calibrating the geometric and radiometric parameters of the prototype. The calibration procedure fed with specific calibration measurements will be able to determine the parameters embedded in the coordinate transformation included in Equation 15.2. Moreover, the calibration will allow us to measure the radiometric response of the instrument to different reflectance levels impressed onto the SLM device.

The origin shift is controlled by the optical configuration of the instrument as well as by the rotation of the spectrometer grating, that is, the selected central wavelength for the current measurement. In other words, the term (ξ_0, η_0) changes when changing a different nominal wavelength for the spectrometer.

The realized experimental setup was used to perform a preliminary realistic set of CS acquisitions. Those results were used to test the performance of the CS architecture and algorithms. A first acquisition was executed, acquiring a set of 1300 images obtained by superimposing on the image of the laser source a random Gaussian distribution produced on the Liquid Crystal on Silicon (LCoS) display (Figure 15.5).

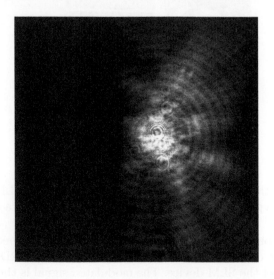

FIGURE 15.5
Image extracted from the CS acquisition performed with a random Gaussian distribution on LCoS.

The temporal behavior of the laser source during CS measurements shows a significant variability of the light reaching the LCoS display and DALSA CCD. The reason for this variability is strongly related to the multimode nature of the diode laser module. Another cause is the relevant distance between laser source and slit spectrometer. In order to reduce these effects, the experimental setup was improved by using a stabilized high-power diode laser operating at 670 nm coupled with a monomode optical fiber. Laser power can be adjusted and stabilized with a precision of 0.1 mW, while the monomode optical fiber removes the effects of mode-hopping. Moreover, the laser source was placed close to the spectrometer entrance slit thanks to the use of the optical fiber. The acquisition made with the described setup shows a better temporal stability, both for mean and standard deviation, and for both black and white measurements (obtained by programming the LCoS with values all equal to zero and 255, respectively).

15.2.3 CS acquisitions

The control of the instrument prototype performing acquisition of CS data is obtained by two independent computers, one devoted to the acquisition of digital images captured by the DALSA camera and the other dedicated to generation of the pseudorandom images displayed by the LCoS modulator. The computers operate according to a master–slave model in which the LCoS computer acts as the master and the DALSA computer as a slave. The master computer manages the acquisition sequence that is composed by two main steps: the generation and display of a new pseudo-random image (LCoS) and the acquisition of a new image (DALSA). Usually, DALSA images are the result of a user-programmable temporal average of successive images digitized by the camera. Temporal averaging, often between 4 and 16 frames, helps to mitigate random experimental noise affecting the measurements. Synchronization of the activity is guaranteed by a serial communication link (RS232) that implements a custom communication protocol. Therefore, the CS acquisition software is split in two applications installed on the aforementioned computers. Both the computers run on the MS Windows operating system. Figure 15.6 shows the layout of the entire acquisition procedure with its main steps.

Ten CS acquisitions were performed in the IFAC laboratories in order to fine tune and test the prototype, employing a binomial distribution for the random light modulation. The number of black-and-white calibration frames were set to one every three random binomial frames, for a better compensation of the residual source instability. Hereafter, we present one of the last acquisitions; this sequence consists of 7635 frames and was chosen for the good stability of the source during the entire acquisition. In Figure 15.7, a subimage of 128×128 pixels, cropped around the LCoS center, projected on the LCoS, and the respective images acquired by the DALSA camera are shown.

15.2.4 Pre-elaboration of acquired images

The relevant steps were performed for calibrating the CS setup, whose acquisitions need to be radiometrically calibrated and corrected for geometric and radial distortions. In order to correct all these effects, a calibration procedure was implemented and performed before every CS acquisition aiming to recover the one-to-one pixel correspondence between the LCoS and DALSA images. Such correspondence is essential for good-quality reconstructed images.

First, the procedure considered the acquisition of two different images: one for correcting radial aberration and one for fixing geometric misalignments (Figure 15.8d) between the LCoS and DALSA images. Figure 15.8b and d shows the images acquired by the DALSA camera when the LCoS was programmed with a homogenous grid of lines (Figure 15.8a),

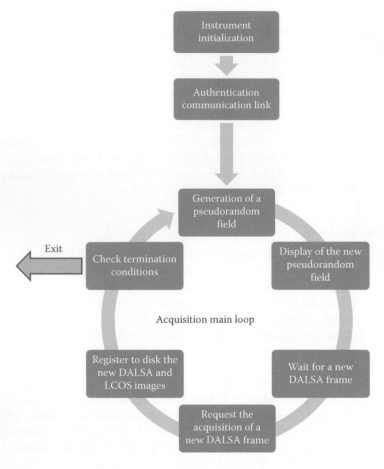

FIGURE 15.6
Layout of the CS acquisition procedure.

FIGURE 15.7
(a) Image projected on the LCoS and (b) image acquired by the DALSA camera.

whereas Figure 15.8d shows the image acquired by the DALSA camera when the LCoS projected an image composed of squares of different dimensions symmetrically disposed inside the scene (Figure 15.8c).

In Figure 15.8, the effect of pincushion distortion caused by the optical system is clearly evident. Figure 15.8b shows how the radial distortion strongly changes the shape of the squares projected by the LCoS display.

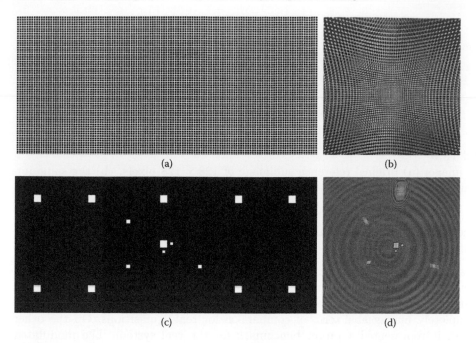

FIGURE 15.8
Acquisitions for fixing the effect of radial and geometric distortions. (a, c) Images projected on the LCoS and (b, d) images acquired by the DALSA camera.

The pincushion distortion was corrected with a procedure that, after a coordinate transformation from Cartesian to polar, applied a change of variable to the radius in the following form:

$$s = r + a \cdot r^3,$$

where s is the new radius and $a = 0.000008$; it then transformed the coordinate back from polar to Cartesian. For good correction of the pincushion distortion, correct evaluation of the origin of the distortion itself is crucial.

The following step of the data pre-elaboration consisted of trying to restore the one-to-one pixel correspondence between the LCoS and DALSA pixels. A reference image was used to correct the roto-translation and zoom factor that is observed between the LCoS and DALSA images. Taking the coordinates of the reference points in the LCoS and DALSA images, it is possible to build a linear system where the unknown variables are the parameters of the affine transformation between the images. Such an operation is performed by an appropriate routine on a subimage of 128×128 pixels cropped around the central square. In Figure 15.9, it is possible to evaluate the effect of the two steps of the correction procedure, both on the reference image and on one of the frames of the informative sequence.

The procedure described so far is applied to the whole acquired sequence; the corrected images can then be passed to the typical reconstruction step of the CS theory.

15.2.5 Reconstruction results

This section describes the reconstruction results from measurements acquired by the instrumentation described in the Section 15.2.1 of this chapter. In particular, each dataset underwent the following processing procedures.

1. The behavior over time of the acquired images was analyzed, both in the sampling domain and in the Discrete Cosine Transform (DCT) domain (white image only).

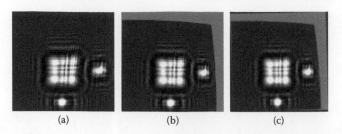

(a)   (b)   (c)

FIGURE 15.9
Reference image: (a) original reference subimage of 128×128 pixels, (b) the same image after pincushion correction, and (c) the final image after affine transformation.

The purpose was to detect abnormal behaviors of the source, for example, energy drifts.

2. A set of synthetically generated random images was created, using the patterns driving the modulator as sensing matrices and the white images as the acquired signal. The purpose of this operation was twofold. On one hand, it served as a completely manageable term of comparison for real acquisitions. On the other hand, it represented a target benchmark for the real system. The modulated synthetic measurements were then added up together, simulating the detector in order to obtain a set of "single-pixel" measurements.

3. A calibration procedure was performed in order to compensate for the fluctuations of the white image and of the black image. Both were interpolated in order to have a white image and a black image available for every random image. The difference of each image with respect to the first temporal frame (taken as *ground truth*) was evaluated. Then, these fluctuations were multiplied pixel-wise by the sensing matrices, summed up, and subtracted from the real and synthetic measurements. Finally, the average of both sets was subtracted to obtain two zero-mean datasets.

4. The reconstruction was performed as follows. First, the synthetic measurements were reconstructed to be used as a benchmark, using the basis pursuit with denoising method of Equation 15.1. Then, smaller and smaller sets of real measurements (random projections) were reconstructed. We started with the full set of measurements, and we obtained smaller sets by picking the measurements that are most similar to the synthetic measurements; that is, its indexes were selected in order to minimize the mean square error with respect to the corresponding set of synthetic measurements. Each set was reconstructed using the basis pursuit with denoising method of Equation 15.1. The reason behind this test is that the real measurements were expected to be affected by errors, and we wished to estimate the maximum quality that could be achieved when the real measurements were of good quality (i.e., as similar as possible to the synthetic ones).

The following figures had the purpose of a visual inspection of the behavior of the source image over time, in order to detect anomalies like unexpected energy drifts. Figure 15.10 shows the white image at nine time indexes uniformly distributed over the entire acquisition interval. Figure 15.11a shows the evolution over time of the most significant DCT coefficients of the first frame of the white image. By *most significant DCT coefficients*, we mean the DCT coefficients encompassing 99% of the frame energy. Finally, Figure 15.11b shows the pixel

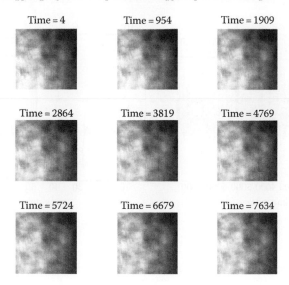

FIGURE 15.10
White image at different time indexes.

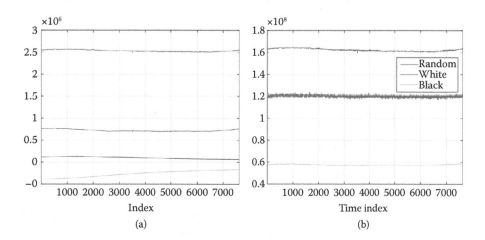

FIGURE 15.11
(a) Value of the most significant DCT coefficients of the first frame of the white image evolution over time and (b) integrated sum of white/random/black image evolution over time.

sum evolution over time of the three acquisitions, namely the white image, random image, and black image. These figures show that the experimental conditions did indeed vary over the scene acquisition time interval.

15.2.5.1 Simulated measurement generation and comparison with real acquisitions

First, a set of CS-like acquisitions from real measurements was generated. To this purpose, the pixels of each frame of random images were summed up to obtain a sequence of

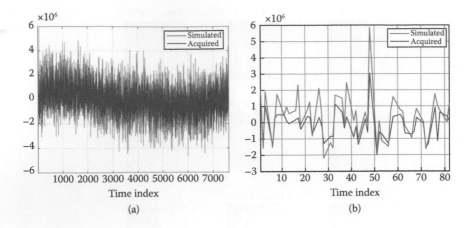

FIGURE 15.12
Synthetic vs. real measurement vectors: (a) entire vector and (b) first 50 samples.

"single pixel" measurements (i.e., the vector y). Then, the mean value of the sequence was subtracted to get a zero-mean measurement vector. In parallel, a set of synthetically generated measurements was produced in the following way. First, the white image (along with the black image, used later) was interpolated to simulate acquisition of those quantities for each random acquisition. Then, the modulation with the random patterns was simulated by a pixel-wise product of the interpolated white image with the random patterns used to drive the modulator when acquiring the real random images. The pixels of these frames were summed together to obtain the simulated measurement vector; its mean value was then subtracted. The result is depicted in Figure 15.12, where the synthetically generated measurements are compared with the real ones.

15.2.5.2   Calibration procedure

This section describes how the measurement vectors were calibrated for black and white fluctuations. It has to be underscored that the following procedure was applied *a posteriori* on the random measurements. Hence, it represents a realistic calibration procedure that could be applied to a CS detector. The procedure applies both to white and black images. First, the frame-by-frame difference with the first frame of the sequence was computed. This represents what is considered the *fluctuation*. Then, the sensing matrix was multiplied to the fluctuation and the pixels of the product were summed up together. This represents the "measurement vector" of the fluctuation that can be subtracted from the real and synthetic measurement vectors. In a realistic scenario, this procedure would be obtained by periodically modulating the source with a completely black/white pattern and subtracting the measurement to the first acquisition.

After subtraction of the fluctuations and of the mean value, the real and synthetically generated sequences are depicted in Figure 15.13.

The calibration procedure is crucial for the reconstruction. Figure 15.14 shows a comparison between the reconstruction from uncalibrated measurements and from calibrated ones (synthetic set). The figure shows the ground truth, as well. It can be easily observed that the reconstruction from uncalibrated measurements leads to a completely random result, due to the fact that fluctuations, albeit barely noticeable to the naked eye (see Figure 15.10), have a significant impact on the reconstruction quality.

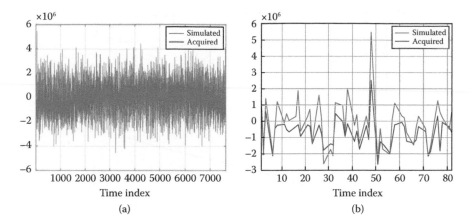

Time index | Time index
(a) | (b)

FIGURE 15.13
Synthetic vs. real calibrated measurement vectors: (a) entire vector and (b) first 50 samples.

(a)   (b)   (c)

FIGURE 15.14
Reconstruction from synthetic measurements: (a) uncalibrated, (b) calibrated, and (c) ground truth.

15.2.5.3 Reconstruction from acquired measurements: Visual results

In this section, we show some visual results concerning the reconstruction from real measurements. We used the synthetic measurements as a benchmark, and we selected from the real acquisitions the M measurements most similar to their synthetic counterparts, with M ranging from 10% to 90% of the total available, which is 4581. The purpose of this operation was to show that, despite all the issues involved in realization of such a complicated hardware, causing significant differences between the ideal and real acquired signals, it is still possible to reconstruct an estimate of the source from real acquisitions. This can be observed in Figure 15.15, which shows the reconstruction Peak Signal-to-Noise Ratio (PSNR) as a function of discarded measurements. It can be clearly observed that, discarding the measurements with the biggest difference with respect to their synthetic counterpart, the PSNRs of the reconstructions tend to converge. This is due to the fact that for synthetically generated measurements, each of which is *perfect*—in the sense that the product between the source and the random matrix is perfectly computed—the fewer the measurements that are used for the reconstruction, the lower the quality. By contrast, for real acquisitions, removing the *worst quality* measurements has a positive effect on the reconstruction. This means

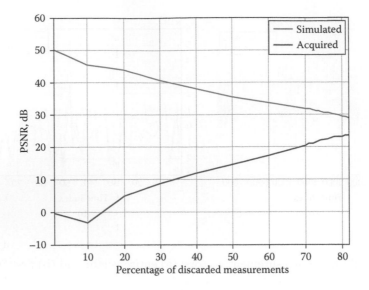

FIGURE 15.15
PSNR of the reconstruction as a function of the percentage of discarded measurements: synthetic vs. real acquisitions.

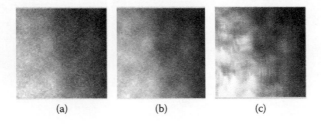

FIGURE 15.16
(a) Reconstruction from the best 12% of real measurements, (b) reconstruction from their synthetic counterparts, and (c) ground truth.

that, refining the acquisition hardware to mitigate the effects of the involved distortions, it would be possible to obtain a reconstruction quality comparable to the one obtained from synthetically generated measurements.

Choosing as the best number of measurements the one corresponding to the peak in the PSNR performance of the reconstruction, that is, $M = 12\% = 550$, we show in Figure 15.15 a comparison between the *clean* measurement vector and its synthetically generated counterparts and in Figure 15.16 the visual quality of the reconstruction.

15.3   Lessons learned and conclusions

The research activity leading to the implementation of this prototype of compressive hyperspectral imaging system gave us the opportunity to gain new insights regarding the performance expected from the hardware utilized for implementing the CS scheme. As shown in the previous sections, signal reconstruction gave partially surprising results. When attempting to reconstruct the signal using all the data collected during an experiment, we usually got poor results, characterized by a small Signal-to-Noise Ratio (SNR). A deep analysis of this phenomenon showed a weak correlation between the modulated signal observed by the DALSA camera (CS modulated image) and the simulated modulated image (i.e., the image computed stemming from the corresponding pseudorandom field and the known source image). We noted that the result of simulation would originate a modulated image equal to the one registered by the prototype. Therefore, the lower correlation between images modulated with optical and numerical methods was presumably connected with some unknown degradation of the gathered images, probably due to poor optical quality of the developed prototype.

Validation of the above hypothesis was investigated by attempting a signal reconstruction strategy in which only those experimental random projections that showed the best correlation with the (numerically) simulated data were utilized for reconstruction. This kind of analysis was described in the previous sections and produced a reconstructed signal of higher quality. It was even possible to show that the SNR of the signal estimated with this method tends toward that obtained with an equal number of simulated (modulated) random projections. This was the first and most important confirmation of the shortcomings affecting the developed prototype. A second confirmation of the poor quality of the prototype comes from analysis of the acquired modulated images. We noted that the image available in the focal plane of the DALSA camera was affected by the following disturbances:

- *Multiple paths*: The occurrence of partially shifted multiple images of the source has been ascribed to two optical elements of the instrument that are well known as possible sources of unwanted light reflection: the Beam Splitter (BS) film and the polarizing filter. As an example, the reflection of the incoming radiation field on the two surfaces of the polarizer gives rise to small-amplitude ghost images in the focal plane of the camera that are seen as multiple images.

- *Interference*: Interference fringes are visible in the images collected by the prototype. This phenomenon is partially connected to the occurrence of multipath when utilizing a radiation source of long coherence time, such as the laser unit we adopted for the experimental activity.

- *Distortion*: Image distortion is always due to poor optical quality of the focusing elements (e.g., lenses) included in the prototype. The available lens produced a typical pincushion deformation of the collected images that was partially removed offline by standard numerical procedures.

- *Source stability*: Source stability has been an important issue with a significant effect on the acquired modulated images. Source stability was affected by temperature changes, air flows inside the laboratory, and instability of the utilized laser unit. The last of these issues was mitigated by substituting the original laser unit with an element of higher quality. Even adopting a monomode optical-fiber laser unit, mechanical vibrations could have some impact on the intensity (brightness) of the DALSA image. The extended time required by the experiment was an additional cause that limited the availability of high-quality data. This issue reinforces the relevance to CS of SLMs with high frame rate.

- *Dynamic range*: The radiometric dynamic range achieved by the prototype was insufficient to obtain acceptable modulation levels in the acquired images. This problem is partially connected to the modulation ability of the selected SLM unit. When imposing the black value in a certain pixel of the modulating unit, significant power is really flowing toward the corresponding pixel of the DALSA camera. This behavior may be connected with a number of causes that are partially recapped in the list below.

- *Stray light*: Parasitic or scattered light originates everywhere inside the prototype, along the optical path. Light is scattered mainly over the surface of several optical elements composing the instruments, such as the polarizing filter, the BS film, the lenses, and so forth. Parasitic light may be due to light reflection over the inner surface of the cylinder that shields the optical path. In order to mitigate this problem, the (cylindrical) carter used for shielding environment light was coated with black paint. An effect of stray light is the cross-talk, generated between adjacent image locations (pixels) in the DALSA focal plane. Due to this cross-talk, a fraction of the light reflected by an SLM high-reflectance pixel may be redirected over the sensitive surface of a nearby picture element onto the DALSA focal plane. This circumstance transforms a theoretically black pixel into a gray pixel. Usually a high-quality optical element (e.g., a lens) originates not less than 0.5% of stray light. The whole instrument might originate up to several percent of stray light.

- *Residual reflection over the SLM*: Residual reflection by the surface of the SLM means that when a null reflectance value is programmed in a certain pixel of the SLM the actual physical reflectance rate is greater than zero, generating the so-called residual reflection. An important physical phenomenon that gives rise to this "reflectance offset" is connected with the mirror-like reflection induced by the refractive index step between the air and the surface of the SLM. This phenomenon is governed by the well-known Fresnel's equations, for the two possible polarization states (orthogonal and parallel polarizations), which predict a mirror-like reflection coefficient that is proportional to the refractive index gap between the two considered materials. It is evident that when ordinary materials are considered for the surface of the SLM (e.g., a crown glass or a polymer that would have a refractive index greater than 1.4 in the visible spectral range), the mirror-like reflectance is never less than 4%. Very often this contribution, when averaged for various propagation directions, amounts to 6%–8%.

- *Volume backscattering*: Volume backscattering takes place inside the SLM, depending on the internal structure of the latter. For example, the volume backscattering originated inside LCoS by the liquid crystal molecules is very difficult to estimate since it depends on the exact composition of the LCoS.

The discussion above clearly demonstrates the need for a hardware prototype of higher optical quality in order to improve the performance of the entire CS system and of the reconstructed signals. An additional issue that should be addressed by prospective research is the impact of this sensitivity of the CS scheme to the quality of the hardware. A question to be addressed is whether this kind of requirement can be fulfilled by a space-borne instrument or whether it is too restrictive. Moreover, other important technological issues should be highlighted, namely (1) the lack of SLMs of high frame rate that ideally would be necessary for any CS instruments and (2) the possible need for small matrix array detectors to be adopted for parallelizing the CS scheme in adjacent image regions (blocks) covering the entire field of view of the instrument.

The highest frame rate attainable with an SLM is about 32 kHz (TI DLP7000), working in binary mode with dedicated Field Programmable Gate Array (FPGA). The final data quality is directly related to the percentage of utilized measurements. A state-of-the-art SLM allows the development of a CS instrument working with block dimension of 16×16 pixels and different levels of reconstructed image quality. Improved reconstruction performance would be obtained using larger blocks. A CS system working with 16×16 pixel blocks allows an overall reduction of detector dimensions and data rate quantifiable with a scale factor of 256. As an example, a 1024×1024 pixel detector can be substituted with a 64×64 pixel one. In practice, this allows a overall reduction in system dimensions, memory requirements, and volume of data to be transmitted.

Acknowledgment

This work has been performed under ESA ITI-B project (ESA contract n.4000106941/12/NL/CO).

References

R. G. Baraniuk, M. Davenport, R. DeVore, and M. Wakin, A Simple Proof of the Restricted Isometry Property for Random Matrices, *Constructive Approximation*, vol. 28, no. 3, pp. 253–263, 2008.

A. Barducci, D. Guzzi, C. Lastri, V. Nardino, P. Marcoionni, and I. Pippi, Development of a Compressive Sampling Hyperspectral Imager Prototype, in *SPIE Remote Sensing*, 2013.

A. Barducci, D. Guzzi, C. Lastri, and I. Pippi, Compressive Sensing and Hyperspectral Imaging, in *Proceedings of the 9th International Conference on Space Optics (ICSO 2012), Topics*, 2012.

A. Barducci, D. Guzzi, C. Lastri, and V. Raimondi, Compressive Sensing for Hyperspectral Earth Observation from Space, in *International Conference on Space Optics*, 2014, p. 10.

D. J. Brady, M. Feldman, N. Pitsianis, J.P. Guo, A. Portnoy, and M. Fiddy, Compressive Optical MONTAGE Photography, in *Optics & Photonics 2005*, 2005, pp. 590708–590708.

D. J. Brady, M. E. Gehm, N. Pitsianis, and X. Sun, Compressive Sampling Strategies for Integrated Microspectrometers, in *Defense and Security Symposium*, 2006, pp. 62320C–62320C.

E. J. Candes and T. Tao, Near-Optimal Signal Recovery from Random Projections: Universal Encoding Strategies? *IEEE Transactions on Information Theory*, vol. 52, no. 12, pp. 5406–5425, 2006.

H. Chen, M. S. Asif, A. C. Sankaranarayanan, and A. Veeraraghavan, FPA-CS: Focal Plane Array-Based Compressive Imaging in Short-Wave Infrared, in *2015 IEEE Conference on Computer Vision and Pattern Recognition (CVPR)*, 2015.

G. Coluccia, S. K. Kuiteing, A. Abrardo, M. Barni, and E. Magli, Progressive Compressed Sensing and Reconstruction of Multidimensional Signals Using Hybrid

Transform/Prediction Sparsity Model, *IEEE Journal on Emerging and Selected Topics in Circuits and Systems*, vol. 2, no. 3, pp. 340–352, 2012.

COMP-I. http://www.disp.duke.edu/history/archives/imaging/index.ptml.

Compressive sensing and sparse approximation. http://dsp.rice.edu/research/compressive-sensing.

J. R. Dupuis, M. Kirby, and B. R. Cosofret, Longwave Infrared Compressive Hyperspectral Imager, in *SPIE Sensing Technology+ Applications*, 2015, pp. 94820Z–94820Z.

M. P. Edgar, G. M. Gibson, R. W. Bowman, B. Sun, N. Radwell, K. J. Mitchell, S. S. Welsh, and M. J. Padgett, *Simultaneous Real-Time Visible and Infrared Video with Single-Pixel Detectors*. Scientific Reports, Nature Publishing Group, 2015.

M. E. Gehm, R. John, D. J. Brady, R. M. Willett, and T. J. Schulz, Single-Shot Compressive Spectral Imaging with a Dual-Disperser Architecture, *Optics Express*, vol. 15, no. 21, pp. 14013–14027, 2007.

M. A. Golub, M. Nathan, A. Averbuch, E. Lavi, V. A. Zheludev, and A. Schlar, Spectral Multiplexing Method for Digital Snapshot Spectral Imaging, *Applied Optics*, vol. 48, pp. 1520–1526, 2009.

Q. Guan, X. Yin, X. Guo, and G. Wang, A Novel Infrared Motion Sensing System for Compressive Classification of Physical Activity, *IEEE Sensors Journal*, vol. 16, no. 8, pp. 2251–2259, 2016.

A. B. Hamza and D. J. Brady, Reconstruction of Reflectance Spectra Using Robust Nonnegative Matrix Factorization, *IEEE Transactions on Signal Processing*, vol. 54, no. 9, pp. 3637–3642, 2006.

G. Huang, H. Jiang, K. Matthews, and P. Wilford, Lensless Imaging by Compressive Sensing, in *20th IEEE International Conference on Image Processing (ICIP)*, 2013.

Inview Technology Portfolio. http://inviewcorp.com/inview/wp-content/uploads/2015/12/InViewTechnologyCorp-Patents.pdf.

M. H. Kim, T. A. Harvey, D. S. Kittle, H. Rushmeier, J. Dorsey, R. O. Prum, and D. J. Brady, 3D Imaging Spectroscopy for Measuring Hyperspectral Patterns on Solid Objects, *ACM Transactions on Graphics*, vol. 31, no. 4, p. 38, 2012.

S. K. Kuiteing, G. Coluccia, A. Barducci, M. Barni, and E. Magli, Compressive Hyperspectral Imaging Using Progressive Total Variation, in *2014 IEEE International Conference on Acoustics, Speech and Signal Processing (ICASSP)*, 2014, pp. 7794–7798.

K. W. C. Lai, N. Xi, H. Chen, L. Chen, and B. Song, Development of 3D Hyperspectral Camera Using Compressive Sensing, in *2012 IEEE Sensors*, 2012.

E. Magli, M. Barni, A. Barducci, D. Guzzi, and I. Pippi, Technological Issues in Compressive Sensing, in *Proceedings of 2012 ESA Workshop on Onboard Payload Data Compression (OBPDC)*, 2012.

P. Potuluri, M. Gehm, M. Sullivan, and D. Brady, Measurement-Efficient Optical Wavemeters, *Optics Express*, vol. 12, no. 25, pp. 6219–6229, 2004.

N. Radwell, K. J. Mitchell, G. M. Gibson, M. P. Edgar, R. Bowman, and M. J. Padgett, Single-Pixel Infrared and Visible Microscope, *Optica*, vol. 1, no. 5, pp. 285–289, 2014.

R. Robucci et al., Compressive Sensing on a CMOS Separable Transform Image Sensor, in *IEEE International Conference on Acoustics, Speech and Signal Processing (ICASSP 2008)*, 2008.

Shortwave Infrared (SWIR) Cameras. [Online]. http://inviewcorp.com/products/shortwave-infrared-swir-cameras.

F. Soldevila, E. Irles, V. Durán, P. Clemente, M. Fernández-Alonso, E. Tajahuerce, J. Lancis, Single-Pixel Polarimetric Imaging Spectrometer by Compressive Sensing, *Applied Physics B*, vol. 113, no. 4, pp. 551–558, 2013.

T. Sun and K. Kelly, Compressive Sensing Hyperspectral Imager, *Computational Optical Sensing and Imaging*, p. CTuA5, 2009.

H. Uzeler, S. Cakir, and T. Aytaç, Compressive Sensing Applications for Single Detector Rosette Scanning Infrared Seekers, *SPIE Security+ Defence*, pp. 964809–964809, 2015.

A. Wagadarikar, R. John, R. Willett, and D. Brady, Single Disperser Design for Coded Aperture Snapshot Spectral Imaging, *Applied Optics*, vol. 47, no. 10, pp. B44–B51, 2008.

R. M. Willett, R. F. Marcia, and J. M. Nichols, Compressed Sensing for Practical Optical Imaging Systems: A Tutorial, *Optical Engineering*, vol. 50, no. 7, pp. 072601–072601, 2011.

Index

Printed and bound by CPI Group (UK) Ltd, Croydon, CR0 4YY

01/11/2024

01782603-0006